IUTAM SYMPOSIUM ON COMPUTATIONAL METHODS IN CONTACT MECHANICS

IUTAM BOOKSERIES

Volume 3

Aims and Scope of the Series

The IUTAM Bookseries publishes the proceedings of IUTAM symposia under the auspices of the IUTAM Board.

For a list of related mechanics titles, see final pages.

IUTAM Symposium on Computational Methods in Contact Mechanics

Proceedings of the IUTAM Symposium held in Hannover, Germany, November 5–8, 2006

Edited by

PETER WRIGGERS

Leibniz Universität Hannover, Germany

and

UDO NACKENHORST

Leibniz Universität Hannover, Germany

Springer

A C.I.P. Catalogue record for this book is available from the Library of Congress.

ISBN 978-1-4020-6404-3 (HB)
ISBN 978-1-4020-6405-0 (e-book)

Published by Springer,
P.O. Box 17, 3300 AA Dordrecht, The Netherlands.

www.springer.com

Printed on acid-free paper

Table of Contents

Preface

Computational methods for the analysis of contact problems have been under development for more than three decades. One could think that it is a mature field which does not necessarily need new input from scientists working in the area of numerical methods. But even now there is still room for research in computational contact mechanics, hence engineering analysis includes unilateral constraints when a detailed modelling of a problem is needed. This detailed modelling of real structures includes e.g. rolling wheels, metal forming processes and car crash analysis, and is related to the increasing power of modern computering equipment. But also multi-physics modelling of thermo-mechanical or electro-mechanical contact as well as microscopic modelling are used in order to get a deeper insight into the mechanism of friction and wear. Besides this also new numerical simulation techniques were developed within the finite element or boundary element method yielding better discretization tools and more efficient algorithms for contact. But up to now there exists no approach which solves all problems of the above mentioned areas in a robust and reliable way, hence investigations and developments are still needed. Even today, scientists try to construct numerically more efficient and stable algorithms for the treatment of nonlinear constraint optimization problems and to develop better and more stable discretization schemes for finite deformation contact problems.

The aim of the symposium was to bring scientists together who work on the fron tier of research in computational contact mechanics described above, or in closely related subjects. During the symposium 37 scientist from more than 11 countries (Australia, Brasil, China, France, Germany, Greece, Italy, Poland, Spain, Russia and USA) presented their newest findings in the area of discretization techniques for contact, algorithms and interesting numerical simulations of contact problems. The symposium gave detailed insight in the newest development in the area of simulation techniques for contact problems. Presentations included new mathematical techniques like multi-level approaches. New discretization techniques like the mortar-method were considered for finite deformations and used within primal-dual active set strategies. But also advanced applications of unilateral contact to masonry structures, decohesion analysis and tractive rolling of tires were among many interesting presentations which showed the wide applicability of the newest algorithms and formulations but also their limitations.

This book contains the contributions which are related to the findings presented throughout the symposium. It provides a good overview on modern techniques and state-of-the-art discretizations schemes applied in contact mechanics.

During the symposium scientists from leading research groups in computational contact mechanics, including the disciplines of mechanics and applied mathematics, had a fruitful exchange of experience. Many still open questions were discussed after the presentations, in the coffee breaks and during the social events. New friendships were established and will lead to future joint collaboration in science related to computational contact mechanics and in the organization of next minisymposia and workshops in the area contact mechanics.

Peter Wriggers and Udo Nackenhorst
Leibniz Universität Hannover
April 2007

Multigrid Methods for Unilateral Contact Problems with Friction

Frédéric Lebon[1,2], Michel Raous[1] and Iulian Rosu[1]

[1]*Laboratoire de Mécanique et d'Acoustique – CNRS, 13402 Marseille Cedex 20, France*
E-mail: {lebon, raous, rosu}@lma.cnrs-mrs.fr
[2]*UFR Mathématiques, Informatique, Mécanique, Université de Provence, France*

Abstract. In this paper, multigrid methods are tested on unilateral problems with friction. An optimal strategy is presented and efficiency of the solver is discussed on several examples.

Key words: Unilateral contact, friction, multigrid.

1 Introduction

Multigrid methods have been widely used in fluids mechanics when large numbers of degrees of freedom are involved. Usually the geometries are sufficiently simple to enable the generation of multiple overlapped meshes in an easy way (essentially in the context of finite difference methods). In nonlinear structure mechanics, the computational costs increase because of the treatment of nonlinearities and finite elements methods are dominant because of the complexity of the geometries. The present work investigates the ability of multigrid methods to reduce the computational times and analyzes the specific problems of formulation and implementation related to the treatment of nonlinearities in the context of finite element methods. This work is conducted on contact problems involving unilateral contact and friction between an elastic body and a rigid obstacle. The nonlinearities are stiff because the contact behavior laws are nonsmooth (the nonpenetration is characterized by the nonregularized Signorini conditions) and nondifferentiable because of the use of the nonregularized Coulomb law.

2 The Frictional Contact Problem

2.1 The Contact Model

Details of the formulation are to be found in [21, 29, 30]. Considering 2D problems, let Ω be an open bounded set of $\mathbb{R}^2$, which is the interior of an elastic obstacle, with

Peter Wriggers and Udo Nackenhorst (eds.), IUTAM Symposium on Computational Contact Mechanics, 1–16.

a sufficiently smooth boundary $\Gamma = \Gamma_F \cup \Gamma_D \cup \Gamma_C$, in contact with a rigid body. The boundary Γ_D (respectively Γ_F) is the part of Γ on which the displacements (respectively the forces) are prescribed. Part Γ_C is that on which the contact is possible with the rigid obstacle. Let σ be the stress tensor, K the elasticity tensor, ε the strain tensor and u the displacement.

- Unilateral contact conditions
  On the contact boundary Γ_C, the displacement u and the unknown contact force F are written in local coordinates (n,t) where n denotes the outward normal unit vector to Γ_C :

$$\sigma.n = F \text{ on } \Gamma_C \tag{1}$$

$$u = u_N\, n + u_T \tag{2}$$

$$F = F_N\, n + F_T \tag{3}$$

  On Γ_C, the unilateral conditions are written as follows where d denotes the initial gap:

$$u_N \leq d \tag{4}$$

$$F_N \leq 0 \tag{5}$$

$$(u_N - d).F_N = 0 \tag{6}$$

  This is a Signorini problem. Neither regularization nor penalty are used.

- The Coulomb friction law
  It is written as follows where μ denotes the friction coefficient:

$$\|F_T\| \leq \mu|F_N| \text{ with } \text{ if } \|F_T\| < \mu|F_N| \text{ then } \dot{u}_T = 0 \tag{7}$$

$$\text{if } \|F_T\| = \mu|F_N| \text{ then } \dot{u}_T \text{ is colinear and opposite to } F_T \tag{8}$$

  As presented in [6, 29, 30], by introducing an incremental formulation, it was established that the problem to be solved at each step, is subsequently very similar to the static one. In that case, the unknowns are the displacement increments and there are extra terms both in the functional which has to be minimized and in the convex characterizing the constraints. In this paper, we consider only the static problem using the following displacement formulation for the friction ((9) and (10)). Formulation and numerical algorithms can easily be extended to quasistatic evolutive problems [6, 29]. As for the unilateral contact, no regularization (compliance) is used.

$$\|F_T\| \leq \mu|F_N| \text{ with if } \|F_T\| < \mu|F_N| \text{ then } u_T = 0 \tag{9}$$

$$\text{if } \|F_T\| = \mu|F_N| \text{ then } u_T \text{ is colinear and opposite to } F_T \tag{10}$$

2.2 Variational Formulation

The variational formulation of the Coulomb problem can be written as an implicit variational inequality [8]. By introducing a fixed point problem P_1 on the sliding limit g [27, 29], the problem can be solved as a sequence of classical variational inequalities including nevertheless a nondifferentiable term. Because of the symmetry of the elasticity mapping, these problems can be solved as minimization problems under constraints (problem P_2).

Problem P_1: Find the sliding limit function G, fixed point of the application $g \to \mu|F_N(u_g)|$ where u_g is solution to the problem P_2 depending on g.

Problem P_2: Let ϕ_1 be the load applied in Ω, ϕ_2 be the load applied to part Γ_F of the boundary, and g be given,

find $u_g \in \mathbb{K} = \{v, v = 0 \text{ on } \Gamma_D \text{ and } v_N \leq d \text{ on } \Gamma_C\}$ such that:

$$J(u_g) \leq J(v) \qquad \forall v \in \mathbb{K} \tag{11}$$

with

$$J(v) = \frac{1}{2}a(v, v) - (f, v) + j(v) \tag{12}$$

$$\text{where: } a(u, v) = \int_\Omega \text{grad}_s u \; K \; \text{grad}_s v \, dx \tag{13}$$

$$j(v) = \int_{\Gamma_C} g \|v_T\| \, dl \tag{14}$$

$$(f, v) - \int_\Omega \phi_1 v \, dx + \int_{\Gamma_F} \phi_2 v \, dl \tag{15}$$

3 Preconditioner for Mixed Formulation

In this section, we adopt the mixed formulation proposed in [1]. The problem is split into two parts involving the pair $x = (u, \lambda)$, where u denotes the displacement vector and λ the contact nodal forces, say a differentiable part G and a nondifferentiable one F:

$$G(x) + F(x) = 0 \tag{16}$$

This last problem is solved using the Generalized Newton Method (GNM):

$$x^{i+1} = x^1 - (K_i + J_i)^{-1}(G(x^i) + F(x^i)) \tag{17}$$

where $K_i = \partial G(x^i)$ and $J_i \in \partial F(x^i)$. The tangent matrix $A = K_i + J_i$ is nonsymmetric (with zeros on the diagonal) and the previous linear system can be solved by the Conjugate Gradient Squared Method (CGS) associated with the coarse/fine preconditioner proposed in [2]. The basic idea of the preconditioner is to assume that the tangent matrix A has been obtained through the refinement of a coarser mesh. The matrix A is split into four parts (index c refers to coarse level and f to the fine one):

$$A = \begin{bmatrix} A_{ff} & A_{fc} \\ A_{cf} & A_{cc} \end{bmatrix} \tag{18}$$

A block factorization of matrix A gives:

$$A = \begin{bmatrix} A_{ff} & 0 \\ A_{cf} & A_{cc} - A_{cf} A_{ff}^{-1} A_{fc} \end{bmatrix} \begin{bmatrix} I & A_{ff}^{-1} A_{fc} \\ 0 & I \end{bmatrix} \tag{19}$$

We postulate the following preconditioner:

$$C = \begin{bmatrix} A_{ff} & 0 \\ A_{cf} & A_{cc}^{*} \end{bmatrix} \begin{bmatrix} I & A_{ff}^{-1} A_{fc} \\ 0 & I \end{bmatrix} \tag{20}$$

where A_{cc}^{*} is the tangent matrix on the coarse grid. The coarse/fine preconditioner requires to solve two sub-systems with the matrix A_{ff}. It is possible to define an approximated LDU factorization of this matrix associated to each contact status [2]. This preconditioner is very efficient because it is not necessary to find an optimal parameter.

4 Formulation and Implementation of the Multigrid Method for Contact Problems

Classically, in the linear case, the efficiency of multigrid methods can be explained in terms of the spatial frequencies of the error between the solution and the iterates when this error is written in the basis of the eigenvectors of the discretized mapping. The smoothers (Gauss–Seidel method for example) are known to quickly reduce the high frequencies of the error, but a large number of iterations is required to reduce the low components. Multigrid methods increase the convergence on the low frequencies of the error. They correct the solution obtained after only a few smoothings on the fine grid (where the solution is needed) by a correcting error computed on a coarse grid with the current defect of equilibrium. The sequence is then composed of a few smoothings of the solution on the fine grid (large system) and of complete resolution to determine a correction error on a coarse grid (small system). Appropriate interpolation and restriction have to be carried out between the grids.

With linear problems, the problem to be solved on the coarse grid is similar to the initial one defined on the fine grid. With nonlinear problems, one of the difficulties is to write an appropriate problem on the coarse grid, which is defined on the error and associated with the defect of equilibrium computed on the fine grid and carried onto the coarse one. Here, several alternatives arise. Section 4.2.1 is devoted to the problem written on the error. In Section 4.2.2 different methods of carrying the non-linearity information needed onto the coarse grid from the fine one are proposed and tested.

4.1 Basic Solvers

4.1.1 The Discrete Problem

Finite element discretization of problems P_1 and P_2 gives the following problems P_3 and P_4 written in $\mathbb{R}^{2N_h}$, where N_h is the number of fine mesh nodes. Let $\mathbb{K}_h \subset \mathbb{R}^{2N_h}$ be the cone defined by:

$$\mathbb{K}_h = K_{h,1} \times K_{h,2} \times \ldots \times K_{h,2N_h}$$

Let I_h be the set of indexes of the equations relating to the normal components of the contact nodes. Let L_h be the set of suffixes of the tangential components of the nodes of Γ_C.

$$K_{h,i} =]-\infty, d\,] \text{ if } i \in I_h$$
$$K_{h,i} = \mathbb{R} \qquad\qquad \text{otherwise}$$

Problem P_3: find the sliding limit G_h fixed point of the application $g_h \rightarrow \mu|F_{h,N}(u_h)|$ where u_h is the solution to the problem P_4 depending on g_h.

Problem P_4. find $u_h \in \mathbb{K}_h$ such that:

$$J_h(u_h) < J_h(v_h) \qquad \forall\, v_h \in \mathbb{K}_h \tag{21}$$

where:
$$J_h(v_h) = \frac{1}{2} v_h^T A_h v_h - v_h^T f_h + |v_h|^T g_h \tag{22}$$

with

$$|v_h|^T = (|v_{h,1}|,\ |v_{h,2}|,\ \ldots,\ |v_{h,2N_h}|) \tag{23}$$

$$|v_h|^T g_h = \sum_{i \in L_h} |v_{h,i}| g_{h,i} \tag{24}$$

and

– $A_h = (a_{h,ij})$ is the matrix $(2N_h \times 2N_h)$ associated to the elasticity operator,

- $f_h = (f_{h,ij})$ is the load vector,
- g_h is the vector associated to the sliding limit g ($g_h \in \mathbb{R}^{\mathrm{card}(L_h)}$).

The discrete problem P_4 on the fine grid has been derived from an appropriate form of the approximation problem given in [23].

The problem P_3 is solved using a fixed point algorithm (successive approximations). Starting with g_h^0 equal to zero (frictionless case), the solution to problem P_4 gives the displacement u_h^ℓ and the contact force $F_{h,N}^\ell$ at each step ℓ ($F_{h,N}^\ell$ is computed directly from the defect of equilibrium $A_h u_h^\ell - f_h$). We then define the iterative process as $g_h^{\ell+1} = \mu |F_{h,N}^\ell|$.

4.1.2 Gauss–Seidel Algorithm for the Contact

Problem P_4, where the sliding limit g_h^ℓ is given by the fixed point method, is solved by an internal multigrid process. The first step in the multigrid process is the smoothing step for which a Gauss–Seidel method is often used. Here we consider a Cryer–Christopherson method [7] adapted to deal with the nondifferential term due to friction [21,29,30]. It is a Successive Over-Relaxation method with Projection (SORP):

$$u_{h,i}^{k+1} = P_{K_{h,i}}((1-\omega)u_{h,i}^k + \omega m) \tag{25}$$

where

$$\begin{cases} m = \dfrac{1}{a_{h,ii}}\left(f_{h,i} - \displaystyle\sum_{j=1}^{i-1} a_{h,ij} u_{h,j}^{k+1} - \sum_{j=i+1}^{2N_h} a_{h,ij} u_{h,j}^k - sgn(m) g_{h,i}^\ell\right) & \text{if } m \in \mathbb{R}^* \\ m = 0 & \text{else} \end{cases} \tag{26}$$

with $g_{h,i}^\ell = 0$ if $i \notin L_h$. The second alternative of (26) occurs when there is no compatibility between the calculated m and its assumed sign, i.e., we obtain a negative (or positive) value of m when its sign (denoted sgn(m)) was taken to be positive (or negative).

$P_{K_{h,i}}$ is the projection operator on $K_{h,i}$ and ω is the relaxation coefficient ($\omega \in [1, 2[$). Because of the lack of theoretical estimates, an optimal value of the relaxation coefficient can be only obtained through dichotomic research and this Over-Relaxation method will be used only when the method acts as a solver on the coarse grid. When the method acts as a smoother on the fine grid, ω is set equal to 1 and the smoother is then the Symmetric Gauss–Seidel method with Projection (SGSP) [9].

4.1.3 Mathematical Programming Solver and Complementary Formulation

The frictional contact problem introduced in Section 2 can also be formulated, after discretization, as a complementarity problem [5,9,16,29,30]).

Problem P_5: let f be given, find the displacement u and the contact force F such that:

$$\left\{\begin{array}{l} A_h^* u_h = f_h^* + F_h \\ \\ \text{for } i \in I_h (\text{normal components}) \\ \left\{\begin{array}{l} u_{h,i} \leq 0 \\ F_{h,i} \leq 0 \\ u_{h,i} F_{h,i} = 0 \end{array}\right. \\ \\ \text{for } i \in L_h (\text{tangential components}) \\ \left\{\begin{array}{l} u_{h,i} = \lambda_2 - \lambda_1 \\ F_{h,i} = \frac{1}{2}(\varphi_1 - \varphi_2) \\ \lambda_i \leq 0 \\ \varphi_i \leq 0 \quad i = 1,2 \\ \lambda_i \varphi_i = 0 \end{array}\right. \end{array}\right. \tag{27}$$

Problem P_5 has been reduced to the contact variables by using a condensation procedure described in [29]. Matrix A_h^* is the full sized condensed stiffness matrix which turns out to be nonsymmetric because of the friction, and f_h^* is the condensed loading force vector. The tangential contact displacement has been separated into positive (forward) sliding λ_1 and negative (backward) sliding λ_2 relatively to the unit vector t. This problem is known as a linear complementary problem, the general form of which is written:

$$w - Mz = q \tag{28}$$

$$w \geq 0 \tag{29}$$

$$z \geq 0 \tag{30}$$

$$wz = 0 \tag{31}$$

We solve this problem by using the Lemke method which is a direct pivoting algorithm. Details can be found in [5, 9, 16, 29].

4.2 The Two Grid Method

4.2.1 The Algorithm

We consider a fine mesh (with N_h nodes) and a coarse mesh (with N_H nodes). We consider the operator (prolongation) p_H^h: $\mathbb{R}^{2N_H} \rightarrow \mathbb{R}^{2N_h}$ and the operator (restriction) $p_h^H : \mathbb{R}^{2N_h} \rightarrow \mathbb{R}^{2N_H}$.

Starting with $u_h^0 \in \mathbb{K}_h$, we define an iterative process:

$$u_h^k \rightarrow u_h^{k+1} \tag{32}$$

- *Smoothing on the fine grid*
We carry out ν smoothing iterates (two or three iterates) of the SGSP$^\nu$ procedure (Symmetric Gauss–Seidel with Projection). We denote:

$$u_h^{k+1/2} = \mathrm{SGSP}_\omega^\nu(u_h^k, f_h, A_h) \tag{33}$$

We introduce the variable $e_h^k = u_h - u_h^{k+1/2}$, u_h being the (unknown) exact solution of problem P_4. Now, in order to evaluate the correction, we have to solve the new problem:

Problem P_6: find e_h^k (such that $e_h^k + u_h^{k+1/2} \in \mathbb{K}_h$) such that:

$$J_h^C(e_h^k) \leq J_h^C(w_h) \qquad \forall\, w_h \text{ with } w_h + u_h^{k+1/2} \in \mathbb{K}_h \tag{34}$$

where

$$J_h^C(w_h) = \frac{1}{2} w_h^T A_h w_h - w_h^T(f_h - A_h u_h^{k+1/2}) + |w_h + u_h^{k+1/2}|^T g_h^\ell \tag{35}$$

Problem P_6 now has to be set on the coarse grid. Let $r_h^k = f_h - A_h u_h^{k+1/2}$ and $r_H^k = p_h^H r_h^k$.

- *Coarse grid correction*
Three difficulties occur in treating the nonlinearities of contact problems with friction:
  - how to carry the nonpenetration condition from the fine onto the coarse grid,
  - how to treat the absolute value on the coarse grid,
  - how to transfer the sliding limit g_h^ℓ from the fine grid onto the coarse one?

These points will be discussed in Section 4.2.2.
We take U_H^k a vector chosen in $\mathbb{R}^{2N_H}$ and we set:

$$\mathbb{K}_H^C = K_{H,1}^C \times K_{H,2}^C \times \ldots \times K_{H,2N_H}^C$$

with

$$\begin{cases} K_{H,i}^C = \mathbb{R} & \text{if } i \notin I_H \\ K_{H,i}^C =]-\infty, d - U_{H,i}^k\,] & \text{if } i \in I_H \end{cases} \tag{36}$$

We define two vectors $G_H^\ell \in \mathbb{R}^{\mathrm{card}(L_H)}$ and $V_H \in \mathbb{R}^{2N_H}$. As with classical nonlinear multigrid methods, these vectors are chosen to approximate the values on the coarse grid. A new problem (similar to eq. (35)) is written on the coarse grid:

Problem P_7: find $e_H^k \in \mathbb{K}_H^C$ such that:

$$J_H^C(e_H^k) \leq J_H^C(w_H) \qquad \forall\, w_H \in \mathbb{K}_H^C \tag{37}$$

where $J_H^C(w_H) = \frac{1}{2} w_H^T A_H w_H - w_H^T r_H + |w_H + V_H^k|^T G_H^\ell$.
How to choose U_H^k, V_H^k and G_H^ℓ will be discussed in Section 4.2.2.
Problem P_7 involving the error e_H^k has to be solved (the loads are replaced by the defect r_H^k in the second member). With linear problems, it is natural to use a direct method (Cholesky). With contact problems, the SORP method or the Lemke method (linear programming method) will be used. When the Lemke method is used, P_7 must be changed into a complementary problem (P_7^b) similar to P_5 [9, 29]. The solution e_H is interpolated onto the fine grid:

$$e_h^k = p_h^H e_H^k \tag{38}$$

The previous solution $u_h^{k+1/2}$ is corrected (we must ensure that u_h^{k+1} stays within the convex $\mathbb{K}_h$):

$$u_h^{k+1} = P_{\mathbb{K}_h}(u_h^{k+1/2} + e_h^k) \tag{39}$$

and we return to Equation (33) until a given degree of precision is achieved.

- *Prolongation and restriction operators*
The prolongation p_H^h is a linear interpolation. Consistence of the formulation is insured by choosing $p_H^h = (p_h^H)^T$. With three-node triangles when each element is divided into four triangles, we have

$$r_H^k(x) = p_h^H r_h^k(x) = r_h^k(x) + \frac{1}{2} \sum_{y \in \mathcal{C}_x} r_h^k(y).$$

$\mathcal{C}_x$ is the set of nodes connected to x [20].

- *The initial condition (full multigrid)*
The two grid process will be initiated by solving first the problem on the coarse grid (problem P_4) and then using the prolongation of this solution u_H^0 on the fine grid as the initial condition $u_h^0 = p_H^h u_H^0$ to start the first smoothings. This is known as the full multigrid method. Numerical tests have confirmed that the choice of the full multigrid method turns out to be essential for our problem, as generally observed for nonlinear problems [9, 19].

4.2.2 *Transfer of the Nonlinearities*

- *Convex definition*
The nonpenetration condition is satisfied by the condition that the solution must be within the convex $\mathbb{K}_H^k$ in problem P_5. On the coarse grid, the convex $\mathbb{K}_H^k$, associated with the error e_H^k depends on the solution after the smoothing: it is

characterized by the condition $(U_H^k + e_H^k)_N \leq d$ on Γ_C. Therefore U_H^k, which is not needed in linear classical case, has to be evaluated on the coarse grid. There exist various means of evaluating U_H^k from the computed solution $u_h^{k+1/2}$ on the fine grid. Four of them are considered here:

(a) $U_H^k = q_h^H u_h^{k+1/2}$ where q_h^H is a restriction. We can use:
   - the *canonic injection* $U_H^k = u_h^{k+1/2}$,
   - the *standard operator* defined in Section 4.2.1, $U_H^k = \frac{1}{2} p_h^H u_h^{k+1/2}$. Operator p_h^H has been defined for the forces: it is used here on the displacements, where a coefficient $1/2$ has to be introduced,
   - a more sophisticate operator which involves five nodes of the fine grid.

(b) U_H^k is defined as the largest normal displacement through the node under consideration and its two neighbors: $U_H^k(x) = u_h^{k+1/2}(z)$ with $z = -\text{argmax}\,\{u_h^{k+1/2}(y),\ y \in \mathcal{C}_x\}$. This is a procedure developed by *Mandel* [25]. It is a very good choice because it ensures that $u_h^{k+1/2} + e_h$ stays within the convex. This is convenient for theoretical reasons.

(c) $U_H^k = u_H^0$ where u_H^0 is the solution initially computed on the coarse grid for the full multigrid initialization defined in Section 4.3. This is derived from the *Hackbusch* [10] choice for nonlinear problems.

(d) $U_H^k = U_H^{k-1} + e_H^{k-1}$ (with the initial condition $U_H^1 = u_H^0$). This is a variant of (c), where the computed correction e_H^k is applied at each iteration (*Hackbusch with updating*).

The cases (c) and (d) need the use of the nested iteration or the full multigrid method [10]. The advantage of the full multigrid method is that it provides a good starting solution at each level. The algorithm begins at the coarsest level.

- *The undifferentiable term*

  The problem of the absolute value is very important from the mathematical point of view, and the term j(v) is undifferentiable. The term $|e_H + V_H^k|$ depends on the choice of vector V_H^k. An approximation of the displacement is needed on the coarse grid. Various choices of V_H^k are given in [19].

  (a) $V_H^k = q_h^H u_h^{k+1/2}$ where q_h^H is a restriction. We can use:
     - the *standard injection* $V_H^k = \frac{1}{2} p_h^H u_h^{k+1/2}$ (tangential components),
     - the *canonic injection*.

  (b) the *Hackbusch choice*: $V_H^k = u_H^0$ where u_H^0 is the solution initially computed on the coarse grid for the full multigrid initialization. The use of a full multigrid is necessary in that case.

  (c) the *Hackbusch with updating*: $V_H^k = V_H^{k-1} + e_H^{k-1}$ (with the initial condition $V_H^1 = u_H^0$).

- *The sliding limit*

  G_H^ℓ is an approximation of the sliding limit on the coarse grid. Two cases are proposed:

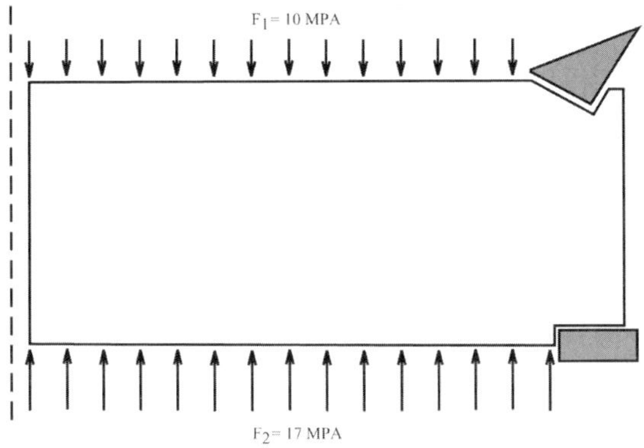

Fig. 1. Plate in contact with two rigid bodies.

(a) the *restriction* of the fine sliding limit at each multigrid iteration,
(b) the *initial value* G_H^0 computed with the initial resolution on the coarse grid of the full multigrid method.

4.3 Multigrid Method

With very large sized problems, introducing several levels of grids can keep the coarse problem small sized. This is of course important because total resolutions are conducted on the coarse grid. It is easy to generalize the previous algorithm to a MultiGrid Method (MGM) and to implement either V-cycle or W-cycle.

5 Optimal Strategy: Tests on the Various Alternatives

The multigrid methods have been developed in our own finite element code EUXENE which is a specific version of GYPTIS dedicated to the contact problems at the LMA [18].

The optimal strategy for dealing with the nonlinearities is discussed in the case of a plate which is part of a pressure vessel used in nuclear engineering [21] (see Figure 1). This is an axisymmetrical problem. The structure is in contact with two rigid solids on an oblique zone (30°) and on a horizontal zone. The plate is squeezed between the two rigid obstacles by prescribing a vertical displacement (0.05 mm) of the lower one. Primary and secondary pressures are applied on the two sides of the plate. The numerical tests have been conducted by using five meshes, with 51617 nodes for the finest one and 227 nodes for the coarsest one. Of course, the solution is always computed on the finest one. The meshes are unstructured with local refinements in the contact zone (194 contact nodes on the finest mesh).

Numerical tests have been conducted on this example in order to compare the various alternatives presented in Section 4 for the implementation of multigrid methods for that nonlinear problem. The main results are given and details of the numerical results could be found in [17].

5.1 Convex Definition

In this case, the tests are conducted on a frictionless case. The different possible choices for the convex definition turn out to be equivalent. For mathematical reasons, the Mandel process is suitable. This is confirmed by other numerical results obtained on other examples. The Mandel procedure will be used in the computations.

5.2 The Undifferentiable Term

The same example is now treated with friction. The convex is defined as chosen in the previous paragraph. We test the various ways of choosing V_H^k in (P_7) as described in Section 4.2.2.

In this case, results have shown that the use of informations from the fine grid can be very unsatisfactory when three grids are implemented. The canonic and standard injection should be avoided. The computations are conducted with 3 pre-smoothings, 2 post-smoothings and by using V-cycle. The "Hackbusch" procedure is always efficient and will be chosen in what follows.

5.3 The Sliding Limit

We tested the two alternatives given in Section 4.2.2 for G_H^ℓ in (P_7). The computations are conducted with 3 pre-smoothings and 2 post-smoothings. The results have shown that the Coulomb's condition $|F_T| = \mu|F_N|$ is verified with a good accuracy when the "restriction" is used to define the sliding threshold G_H^l. When the "initial value" is used, the accuracy varies from 5 to 10%. The choice for the coarse sliding limit is therefore the restriction of the fine sliding limit.

5.4 Number of Smoothings

Taking the same example, we study the effects of the number of pre- and post-smoothings on the convergence. The results have shown that it is convenient to perform both a few pre-smoothings and a few post-smoothings. We will choose in the following 3 pre-smoothings and 2 post-smoothings which turned out to be convenient choices in the numerical tests.

Table 1. Direct and multigrid resolution for two meshes of the plate with two contact zones (Lemke solver).

Number of d.o.f	Direct method	2-grid method		
	CPU	Iterates	Smoothings	CPU
26018	1′43″	6	33	15.7″
103234	28′00″	6	33	2′16″

5.5 Type of Multigrid Cycle

Taking the three grid example, we have compared the results using V-cycle and W-cycle on the same example of contact with friction. The W-cycle turns out to be slightly better than the V-cycle when considering the number of resolutions on the coarse grid (number of 3-grid iterates) and the total number of smoothings. In the following W-cycles are used for the case 3-grid or more.

5.6 Number of Grids

The number of grids is of course an important parameter. A large number of grids is time and memory size consuming. The tests have shown that in the case of large number of degrees of freedom (very fine mesh) the choice 3 or 4 grids is efficient. For smaller sized problems 2 grids will be sufficient.

6 Global Efficiency of the Multigrid Method on a Contact Problem

The efficiency of the method is presented here by comparing the multigrid method to the direct resolution on two examples (other ones could be found in [17]. The example is still that of the axisymmetrical plate in contact with two rigid obstacles (see Figure 1). Results for 26018 and 103234 d.o.f. meshes are given in Table 1. The Lemke solver is used both for the direct resolution and for the resolutions on the coarse grids. The multigrid method divides the total CPU time by 8 to 10. In Figure 2, we have plotted the evolution of the CPU time relative to the number of degrees of freedom.

It has to be noted first that control of the efficiency of the multigrid methods for that nonlinear problem was not as simple as for the linear case, because among other features it depends on the contact status. So only general and qualitative comments can be given.

It is interesting to note that in nonlinear structure mechanics such as the present frictional contact analysis, the multigrid method was found to be efficient, even with medium sized discretization instead that for linear problems they are proved to be efficient only for very large sized problems. In the case of a plate with 13009 nodes,

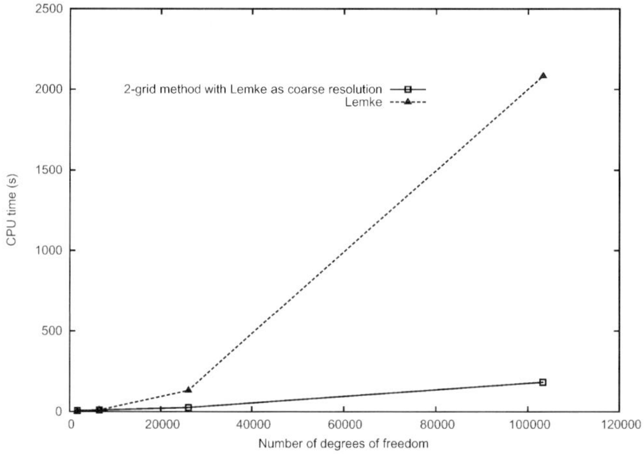

Fig. 2. CPU time versus the number of degrees of freedom for the vessel pressure plate.

the total CPU time was already divided by 8 (Table 1) when the Lemke solver is used.

Figure 2 shows that the rate of convergence does not seem to be strongly modified by the nonlinearity of the problem. In our case no theoretical results exist. In Figure 2, the rate of convergence for the two grid method is quasi linear.

We would like to make some comments on the influence of the computer processor properties on the multigrid efficiency. When we implemented first the multigrid methods, we have tested them on a Dec Alpha processor and also on a CISC processor. Of course, the CPU times are drastically reduced by using the very powerful new processors, but it turned out that multigrid methods were more efficient with the CISC processor than with the recent ones when the multigrid resolution is compared with the direct one. It should be noted that the new generation of processors are highly efficient at computation but still relatively slow in performing addressings and transfers. So simple solvers such as the Gauss–Seidel and SORP ones turn out to be more and more fast because they only do a large number of multiplications and only a small number of transfers. That kind of solvers becomes more and more competitive regarding sophisticated methods such as the Lemke's one and multigrid involving a lot of transfers. So when one refers the direct computation of the solution, it can be observed that efficiency of multigrid decreases when the processor power increases.

References

1. P. Alart and A. Curnier. A mixed formulation for frictional contact problems prone to newton like solution methods. *Comp. Meth. Appl. Mech. Eng.*, 92:353–375, 1991.

2. P. Alart and F. Lebon. Solution of frictional contact problem using ilu and coarse/fine preconditioners. *Comp. Mech.*, 16:98–105, 1995.
3. A. Boersma and P. Wriggers. An algebraic multigrid solver for finite element computations in solids mechanics. *Engng. Comp.*, 14(2):202–215, 1997.
4. A. Brandt and C. W. Cryer. Multigrid algorithms for the solutions of linear complementarity problems arising from free boundary problems. *SIAM J. Sci. Statist. Computational*, 4:655–684, 1983.
5. P. Chabrand, F. Dubois, and M. Raous. Numerical methods for solving unilateral contact problems with friction. *Math. Comp. Mod.*, 28(4-8):97–108, 1998.
6. M. Cocu, E. Pratt, and M. Raous. Formulation and approximation of quasi-static frictional contact. *Int. J. Engng. Sci.*, 34(7):783–798, 1996.
7. C. W. Cryer. The solution of a quadratic programming problem using systematic overrelaxation. *SIAM J. Control*, 9(3):385–392, 1971.
8. G. Duvaut and J. L. Lions. *Inequalities in Mechanics and Physics*. Springer-Verlag, Berlin, 1976.
9. L. Grego. *Méthodes multiniveaux pour des problèmes de contact unilatéral avec frottement*. PhD thesis, Provence University, Marseille, 1995.
10. W. Hackbusch. Multigrid method of the second kind. In *Multigrid Methods*, W. Hackbush and U. Trottenberg (Eds), Lectures Notes in Math, Vol. 960. Springer-Verlag, Berlin, 1982.
11. W. Hackbusch. *Multi-Grid Methods and Applications*. Springer, Berlin, 1985.
12. W. Hackbusch and H. D. Mittelman. On multigrid methods for variational inequalities. *Numer. Math.*, 42:65–76, 1983.
13. W. Hackbusch and U. Trottenberg (Eds). *Multigrid Methods*, Lecture Notes in Math., Vol. 960. Springer-Verlag, Berlin, 1982.
14. P. W. Hemker and P. M. De Zeeuw. Some implementations of multigrid linear system solvers. In *Multigrid Methods for Integral and Differential Equations*, D. J. Paddon and H. Holstein (Eds). Clarendon Press, 1985.
15. R. H. W. Hoppe. Multigrid algorithms for variational inequalities. *SIAM J. Numer. Anal.*, 24(5):1046–1065, 1987.
16. A. Klarbring and G. Bjorkman. The treatment of problems in contact mechanics by mathematical programming. *J. Th. Appl. Mech.*, 7(1):111–128, 1988.
17. F. Lebon M. Raous L. Grego, J.-C. Latil and I. Rosu. Multigrid methods for unilateral contact problems with friction. submitted.
18. J.-C. Latil and M. Raous. Module gyptis version 1.0: Contact unilatéral avec frottement en mécanique des structures – inéquations variationnelles. *LMA Publications*, 132, 1991. Also available in INRIA-Modulef publications.
19. F. Lebon. *Résolution numérique de problème de frottement de Coulomb. Accélération de convergence par une méthode multigrilles interne*. PhD thesis, Provence University, France, 1989.
20. F. Lebon, M. A. Ledain, M. Bellet, J. L. Chenot, and M. Raous. *Introduction aux méthodes multigrilles en mécanique des structures linéaires et non linéaires*, Vol. 3. Pluralis, 1989.
21. F. Lebon and M. Raous. Multibody contact problem including friction in structure assembling. *Comp. & Struct.*, 43(5):925–934, 1992.
22. M. A. Ledain. *Méthodes multigrilles et h-adaptation: Application au calcul de structures élastiques bidimensinnelles*. PhD thesis, ENSM, Paris, 1991.

23. C. Licht, E. Pratt, and M. Raous. Remarks on a numerical method for unilateral contact including friction. *Int. Series Numer. Math.*, 101:129–144, 1991.
24. J. Mandel. Etude algébrique d'une méthode multigrille pour quelques problemes de frontière libre. *CRAS Paris*, 298(18):469–472, 1984. Série I.
25. J. Mandel. A multilevel iterative method for symmetric, positive definite linear complementarity problems. *Appl. Math. Optim.*, 11:77–95, 1984.
26. D. J. Paddon and H. Holstein (Eds). *Multigrid Methods for Integral and Differential Equations*. Clarendon Press, 1985.
27. P. D. Panagiotopoulos. *Inequality Problems in Mechanics and Application*. Birkhauser, Boston, MA, 1985.
28. Y. Pressburger and R. Perucchio. A hierarchical two-level multigrid solver. *Comp. & Struct.*, 55(3):471–483, 1995.
29. M. Raous. Quasistatic signorini problem with coulomb friction and coupling to adhesion. In P. Wriggers and P.D. Panagiotopoulos (Eds), *New Developments on Contact Mechanics*, Vol 384, CISM, Courses and Lectures, Springer, Wien, 1999, pp. 101–178.
30. M. Raous, P. Chabrand and F. Lebon. Numerical methods for frictional contact problems and applications. *J. Th. Appl. Mech.*, 7(Suppl. to 1):111–128, 1988.
31. P. Sonneveld, P. Wesseling and P. M. De Zeeuw. Multigrid and conjugate gradient methods as convergence acceleration techniques. In *Multigrid Methods for Integral and Differential Equations*, D. J. Paddon and H. Holstein (Eds). Clarendon Press, 1985.
32. K. Stuben and U. Trottenberg. Multigrid methods: Fundamental algorithms, model problem analysis and applications. In *Multigrid Methods*, W. Hackbush and U. Trottenberg (Eds). Springer-Verlag, Berlin, 1982.

Contact Dynamics with Lagrange Multipliers

Stephan Brunßen, Stefan Hüeber and Barbara Wohlmuth

IANS, University of Stuttgart, Pfaffenwaldring 57, 70569 Stuttgart, Germany
E-mail: {brunssen,hueeber,wohlmuth}@ians.uni-stuttgart.de

Abstract. The efficient modeling of dynamical contact problems with friction is still a callenge in non-linear implicit structural analysis. We employ a mixed formulation in space with the displacement as primal variable and the contact stress as dual variable. For the discretization of the latter we use a discrete Lagrange multiplier space with biorthogonal basis functions. For the treatment of the nonlinear frictional contact conditions semi-smooth Newton methods are applied. To avoid oscillations in the Lagrange multiplier during the solution of dynamical contact problems with mass, we locally under-integrate the mass matrix. We also show the applicability of the mixed formulation to a velocity driven rigid-plastic problem.

Key words: Coulomb friction, semi-smooth Newton methods, non-oscillating Lagrange multiplier, energy conservating time integration.

1 Introduction

Dynamical contact problems with friction play an important role in many applications in structural mechanics. The numerical simulation of such problems is still a challenging task. Here, we focus on a mixed formulation in space. More precisely, the displacement enters as primal variable, and the contact stress is the dual variable. From the mathematical point of view, the dual variable stands for the Lagrange multiplier (LM) which takes care of the non-penetration constraints. In the case of two deformable bodies, we have to face the additional difficulty that, in general, the meshes are non-matching at the contact zone. This situation can be handled by mortar methods [6] which have been successfully adapted to contact problems in, e.g. [4,5,10,18,20,23]. The introduction of the dual variable guarantees that the non-penetration condition is satisfied weakly, and it gives a segment to segment approach which is of optimal approximation order if the primal-dual pairing satisfies a uniform inf-sup condition. Although optimal from the theoretical point of view, standard LM spaces result in unphysical global constraints at the contact interfaces and cannot be

Peter Wriggers and Udo Nackenhorst (eds.), IUTAM Symposium on Computational Contact Mechanics, 17–32.

handled easily from the numerical point of view. Therefore, we replace the standard discrete LM space by one reflecting the duality and use biorthogonal basis functions. As a consequence, the segment to segment coupling reduces to a node to segment one. Moreover, a local basis transformation can be carried out, see [16], and the two-body problem has the same algebraic structure as a one-body problem. Unfortunately, the combination of optimal mixed space discretizations with energy conserving time-stepping methods yields spurious oscillations in the LM. Therefore, we use a locally under-integrated mass matrix, see [11]. To solve the arising nonlinear discrete problems semi-smooth Newton methods are applied, see [14–16]. Of special interest in the numerical examples is the stabilization effect on the LM and the energy consistent time integration. To show the flexibility of the method, the method is adapted to a velocity-driven material model and applied to a rigid-plastic metal forming simulation.

2 Problem Formulation

We consider two bounded bodies given in the reference configuration by $\Omega^s, \Omega^m \subset \mathbb{R}^d$, $d = 2, 3$. The superscript s stands for the slave body and m for the master body; the choice is arbitrary but should be fixed and plays only a role for the discretization in space. On each body, we introduce the displacement $\boldsymbol{u}^i$ depending on the space and time variable as primal unknown and the contact stress $\boldsymbol{\lambda}$ on the possible contact zone as additional dual variable. The local balance of momentum is given by

$$\varrho^i \ddot{\boldsymbol{u}}^i - \mathrm{Div}(\boldsymbol{P}^i) = \boldsymbol{f}^i \quad \text{in} \quad \Omega^i \times (0, T). \tag{1}$$

Here we denote the mass density of the body Ω^i by ϱ^i and the first Piola–Kirchhoff stress tensor by $\boldsymbol{P}^i$. The vector $\boldsymbol{f}^i$ describes the prescribed body forces acting on Ω^i. To obtain a unique solution, we have to specify suitable initial and boundary conditions. The initial conditions are given by

$$\boldsymbol{u}^i(\boldsymbol{x}, 0) = \boldsymbol{0}, \qquad \dot{\boldsymbol{u}}^i(\boldsymbol{x}, 0) = \boldsymbol{v}_0^i \qquad \text{in} \quad \Omega^i, \tag{2}$$

where $\boldsymbol{v}_0^i$ denotes the initial velocity. To formulate the boundary conditions, we decompose the boundary of the reference configuration $\Gamma^i := \partial\Omega^i$ into three open disjoint subsets Γ_D^i, Γ_N^i and Γ_c^i and assume that Γ_D^s is compact embedded in $\Gamma^s \setminus \Gamma_c^s$. Now the boundary conditions are given by

$$\boldsymbol{u}^i = \boldsymbol{u}_D^i \text{ on } \Gamma_D^i \times (0, T), \qquad \boldsymbol{P}^i \boldsymbol{n}_0^i = \boldsymbol{t}_N^i \qquad \text{on} \quad \Gamma_N^i \times (0, T), \tag{3}$$

where $\boldsymbol{n}_0^i$ denotes the outward unit normal vector in the reference configuration on Γ^i and $\boldsymbol{u}_D^i$ and $\boldsymbol{t}_N^i$ are the prescribed displacements and surface forces, respectively. We remark that the boundary segment Γ_c^i refers to the contact boundary. To specify the nonlinear contact conditions, we introduce for each point $\boldsymbol{x} \in \Gamma_c^s$ the normal vector with respect to the actual configuration $\boldsymbol{n} := \boldsymbol{n}^s$. As dual variable, we use the total contact stress $\boldsymbol{\lambda} := -\boldsymbol{P}^s \boldsymbol{n}_0^s$. The normal part of this stress is denoted by $\lambda_n := \boldsymbol{\lambda n}$ and its tangential part by $\boldsymbol{\lambda}_\tau := \boldsymbol{\lambda} - \lambda_n \boldsymbol{n}$. In order to formulate the non-penetration condition of the two bodies, we use a predefined relation between the points of the possible contact zones Γ_c^i, namely a smooth mapping $R_t(\boldsymbol{x}) : \Gamma_c^s \to \Gamma_c^m$ satisfying $R_t(\Gamma_c^s) \subset \Gamma_c^m$ for all $t \in (0, T)$. We assume that the mapping R_t is well defined and that $\boldsymbol{\varphi}^s(\boldsymbol{x}, t) - \boldsymbol{\varphi}^m(R_t(\boldsymbol{x}), t)$, where $\boldsymbol{\varphi}^i(\boldsymbol{y}, t) := \boldsymbol{y} + \boldsymbol{u}^i(\boldsymbol{y}, t)$, is parallel to the actual normal $\boldsymbol{n}(\boldsymbol{x}, t)$ on $\Gamma_{c,t}^s$ at the point $\boldsymbol{\varphi}(\boldsymbol{x}, t)$ and of shortest length. Defining the actual gap at a point $\boldsymbol{x} \in \Gamma_c^s$ by

$$g(\boldsymbol{x}, t) := \big(\boldsymbol{\varphi}^s(\boldsymbol{x}, t) - \boldsymbol{\varphi}^m(R_t(\boldsymbol{x}), t)\big)\boldsymbol{n}(\boldsymbol{x}, t)$$

the contact conditions on Γ_c^s are given by

$$g(\boldsymbol{x}, t) \le 0, \quad \lambda_n(\boldsymbol{x}, t) \ge 0, \quad g(\boldsymbol{x}, t)\lambda_n(\boldsymbol{x}, t) = 0 \tag{4}$$

for all $t \in (0, T)$. The boundary stress in normal direction has to satisfy the condition

$$\boldsymbol{\lambda} = -\boldsymbol{P}^s(\boldsymbol{x}, t)\boldsymbol{n}_0^s = \boldsymbol{P}^m(R_t(\boldsymbol{x}), t)\boldsymbol{n}_0^m \quad \text{on} \quad \Gamma_c^s \tag{5}$$

for all $t \in (0, T)$. Furthermore, we introduce the jump of the displacement by

$$[\boldsymbol{u}(\boldsymbol{x}, t)] := \boldsymbol{u}^s(\boldsymbol{x}, t) - \boldsymbol{u}^m(R_t(\boldsymbol{x}), t) \quad \text{for} \quad (\boldsymbol{x}, t) \in \Gamma_c^s \times (0, T)$$

and define its normal part $[\boldsymbol{u}]_n := [\boldsymbol{u}]\boldsymbol{n}$ and its tangential part $[\boldsymbol{u}]_\tau := [\boldsymbol{u}] - [\boldsymbol{u}]_n\boldsymbol{n}$. In terms of the Euclidean norm $\|\cdot\|$, we can write Coulomb's friction law for a given coefficient of friction $\mathfrak{F} \ge 0$ as

$$\begin{cases} \|\boldsymbol{\lambda}_\tau\| \le \mathfrak{F}|\lambda_n|, \\ \|\boldsymbol{\lambda}_\tau\| < \mathfrak{F}|\lambda_n| \Rightarrow [\dot{\boldsymbol{u}}]_\tau = 0, \\ \|\boldsymbol{\lambda}_\tau\| = \mathfrak{F}|\lambda_n| \Rightarrow \exists\beta : \boldsymbol{\lambda}_\tau = \beta^2[\dot{\boldsymbol{u}}]_\tau, \end{cases} \quad \text{on} \quad \Gamma_c^s \times (0, T). \tag{6}$$

Next, we state the variational formulation of the dynamical contact problem consisting of the field equation (1), the boundary conditions (3) and the initial conditions (2) together with the contact conditions (4) and (6). We start with the introduction of the functional spaces we are going to use. The vector valued spaces $\boldsymbol{V}^i$, $i = s, m$, contain the admissible displacement fields for each subdomain Ω^i

$$\boldsymbol{V}^i := \left\{\boldsymbol{u}^i \in [H^1(\Omega^i)]^d \, : \, \boldsymbol{u}^i|_{\Gamma_D^i} = \boldsymbol{u}_D^i\right\},$$

while the corresponding product space is given by $\boldsymbol{V} := \boldsymbol{V}^s \times \boldsymbol{V}^m$. For the test space, we have to use functions satisfying zero Dirichlet boundary conditions on Γ_D^i, denoted by $\boldsymbol{V}_0 := \boldsymbol{V}_0^s \times \boldsymbol{V}_0^m$. Multiplying equation (1) with a test function $\boldsymbol{v}^i \in \boldsymbol{V}_0^i$, integration over Ω^i, applying partial integration and summation over $i = s, m$, together with the boundary conditions (3) and (5) leads to

$$m(\ddot{\boldsymbol{u}}, \boldsymbol{v}) + a(\boldsymbol{u}, \boldsymbol{v}) + \langle \boldsymbol{\lambda}, [\boldsymbol{v}] \rangle = f(\boldsymbol{v}), \tag{7}$$

where $\boldsymbol{v} = (\boldsymbol{v}^s, \boldsymbol{v}^m)$, $\langle \cdot, \cdot \rangle$ stands for the duality pairing on Γ_c^s, and the bilinear and linear forms are given by

$$m(\boldsymbol{u}, \boldsymbol{v}) := \sum_{i=s,m} \int_{\Omega^i} \varrho^i \boldsymbol{u}^i \boldsymbol{v}^i \,\mathrm{d}\boldsymbol{x}, \quad a(\boldsymbol{u}, \boldsymbol{v}) := \sum_{i=s,m} \int_{\Omega^i} \boldsymbol{P}^i : \nabla \boldsymbol{v}^i \,\mathrm{d}\boldsymbol{x},$$

$$f(\boldsymbol{v}) := \sum_{i=s,m} \int_{\Omega^i} \boldsymbol{f}^i \boldsymbol{v}^i \,\mathrm{d}\boldsymbol{x} + \sum_{i=s,m} \int_{\Gamma_N^i} \boldsymbol{t}_N^i \boldsymbol{v}^i \,\mathrm{d}\boldsymbol{s}.$$

To formulate the weak form of the contact conditions (4) and (6), we define the LM space $\boldsymbol{M}$ as the dual space of the trace space $\boldsymbol{W}$ of $\boldsymbol{V}_0^s$ restricted to Γ_c^s. The LM $\boldsymbol{\lambda} \in \boldsymbol{M}(\boldsymbol{\lambda})$ represents the contact stress, where $\boldsymbol{M}(\boldsymbol{\lambda})$ is a convex subspace of $\boldsymbol{M}$:

$$\boldsymbol{M}(\boldsymbol{\lambda}) := \left\{ \boldsymbol{\mu} \in \boldsymbol{M} \,:\, \langle \boldsymbol{\mu}, \boldsymbol{\eta} \rangle \leq \langle \mathfrak{F} \lambda_n, \|\boldsymbol{\eta}_\tau\| \rangle, \;\; \boldsymbol{\eta} \in \boldsymbol{W} \;\text{ with }\; \eta_n \leq 0 \right\}. \tag{8}$$

Then, the weak form of the mechanical contact conditions is given by, see [11,14,16],

$$\langle g, \mu_n - \lambda_n \rangle + \langle [\dot{\boldsymbol{u}}]_\tau, \boldsymbol{\mu}_\tau - \boldsymbol{\lambda}_\tau \rangle \leq 0, \qquad \boldsymbol{\mu} \in \boldsymbol{M}(\boldsymbol{\lambda}). \tag{9}$$

Summarizing (7) and (9), we end up with the variational formulation of the dynamical contact problem: find $(\boldsymbol{u}, \boldsymbol{\lambda}) \in \boldsymbol{V} \times \boldsymbol{M}(\boldsymbol{\lambda})$ such that for all $t \in (0, T)$

$$\begin{aligned} m(\ddot{\boldsymbol{u}}, \boldsymbol{v}) + a(\boldsymbol{u}, \boldsymbol{v}) + \langle [\boldsymbol{v}], \boldsymbol{\lambda} \rangle &= f(\boldsymbol{v}), && \boldsymbol{v} \in \boldsymbol{V}_0, \\ \langle g(\boldsymbol{u}), \mu_n - \lambda_n \rangle + \langle [\dot{\boldsymbol{u}}]_\tau, \boldsymbol{\mu}_\tau - \boldsymbol{\lambda}_\tau \rangle &\leq 0, && \boldsymbol{\mu} \in \boldsymbol{M}(\boldsymbol{\lambda}), \end{aligned} \tag{10}$$

holds together with the weak form of the initial conditions (2), namely

$$(\boldsymbol{u}(\cdot, 0), \boldsymbol{v}) = 0, \qquad (\dot{\boldsymbol{u}}(\cdot, 0), \boldsymbol{v}) = (\boldsymbol{v}_0, \boldsymbol{v}), \qquad \boldsymbol{v} \in \boldsymbol{V}_0,$$

where $(\cdot, \cdot)$ denotes the standard L^2-scalar product on $\Omega := \Omega^s \cup \Omega^m$.

3 Discretization in Space and Time

The quality of the numerical simulation highly depends on the applied discretization scheme. More precisely, time and space discretization have to go hand in hand to obtain a stable scheme. For the space discretization, we apply techniques well-known from the linear mortar setting [6]. In particular the pair of primal and dual discrete spaces has to satisfy a uniform inf-sup condition. In time, we would like to work with a scheme which shows no numerical dissipation. Unfortunately, combining mixed formulations in space with energy preserving schemes in time yields high numerical oscillations in the dual variable. Here, we apply a very simple modification of the mass matrix which fits into the framework of under-integration. The time discretization of the weak form (10) is based on the midpoint rule. We define $t^k := k\,\Delta t$, $k = 0, \ldots, N_T$, where the time step size is given by $\Delta t := T/N_T$. Furthermore, we denote by $\boldsymbol{u}^k(\boldsymbol{x}) \approx \boldsymbol{u}(\boldsymbol{x}, t^k)$ the approximation of the displacement field at time t^k. Defining $\Delta\boldsymbol{u}^k := \boldsymbol{u}^{k+1} - \boldsymbol{u}^k$, $\Delta\dot{\boldsymbol{u}}^k := \dot{\boldsymbol{u}}^{k+1} - \dot{\boldsymbol{u}}^k$ the relations based on the midpoint rule are given by

$$\ddot{\boldsymbol{u}}^{k+1/2} = \frac{\Delta\dot{\boldsymbol{u}}^k}{\Delta t}, \quad \dot{\boldsymbol{u}}^{k+1/2} = \frac{\dot{\boldsymbol{u}}^{k+1} + \dot{\boldsymbol{u}}^k}{2} = \frac{\Delta\boldsymbol{u}^k}{\Delta t}, \quad \boldsymbol{u}^{k+1/2} = \frac{\boldsymbol{u}^{k+1} + \boldsymbol{u}^k}{2}.$$

The evaluation of the first equation of the variational formulation (10) at time $t^{k+1/2}$ leads for all $\boldsymbol{v} \in \boldsymbol{V}_0$ to

$$f^{k+1/2}(\boldsymbol{v}) = \frac{2}{\Delta t^2} m(\Delta\boldsymbol{u}^k, \boldsymbol{v}) + \frac{1}{2} a(\Delta\boldsymbol{u}^k, \boldsymbol{v}) + \langle [\boldsymbol{v}], \boldsymbol{\lambda}^{k+1/2} \rangle, \tag{11}$$

$$f^{k+1/2}(\boldsymbol{v}) := f_{k+1/2}(\boldsymbol{v}) + \frac{2}{\Delta t} m(\dot{\boldsymbol{u}}^k, \boldsymbol{v}) - a(\boldsymbol{u}^k, \boldsymbol{v}), \tag{12}$$

where $f_{k+1/2}(\boldsymbol{v})$ is $f(\boldsymbol{v})$ evaluated at time $t^{k+1/2} := (k + 1/2)\Delta t$.

For each time interval $[t^k, t^{k+1}]$, we linearize the contact conditions (4) in normal direction and end up with

$$[\Delta u]_n^k - g^k \leq 0, \qquad \lambda_n^{k+1/2} \geq 0, \qquad \lambda_n^{k+1/2}([\Delta u]_n^k - g^k) = 0, \tag{13}$$

where the gap g^k at the time t^k is defined by

$$g^k(\boldsymbol{x}) := -\big(\boldsymbol{\varphi}^{s,k}(\boldsymbol{x}) - \boldsymbol{\varphi}^{m,k}(R_{t^k}(\boldsymbol{x}))\big)\boldsymbol{n}^k(\boldsymbol{x}).$$

Here we denote by $\boldsymbol{\varphi}^{\star,k}(\boldsymbol{x}) = \boldsymbol{x} + \boldsymbol{u}^{\star,k}(\boldsymbol{x}) \approx \boldsymbol{x} + \boldsymbol{u}^\star(\boldsymbol{x}, t^k)$ the approximation of the deformation $\boldsymbol{\varphi}^\star$ at time t^k, and $\boldsymbol{n}^k(\boldsymbol{x}) \approx \boldsymbol{n}(\boldsymbol{x}, t^k)$ is the outward unit normal vector to the approximated actual configuration at time t^k. The jump $[\Delta\boldsymbol{u}]^k$ is given by $[\Delta\boldsymbol{u}^k] := \Delta\boldsymbol{u}^{s,k}(\boldsymbol{x}) - \Delta\boldsymbol{u}^{m,k}(R_{t^k}(\boldsymbol{x}))$. Furthermore we set $\lambda_n^{k+1/2} := \boldsymbol{\lambda}^{k+1/2}\boldsymbol{n}^k$ and $[\Delta u]_n^k := [\Delta\boldsymbol{u}]^k\boldsymbol{n}^k$. The corresponding tangential parts are given by $\boldsymbol{\lambda}_\tau^{k+1/2} := \boldsymbol{\lambda}^{k+1/2} - \lambda_n^{k+1/2}\boldsymbol{n}^k$ and $[\Delta\boldsymbol{u}]_\tau^k := [\Delta\boldsymbol{u}]^k - [\Delta u]_n^k\boldsymbol{n}^k$. Defining the bilinear form $b(\cdot,\cdot) : \boldsymbol{W} \times \boldsymbol{M} \to \mathbb{R}$ by

$$b(\boldsymbol{v}, \boldsymbol{\mu}) := b_n(\boldsymbol{v}, \boldsymbol{\mu}) + b_\tau(\boldsymbol{v}, \boldsymbol{\mu}) := \langle [v]_n, \boldsymbol{\mu}_n \rangle + \langle [\boldsymbol{v}]_\tau, \boldsymbol{\mu}_\tau \rangle,$$

the time discretized version of $(10)_2$ using the linearized contact conditions (13) reads as

$$b_n(\Delta\boldsymbol{u}^k, \boldsymbol{\mu} - \boldsymbol{\lambda}^{k+1/2}) + b_\tau(\Delta\boldsymbol{u}^k, \boldsymbol{\mu} - \boldsymbol{\lambda}^{k+1/2}) \leq \langle g^k, \mu_n - \lambda_n^{k+1/2} \rangle. \tag{14}$$

Summarizing (11) and (14), the time discretized variational formulation at each time step $t^{k+1/2}$ is given by: find $(\Delta\boldsymbol{u}^k, \boldsymbol{\lambda}^{k+1/2}) \in \mathbf{V} \times \boldsymbol{M}(\boldsymbol{\lambda}^{k+1/2})$

$$\begin{aligned}
&\frac{2}{\Delta t^2} m(\Delta \boldsymbol{u}^k, \boldsymbol{v}) + \frac{1}{2} a(\Delta \boldsymbol{u}^k, \boldsymbol{v}) + b(\boldsymbol{v}, \boldsymbol{\lambda}^{k+1/2}) = f^{k+1/2}(\boldsymbol{v}), \\
&b_n(\Delta \boldsymbol{u}^k, \boldsymbol{\mu} - \boldsymbol{\lambda}^{k+1/2}) + b_\tau(\Delta \boldsymbol{u}^k, \boldsymbol{\mu} - \boldsymbol{\lambda}^{k+1/2}) \leq \langle g^k, \mu_n - \lambda_n^{k+1/2} \rangle,
\end{aligned} \tag{15}$$

for all $(\boldsymbol{v}, \boldsymbol{\mu}) \in \boldsymbol{V}_0 \times \boldsymbol{M}(\boldsymbol{\lambda}^{k+1/2})$.

For the space discretization of (15), we use standard conforming finite elements of lowest order on Ω^i. The discrete conforming spaces of $\boldsymbol{V}$ and $\boldsymbol{V}_0$ are denoted by $\boldsymbol{V}^h$ and $\boldsymbol{V}_0^h$, respectively. As it is standard in the mortar framework the LM space inherits its $(d-1)$-dimensional mesh on Γ_c^s from the d-dimensional triangulation on the slave side Γ_c^s. We separate the nodes of the finite element mesh into three disjoint sets. The set $\mathcal{S}$ contains all nodes on the discrete counterpart of Γ_c^s, the set $\mathcal{M}$ all nodes on the discrete counterpart of Γ_c^m and the set $\mathcal{N}$ all remaining nodes. Each function $\boldsymbol{v}_h \in \boldsymbol{V}^h$ can now be written in the form $\boldsymbol{v}_h = \sum_{p \in \mathcal{N} \cup \mathcal{M} \cup \mathcal{S}} \boldsymbol{v}_p \phi_p$, where $\boldsymbol{v}_p \in \mathbb{R}^d$ are the nodal values. Defining by $\boldsymbol{n}_p$ the normal vector at the node $p \in \mathcal{S}$, the normal and the tangential part of the nodal value $\boldsymbol{v}_p$ are given by $v_{pn} := \boldsymbol{v}_p \boldsymbol{n}_p$ and $\boldsymbol{v}_{p\tau} := \boldsymbol{v}_p - v_{pn} \boldsymbol{n}_p$ respectively. We then can express the normal and the tangential part of the discrete function $\boldsymbol{v}_h$ on the contact interface Γ_c^s by

$$v_{hn} := \sum_{p \in \mathcal{S}} v_{pn} \phi_p, \qquad \boldsymbol{v}_{h\tau} := \sum_{p \in \mathcal{S}} \boldsymbol{v}_{p\tau} \phi_p, \qquad p \in \mathcal{S}, \tag{16}$$

where ϕ_p denotes the standard finite element shape function of lowest order. For the discretization of the dual variable which enters as LM in the mixed formulation, we use low order biorthogonal shape functions ψ_p satisfying

$$\int_{\Gamma_c^s} \phi_p \psi_q \, \mathrm{d}s = \delta_{pq} \int_{\Gamma_c^s} \phi_p \, \mathrm{d}s, \tag{17}$$

see, e.g., [22]. A function $\boldsymbol{\mu}_h \in \boldsymbol{M}^h$ can be written as $\boldsymbol{\mu}_h = \sum_{p \in \mathcal{S}} \boldsymbol{\mu}_p \psi_p$, $\boldsymbol{\mu}_p \in \mathbb{R}^d$; its normal and tangential parts are defined according to (16). Now the discrete convex set $\boldsymbol{M}^h(\boldsymbol{\lambda}_h)$ corresponding to (8) is given by

$$\boldsymbol{M}^h(\boldsymbol{\lambda}_h) = \left\{ \boldsymbol{\mu}_h = \sum_{p \in \mathcal{S}} \boldsymbol{\mu}_p \psi_p \in \boldsymbol{M}^h \; : \; \mu_{pn} \geq 0, \; \|\boldsymbol{\mu}_{p\tau}\| \leq \mathfrak{F} \lambda_{pn}, \; p \in \mathcal{S} \right\}.$$

For general inf-sup stable pairings, the contact conditions results in a segment to segment approach, where the inequality constraints can only be resolved by some global optimization procedure on the contact boundary. This is not the case, if we work with a locally defined biorthogonal basis of the LM space. Then the segment to segment approach is algebraically equivalent to a node to segment approach, and the inequality constraints decouple. Additionally, we can rewrite by a local preprocess a two body problem into a one body problem and the same solver can be applied. Mathematically, the preprocess is a basis transformation, physically we glue master and slave side together such that the two bodies form a composite material. In the new constrained basis $\hat{\boldsymbol{\phi}}$, a linear combination of the nodal basis functions on the slave side is attached to the nodal basis functions of the master side. The basis functions in $\hat{\boldsymbol{\phi}}$ on the slave side describe the relative displacement between the two bodies, i.e., the jump in the displacement, denoted by $[\boldsymbol{u}_h]$. To describe this transformation, we define the entries of the coupling matrices $\mathbf{D}$ and $\mathbf{B}$ between the finite element basis functions ϕ_q, $q \in \mathcal{S}, \mathcal{M}$, and the dual basis functions for the LM ψ_p by

$$\begin{aligned} \mathbf{D}[p,q] &:= b(\phi_q, \psi_p)\,\mathrm{Id}_d =: D_p\,\mathrm{Id}_d, \qquad && p,q \in \mathcal{S}, \quad D_p \in \mathbb{R}, \\ \mathbf{B}[p,q] &:= b(\phi_q, \psi_p)\,\mathrm{Id}_d, && p \in \mathcal{S}, \quad q \in \mathcal{M}, \end{aligned}$$

where Id_d denotes the identity matrix in $\mathbb{R}^{d\times d}$. Due to the biorthogonality (17), the matrix $\mathbf{D}$ is diagonal. Defining $\hat{\mathbf{B}} := -\mathbf{D}^{-1}\mathbf{B}$, the constrained basis $\hat{\boldsymbol{\phi}}$ of $\boldsymbol{V}^h$ is obtained from $\boldsymbol{\phi}$ by the transformation

$$\hat{\boldsymbol{\phi}} := \begin{pmatrix} \hat{\phi}_{\mathcal{N}} \\ \hat{\phi}_{\mathcal{M}} \\ \hat{\phi}_{\mathcal{S}} \end{pmatrix} := \begin{pmatrix} \mathrm{Id} & \mathbf{0} & \mathbf{0} \\ \mathbf{0} & \mathrm{Id} & \hat{\mathbf{B}}^\top \\ \mathbf{0} & \mathbf{0} & \mathrm{Id} \end{pmatrix} \begin{pmatrix} \phi_{\mathcal{N}} \\ \phi_{\mathcal{M}} \\ \phi_{\mathcal{S}} \end{pmatrix} =: \mathbf{Q}\boldsymbol{\phi}\,. \tag{18}$$

Thus only basis functions associated with a node $p \in \mathcal{M}$ are modified. The relation (18) is defined such that we have

$$b(\hat{\boldsymbol{u}}_q \hat{\phi}_q, \boldsymbol{\mu}_p \psi_p) = 0\,, \quad \hat{\boldsymbol{u}}_q, \boldsymbol{\mu}_p \in \mathbb{R}^d, \quad p \in \mathcal{S}\,, \ q \in \mathcal{M}\,.$$

This equation states that the jump of the finite element basis functions on the master side vanishes in a weak sense. Therefore the matrix $\mathbf{B}$ in the new constrained basis is equal to zero. Let $\mathbf{M}$ and $\mathbf{K}$ denote the mass matrix and the stiffness matrix resulting from the bilinear forms $m(\cdot,\cdot)$ and $a(\cdot,\cdot)$ with respect to the nodal finite element basis. The corresponding matrices $\hat{\mathbf{M}}$ and $\hat{\mathbf{K}}$ with respect to the new constrained basis are carried out locally by $\hat{\mathbf{M}} = \mathbf{Q}\mathbf{M}\mathbf{Q}^\top$ and $\hat{\mathbf{K}} = \mathbf{Q}\mathbf{K}\mathbf{Q}^\top$. Denoting by $\mathbf{f}$ the corresponding vector for the right hand side (12) with respect to the nodal basis $\boldsymbol{\phi}$, the vector with respect to $\hat{\boldsymbol{\phi}}$ is given by $\hat{\mathbf{f}} = \mathbf{Q}\mathbf{f}$. Using for the vector representation of the LM the same symbol $\boldsymbol{\lambda}$ as for the continuous function, the algebraic form of the discrete version of $(15)_1$ reads as

$$\left(\frac{2}{\Delta t^2}\hat{\mathbf{M}} + \frac{1}{2}\hat{\mathbf{K}}\right)\Delta\hat{\mathbf{u}}^k + \begin{pmatrix} \mathbf{0} \\ \mathbf{D} \end{pmatrix}\boldsymbol{\lambda}^{k+1/2} = \hat{\mathbf{f}}^{k+1/2}. \tag{19}$$

We remark that the zero block in the coupling matrix between the finite element shape functions in the constrained basis and the dual LM refers to the nodes in the set $\mathcal{N} \cup \mathcal{M}$.

Defining $g_p^k := D_p^{-1}\int_{\Gamma_c^s} g^k \psi_p \,\mathrm{d}s$ and the scaled values $\lambda_{pns}^{k+1/2} := D_p \lambda_{pn}^{k+1/2}$ and $\boldsymbol{\lambda}_{p\tau s}^{k+1/2} := D_p \boldsymbol{\lambda}_{p\tau}^{k+1/2} - \lambda_{pns}^{k+1/2}\boldsymbol{n}_p^k$, the inequality in $(15)_2$ is equivalent to the constraints

$$\Delta\hat{u}^k_{pn} \le g^k_p, \quad \lambda^{k+1/2}_{pns} \ge 0, \quad \lambda^{k+1/2}_{pns}(\Delta\hat{u}^k_{pn} - g^k_p) = 0 \tag{20}$$

and

$$\|\boldsymbol{\lambda}^{k+1/2}_{p\tau s}\| \le \mathfrak{F}\lambda^{k+1/2}_{pns}, \quad \Delta\hat{\boldsymbol{u}}^k_{p\tau} = \beta^2\boldsymbol{\lambda}^{k+1/2}_{p\tau s}, \quad \Delta\hat{\boldsymbol{u}}^k_{p\tau}\left(\|\boldsymbol{\lambda}^{k+1/2}_{p\tau s}\| - \mathfrak{F}\lambda^{k+1/2}_{pns}\right) = 0 \tag{21}$$

for all $p \in \mathcal{S}$ with any $\beta \in \mathbb{R}$. We note that the scaling of the normal and the tangential component of the LM with D_p leads to a better convergence behavior of the applied algorithm. This comes from the fact that the natural norms for primal and dual variable have different scaling factors.

Remark 1. To fulfill the contact conditions correctly, we evaluate the contact integrals represented by the bilinear form $b(\cdot,\cdot)$ and the discrete gap g^k_p on the actual configuration at time t^k. Therefore the matrices $\mathbf{D}$ and $\mathbf{B}$ and also the basis transformation (18) depend on t^k and the values for the LM are defined on the actual configuration. For ease of notation, we have neglected all the time indices in these quantities before.

Remark 2. It is well known that due to the use of the midpoint scheme, spurious oscillations in the LM occur. To avoid these oscillations, we remove the mass from the nodes in $\mathcal{S}$. The idea of massless contact nodes has first been described in [17], where the modified mass matrix has to be obtained from the solution of a global minimization problem. Here, we use a simple under-integration scheme in the local neighborhood of the contact boundary. This approach is extremely cheap to realize and still provides an optimal discretization, see [12]. A further approach based on the Newmark scheme to overcome these oscillations can be found in [8].

As explained in [21], the applied midpoint scheme conserves the total energy. To get an energy conserving discrete formulation of the contact, we use the approaches given in [7,19]. Following [19], we modify the contact constraints in normal direction and replace (20) by the constraints

$$\begin{aligned} g^k_p > 0 \quad &\Rightarrow \quad \lambda^{k+1/2}_{pns} = 0, \\ g^k_p \le 0 \quad &\Rightarrow \quad \lambda^{k+1/2}_{pns} \ge 0, \quad \Delta\hat{u}^k_{pn} \le g^k_p, \quad \lambda^{k+1/2}_{pns}(\Delta\hat{u}^k_{pn} - g^k_p) = 0. \end{aligned} \tag{22}$$

Defining the kinetic energy and the internal stress energy at time t^k by

$$\mathcal{K}_k := \frac{1}{2}\sum_{i=s,m}\int_{\Omega^i} \varrho^i \|\dot{\boldsymbol{u}}^{ik}_h\|^2 \, \mathrm{d}\boldsymbol{s}, \qquad \mathcal{E}_k := \frac{1}{2}\sum_{i=s,m}\int_{\Omega^i} \boldsymbol{P}^{ik} : \nabla\boldsymbol{u}^{ik}_h \, \mathrm{d}\boldsymbol{s},$$

the total energy $\mathcal{K}_k + \mathcal{E}_k + \mathcal{E}^c_k$ is constant in time, where the contact work $\mathcal{E}^c_k$ performed up to time t^k is given by

$$\mathcal{E}^c_k := \sum_{j=0}^{k-1}\int_{\Gamma^s_c} \boldsymbol{\lambda}^{j+1/2}_h[\Delta\boldsymbol{u}^j_h] \, \mathrm{d}\boldsymbol{s} = \sum_{j=0}^{k-1}\sum_{p\in\mathcal{S}} \boldsymbol{\lambda}^{j+1/2}_{p\tau s}\Delta\hat{\boldsymbol{u}}^j_{p\tau}.$$

4 Numerical Algorithm

One of the challenges in the numerical simulation of contact problems is to design fast and robust iterative solvers. To handle the inequality efficiently, we rewrite it as an equivalent nonlinear system, [1, 13], and apply a semi-smooth Newton type algorithm which can be implemented as a primal-dual active set strategy. For details of the algorithm we refer to [16] for the frictionless case and to [14, 15] for the frictional case. In these articles, the algorithms are presented for the static case. Due to the different contact conditions in normal direction (22), a small modification of the algorithm compared to [16] has to be done. Here we shortly present the main ideas from [15] for the frictional approach for the case $d = 3$. For some constants $c_n, c_\tau > 0$, we define the nonlinear complementarity function $C(\cdot, \cdot)$ by

$$C(\boldsymbol{\lambda}_p^{k+1/2}, \Delta\hat{\boldsymbol{u}}_p^k) := \begin{pmatrix} \lambda_{pns}^{k+1/2} - \max\{0,\ \lambda_{pns}^{k+1/2} + c_n \tilde{g}_p^k\} \\ \max\{\mathfrak{F}(\lambda_{pns}^{k+1/2} + c_n \tilde{g}_p^k),\ \|\boldsymbol{\lambda}_{p\tau s}^{k+1/2} + c_\tau \Delta\hat{\boldsymbol{u}}_{p\tau}^k\|\}\boldsymbol{\lambda}_{p\tau s}^{k+1/2} \\ -\max\{0,\ \mathfrak{F}(\lambda_{pns}^{k+1/2} + c_n \tilde{g}_p^k)\}(\boldsymbol{\lambda}_{p\tau s}^{k+1/2} + c_\tau \Delta\hat{\boldsymbol{u}}_{p\tau}^k) \end{pmatrix},$$

where we used the notation

$$\tilde{g}_p^k := \begin{cases} -\frac{\lambda_{pns}^{k+1/2}}{c_n} - g_p^k & \text{if} \quad g_p^k > 0, \\ \Delta\hat{u}_{pn}^k & \text{if} \quad g_p^k \le 0. \end{cases}$$

One can show that the contact conditions (22) and (21) are equivalent to the condition $C(\boldsymbol{\lambda}_p^{k+1/2}, \Delta\hat{\boldsymbol{u}}_p^k) = \mathbf{0}$ for all $p \in \mathcal{S}$. Applying a semi-smooth Newton method to this condition gives rise to an iterative solution strategy. To simplify the notation, we neglect the time index k in the rest of this section. Starting from the previous iterates $(\boldsymbol{\lambda}_h^{l-1}, \Delta\hat{\boldsymbol{u}}_h^{l-1})$ (the index l now refers to the Newton iterate), we get in each Newton step the disjoint decomposition of $\mathcal{S}$ into

- $\mathcal{I}_n^l := \{p \in \mathcal{S} : b_p^{l-1} \le 0\}$ (estimate for the nodes not in contact),
- $\mathcal{I}_\tau^l := \{p \in \mathcal{S} : \|\boldsymbol{\lambda}_{p\tau s}^{l-1} + c_\tau \Delta\hat{\boldsymbol{u}}_{p\tau}^{l-1}\| - b_p^{l-1} < 0\}$ (estimated sticky nodes),
- $\mathcal{A}_{\tau n}^l := \{p \in \mathcal{S} : \|\boldsymbol{\lambda}_{p\tau s}^{l-1} + c_\tau \Delta\hat{\boldsymbol{u}}_{p\tau}^{l-1}\| \ge b_p^{l-1} > 0\}$ (estimated slippy nodes),

where we used the abbreviation $b_p^{l-1} := \mathfrak{F}(\lambda_{pns}^{k+1/2} + c_n \tilde{g}_p^k)$. Depending on this decomposition, we compute the next iterate $(\boldsymbol{\lambda}_h^l, \Delta\hat{\boldsymbol{u}}_h^l)$ by solving a linear system where we enforce the Neumann boundary condition $\boldsymbol{\lambda}_p^l = 0$ for the nodes $p \in \mathcal{I}_n^l$, the Dirichlet condition $\Delta\hat{u}_{pn}^l = 0$ in normal direction and the Robin condition $\Delta\hat{\boldsymbol{u}}_p^l + (\mathfrak{F}\Delta\hat{\boldsymbol{u}}_{p\tau}^l / b_p^{l-1})\lambda_{pns}^l = \Delta\hat{\boldsymbol{u}}_{p\tau}^{l-1}$ in tangential direction for the nodes $p \in \mathcal{I}_\tau^l$. On the set $\mathcal{A}_{\tau n}^l$, we impose in normal direction the Dirichlet condition $\Delta\hat{u}_{pn}^l = 0$ and in tangential direction the Robin condition $-\boldsymbol{\lambda}_{p\tau s}^l + L_p^{l-1}\Delta\hat{\boldsymbol{u}}_{p\tau}^l + \mathfrak{F}\boldsymbol{v}_p^{l-1}\lambda_{pns}^l =$

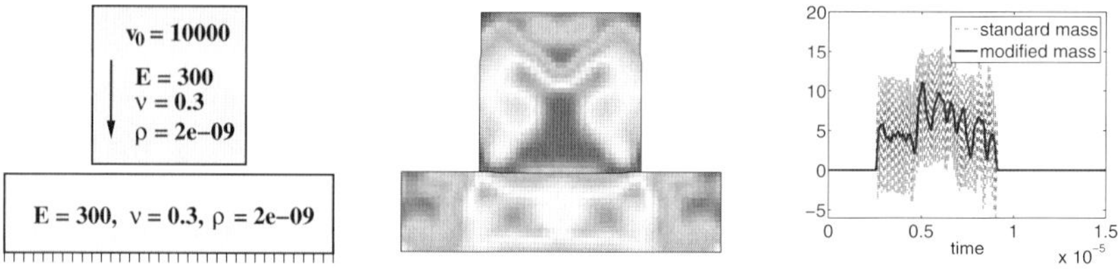

Fig. 1. Problem Setting, effective stress and LM for the frictional contact problem.

$\boldsymbol{r}_p^{l-1} + b_p^{l-1}\boldsymbol{v}_p^{l-1}$. For the definition of the matrix $L_p^{l-1} \in \mathbb{R}^{2\times 2}$ and the vectors $\boldsymbol{v}_p^{l-1}, \boldsymbol{r}_p^{l-1} \in \mathbb{R}^2$ in terms of the previous iterates $(\boldsymbol{\lambda}_h^{l-1},\ \Delta\hat{\boldsymbol{u}}_h^{l-1})$ we refer to [15].

The use of dual LMs allows us to eliminate the dual variable $\boldsymbol{\lambda}$ locally in an efficient way such that only a linear system in the primal variable has to be solved. For details we refer to the cited articles. Numerical examples given in [11] indicate that the inexact Newton approach leads locally to a superlinear convergence rate. We note that we do not have to compute the Jacobian, but have only to switch boundary nodes from Dirichlet to Neumann or to Robin and vice versa.

5 Numerical Examples for Linear Elasticity

In this section, we present two numerical examples. The implementation is based on the finite element toolbox UG, see [3]. For both examples we use linear elasticity. The stress tensor is given by $\boldsymbol{P}^i := \lambda^i \operatorname{tr}(\boldsymbol{\varepsilon}^i)\,\mathrm{Id}_d + 2\mu^i \boldsymbol{\varepsilon}^i$, with the strain tensor $\boldsymbol{\varepsilon}^i = \frac{1}{2}\left(\nabla \boldsymbol{u}^i + (\nabla \boldsymbol{u}^i)^\top\right)$ and the Lamé parameters $\lambda^i := (E^i \nu^i)/(1-(\nu^i)^2)$ and $\mu^i := E^i/(2(1+\nu^i))$ with Young's modulus $E^i > 0$ and Poisson ratio $\nu^i \in (0, 0.5)$. Moreover, tr denotes the matrix trace operator. For the performance of the applied algorithms and its combination with inexact strategies we refer to [14–16].

The first example shows the advantage using the modified mass matrix (see Remark 2). The detailed setting and an extensive numerical discussion about this two-dimensional example can be found in [11]. The problem setting is presented in the left of Figure 1. The upper block moves downwards and impinges the lower block being at rest and fixed at the bottom. The upper block is assumed to be the slave body Ω^s. The middle picture of Figure 1 shows the effective von Mises stress on the distorted blocks during contact. In the right, the normal LM at the midpoint of the lower side of the upper block is presented over time, both for the standard mass matrix and the modified one. In contrast to the use of the standard mass matrix, no oscillations occur using its modified version. In Figure 2, the energy evolution is shown for the frictional (Coulomb friction with $\mathfrak{F} = 0.1$) (left and middle) and the frictionless (right) contact problem. While the total energy is conserved for the frictionless case, we observe dissipation for the frictional problem. However, the middle picture shows that energy is conserved apart from the contact work $\mathcal{E}^c$.

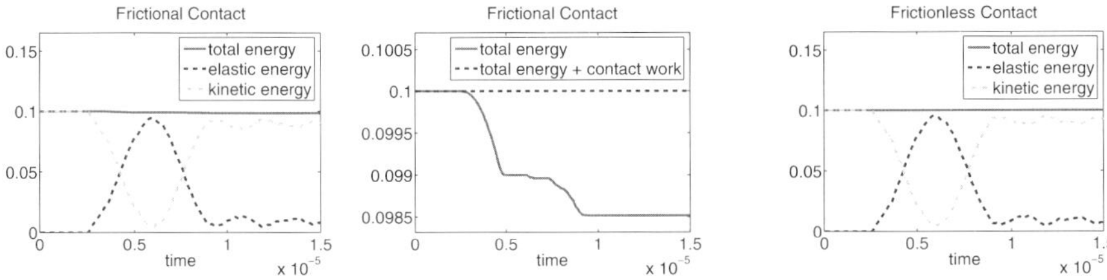

Fig. 2. Energy and contact work for the problem in Figure 1 with and without friction.

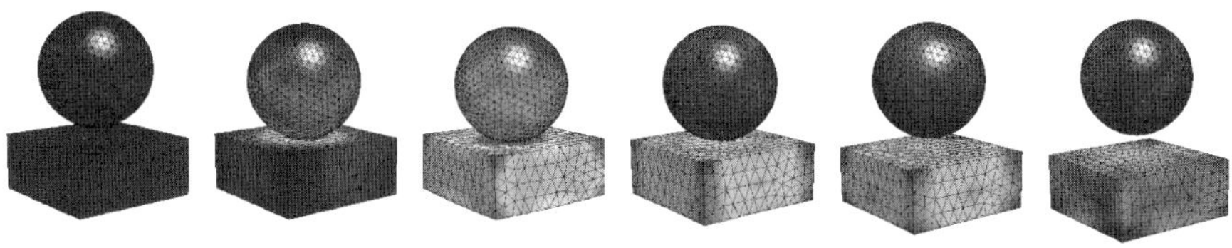

Fig. 3. Situation with effective von Mises stress at different time steps.

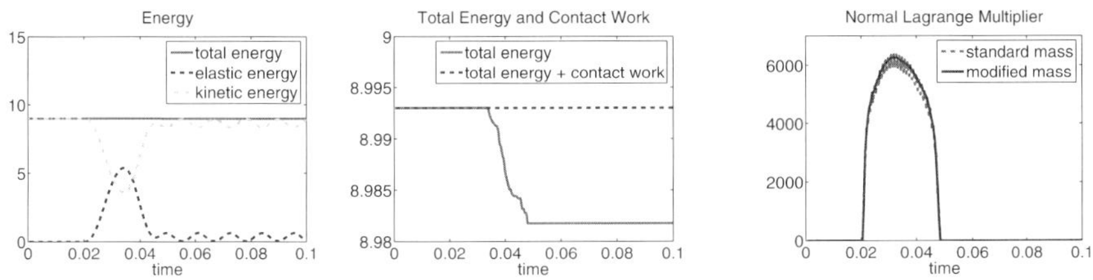

Fig. 4. Energy, contact work and the LM with Coulomb friction.

Our second example is a three dimensional setting. We consider a ball having an initial velocity downwards and impinging on a brick being at rest. The kinetic energy of the ball is transfered to the brick such that the brick moves downwards after the impact. We assume the ball to be the slave body Ω^s and set the coefficient of friction $\mathfrak{F}$ equal to 0.5; see [11] for details. The effective von Mises stress on the distorted bodies for different time steps are shown in Figure 3. The evolution of the contact energy and the normal part of the LM for the standard and the modified mass matrix is shown in Figure 4. As in our first example, we observe that there is no numerical energy dissipation in our algorithm. Moreover, the oscillation in the LM is considerably reduced by our proposed modification of the mass matrix.

6 Rigid-Plastic Problem

In this section, we restrict ourselves to the frictionless one-body contact problem without inertia forces. The body Ω gets into contact with a rigid obstacle Ω_{obs} which

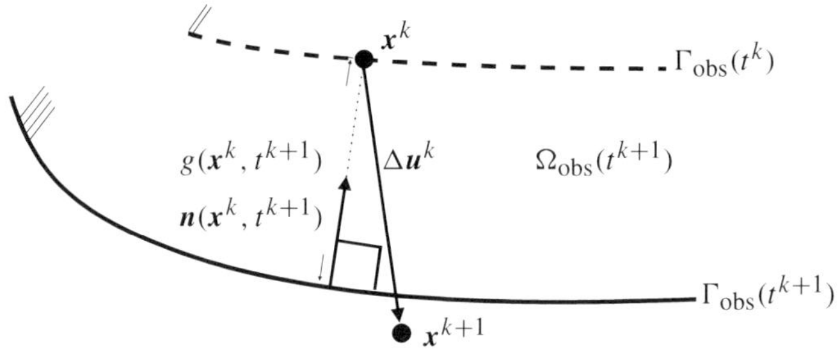

Fig. 5. Gap and normal.

has the boundary Γ_{obs}. The local balance of momentum with respect to the actual configuration is given by

$$-\mathrm{div}(\boldsymbol{\sigma}) = \boldsymbol{f} \quad \text{in} \quad \Omega \times (0, T).$$

By μ^{plas} we denote the viscosity coefficient for the plastic flow and the bulk modulus κ acts as a volume penalty which is usually set to a very high value to nearly guarantee the plastic incompressibility $\mathrm{tr}(\dot{\boldsymbol{\varepsilon}}) \to 0$. By $p := -\frac{1}{d}\mathrm{tr}(\boldsymbol{\sigma})$ we denote the pressure and $\boldsymbol{\sigma}^{dev} := \boldsymbol{\sigma} + p\,\mathrm{Id}_d$ is the deviatoric stress. According to this, the constitutive law [2] formulated in the actual Cauchy stresses reads as

$$\boldsymbol{\sigma}^{dev} = 2\mu^{plas}\dot{\boldsymbol{\varepsilon}}^{dev}, \qquad p = \kappa\ \mathrm{tr}(\dot{\boldsymbol{\varepsilon}}),$$

where $\boldsymbol{\varepsilon}^{dev}$ stands for the deviatoric part of the strain tensor. Since the material law is velocity-driven, the Dirichlet boundary condition according to (3) has to be given for the velocity $\dot{\boldsymbol{u}}$. The weak form of this problem is similar to (7) but without the bilinear form $m(\cdot, \cdot)$ and the jump operator $[\cdot]$ and with the bilinear form $a(\cdot, \cdot)$ replaced by

$$a^{rp}(\dot{\boldsymbol{u}}, \boldsymbol{v}) := \int_{\varphi(\Omega)} 2\mu^{plas}\dot{\boldsymbol{\varepsilon}}^{dev}(\boldsymbol{u}) : \boldsymbol{\varepsilon}^{dev}(\boldsymbol{v}) - \kappa\ \mathrm{tr}(\dot{\boldsymbol{\varepsilon}}(\boldsymbol{u}))\mathrm{tr}(\boldsymbol{\varepsilon}(\boldsymbol{v}))\ \mathrm{d}\boldsymbol{x}.$$

According to the contact conditions (13), the difficulty of the large deformations is tackled in an updated Lagrange manner. Since the structural term $a^{rp}(\cdot, \cdot)$ is velocity-driven whereas the contact conditions are displacement-driven, we have to express (13) in terms of the velocity $\dot{\boldsymbol{u}}^{k+1}$. Therefore we use a generalized trapezoidal rule with the parameter $\zeta \in [0, 1]$

$$\Delta \boldsymbol{u}^k = \boldsymbol{u}^{k+1} - \boldsymbol{u}^k = (1 - \zeta)\ \Delta t\ \dot{\boldsymbol{u}}^k + \zeta\ \Delta t\ \dot{\boldsymbol{u}}^{k+1}. \tag{23}$$

Furthermore, the gap g and the contact normal $\boldsymbol{n}$ are chosen with respect to the last known configuration at time t^k. This means that we denote by $g(\boldsymbol{x}^k, t^{k+1})$ the distance between the position $\boldsymbol{x}^k$ at time t^k and its closest point projection onto $\Gamma_{\text{obs}}(t^{k+1})$. By $\boldsymbol{n}(\boldsymbol{x}^k, t^{k+1})$, we indicate the direction of this closest point projection, see Figure 5. We remark that the actual position of the moving tool boundary $\Gamma_{\text{obs}}(t^{k+1})$ is always known in advance. Now we define a dynamic gap G^{k+1} in such a way that the non-penetration condition (13) in terms of displacements at the actual time step t^{k+1} is fulfilled. Setting

$$G^{k+1} := \frac{1}{\zeta\,\Delta t}\big(g(\boldsymbol{x}^k, t^{k+1}) - (1-\zeta)\,\Delta t\,\dot{\boldsymbol{u}}^k \boldsymbol{n}(\boldsymbol{x}^k, t^{k+1})\big),$$

for the case $\zeta \in (0, 1]$, the velocity-driven form of the non-penetration condition (13) is given by

$$\dot{u}_n^{k+1} - G^{k+1} \le 0, \qquad \lambda_n^{k+1} \ge 0, \qquad \lambda_n^{k+1}(\dot{u}_n^{k+1} - G^{k+1}) = 0, \tag{24}$$

with the definitions $\dot{u}_n^{k+1} := \dot{\boldsymbol{u}}^{k+1}\boldsymbol{n}(\boldsymbol{x}^k, t^{k+1})$ and $\lambda_n^{k+1} := \boldsymbol{\lambda}^{k+1}\boldsymbol{n}(\boldsymbol{x}^k, t^{k+1})$. Using (23), it can be readily verified that we get the displacement-driven non-penetration condition by

$$\begin{aligned}\Delta\boldsymbol{u}^k \boldsymbol{n}(\boldsymbol{x}^k, t^{k+1}) &= \big[(1-\zeta)\,\Delta t\,\dot{\boldsymbol{u}}^k + \zeta\,\Delta t\,\dot{\boldsymbol{u}}^{k+1}\big]\,\boldsymbol{n}(\boldsymbol{x}^k, t^{k+1})\\ &\le (1-\zeta)\,\Delta t\,\dot{\boldsymbol{u}}^k \boldsymbol{n}(\boldsymbol{x}^k, t^{k+1}) + \zeta\,\Delta t\,G^{k+1}\\ &= g(\boldsymbol{x}^k, t^{k+1}).\end{aligned}$$

Remark 3. Since the structural part is velocity-driven, we enforce the equilibrium condition at time t^{k+1} and not at $t^{k+1/2}$ as in the previous sections. Therefore, in contrast to (13), we employ λ_n^{k+1} in (24).

Remark 4. We remark that for the special case $\zeta = 0$ (explicit Euler), the desired non-penetration can only be guaranteed in the next time step t^{k+2} since the actual displacements are completely determined by quantities from the last time step ($\boldsymbol{u}^{k+2} - \boldsymbol{u}^{k+1} = \Delta\boldsymbol{u}^{k+1} = \Delta t\,\dot{\boldsymbol{u}}^{k+1}$). In this case, the choice

$$G^{k+2} := \frac{1}{\Delta t} g(\boldsymbol{x}^{k+1}, t^{k+2}), \qquad \dot{u}_n^{k+1} - G^{k+2} \le 0$$

leads to

$$\Delta\boldsymbol{u}^{k+1}\,\boldsymbol{n}(\boldsymbol{x}^{k+1}, t^{k+2}) = \Delta t\,\dot{\boldsymbol{u}}^{k+1}\boldsymbol{n}(\boldsymbol{x}^{k+1}, t^{k+2}) \le g(\boldsymbol{x}^{k+1}, t^{k+2}).$$

As we consider only the frictionless case, the LM space in (8) simplifies to the space $\boldsymbol{M}^+$ given by

$$\boldsymbol{M}^+ := \{\boldsymbol{\mu} \in \boldsymbol{M} \;:\; \langle \boldsymbol{\mu}, \boldsymbol{\eta}\rangle \le 0, \;\; \boldsymbol{\eta} \in \boldsymbol{W} \;\text{ with }\; \eta_n \le 0\}\,.$$

The time discretized variational formulation for the time step t^{k+1} reads as follows (compare (15)): find $(\dot{\boldsymbol{u}}^{k+1}, \boldsymbol{\lambda}^{k+1}) \in \boldsymbol{V} \times \boldsymbol{M}^+$ such that

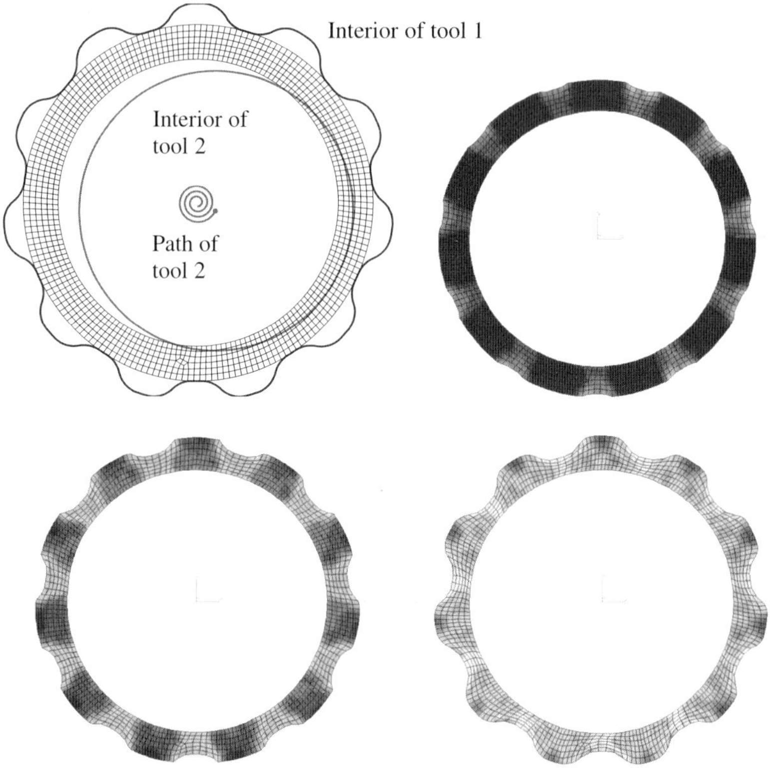

Fig. 6. Ring rolling of external spur gears. For three snapshots of the simulation, the equivalent plastic strain is shown.

$$
\begin{aligned}
a^{\mathrm{rp}}(\dot{\boldsymbol{u}}^{k+1}, \boldsymbol{v}) + \langle \boldsymbol{v}, \boldsymbol{\lambda}^{k+1} \rangle &= f^{k+1}(\boldsymbol{v}), && \boldsymbol{v} \in \boldsymbol{V}_0, \\
\langle \dot{u}_n^{k+1}, \mu_n - \lambda_n^{k+1} \rangle &\le \langle G^{k+1}, \mu_n - \lambda_n^{k+1} \rangle, && \boldsymbol{\mu} \in \boldsymbol{M}^+.
\end{aligned}
\tag{25}
$$

Let $\mathbf{K}^{\mathrm{rp}}$ be the rigid plastic stiffness matrix resulting from the bilinear form $a^{\mathrm{rp}}(\cdot,\cdot)$ after discretization in space. Then, in analogy to (19)–(21), the algebraic form of (25) is given by

$$
\mathbf{K}^{\mathrm{rp}} \dot{\mathbf{u}}^{k+1} + \begin{pmatrix} \mathbf{0} \\ \mathbf{D} \end{pmatrix} \boldsymbol{\lambda}^{k+1} = \mathbf{f}^{k+1},
$$

together with the contact conditions

$$
\dot{u}_{pn}^{k+1} - G_p^{k+1} \le 0, \qquad \lambda_{pn}^{k+1} \ge 0, \qquad (\dot{u}_{pn}^{k+1} - G_p^{k+1})\lambda_{pn}^{k+1} = 0, \qquad \boldsymbol{\lambda}_{p\tau}^{k+1} = 0.
$$

With the definition $\boldsymbol{n}_p^{k+1} := \boldsymbol{n}(\boldsymbol{x}_p^k, t^{k+1})$ we used the abbreviations

$$\dot{u}_{pn}^{k+1} := \dot{\boldsymbol{u}}_p^{k+1} \boldsymbol{n}_p^{k+1}, \qquad \lambda_{pn}^{k+1} := \boldsymbol{\lambda}_p^{k+1} \boldsymbol{n}_p^{k+1}, \qquad \boldsymbol{\lambda}_{p\tau}^{k+1} := \boldsymbol{\lambda}_p^{k+1} - \lambda_{pn}^{k+1} \boldsymbol{n}_p^{k+1}.$$

As a numerical example we consider an incremental metal forming process, see the upper left picture in Figure 6. One mobile tool presses the material into a rigid mold and forms a spur gear within 6000 time steps. The simulation was done in LARSTRAN/SHAPE [9], a FE-package dedicated to large strain plasticity in the context of metal forming simulations. In Figure 6, one can see three snapshots out of the simulation process.

References

1. P. Alart and A. Curnier, A mixed formulation for frictional contact problems prone to Newton like solution methods, *Comp. Meth. Appl. Mech. Engng.*, 92:353–375, 1991.
2. J. Argyris and J. Doltsinis, A primer on superplasticity in natural formulation, *Comp. Meth. Appl. Mech. Engng.*, 46:83–131, 1984.
3. P. Bastian, K. Birken, K. Johannsen, S. Lang, N. Neuß, H. Rentz-Reichert, and C. Wieners, UG – A flexible software toolbox for solving partial differential equations, *Comput. Vis. Sci.*, 1:27–40, 1997.
4. F. B. Belgacem, P. Hild, and P. Laborde, The mortar finite element method for contact problems. Recent advances in contact mechanics, *Math. Comput. Modelling*, 28:263–271, 1998.
5. F. Ben Belgacem and Y. Renard, Hybrid finite element methods for the Signorini problem, *Math. Comp.*, 72:1117–1145, 2003.
6. C. Bernardi, Y. Maday, and A. T. Patera, A new nonconforming approach to domain decomposition: The mortar element method. In *Nonlinear Partial Differential Equations and Their Applications*. Collège de France Seminar, Vol. XI (Paris, 1989–1991), Pitman Res. Notes Math. Ser., Vol. 299, Longman Sci. Tech., Harlow, pp. 13–51, 1994.
7. V. Chawla and T. A. Laursen, Energy consistent algorithms for frictional contact problems, *Internat. J. Numer. Methods Engrg.*, 42:799–827, 1998.
8. P. Deuflhard, R. Krause, and S. Ertel, A contact-stabilized Newmark method for dynamical contact problems, *Internat. J. Numer. Methods Engrg.*, 2007 (to appear).
9. R. Diez, U. Hindenlang, and A. Kurz, Larstran user's manual – Larstran documentation, Tech. Rep., LASSO GmbH, Echterdingen, 1996.
10. K. A. Fischer and P. Wriggers, Mortar based frictional contact formulation for higher order interpolations using the moving friction cone, *Comput. Methods Appl. Mech. Engrg.*, 195:5020–5036, 2006.
11. C. Hager, S. Hüeber, and B. I. Wohlmuth, A stable energy conserving approach for frictional contact problems based on quadrature formulas, Tech. Rep. 011, Universität Stuttgart, SFB 404, 2006.
12. C. Hager and B. I. Wohlmuth, Analysis of a modified mass lumping method for the stabilization of frictional contact problems, Tech. Rep. 003, Universität Stuttgart, IANS, 2007.
13. M. Hintermüller, K. Ito, and K. Kunisch, The primal-dual active set strategy as a semismooth Newton method, *SIAM J. Optim.*, 13:865–888, 2003.
14. S. Hüeber, A. Matei, and B. I. Wohlmuth, Efficient algorithms for problems with friction, *SIAM J. Sci. Comput.*, 29:70–92, 2007.

15. S. Hüeber, G. Stadler, and B. I. Wohlmuth, A primal-dual active set algorithm for three-dimensional contact problems with Coulomb friction, Tech. Rep., Pré-publicações do Departamento de Matemática da Universidade de Coimbra 06-16, 2006.
16. S. Hüeber and B. I. Wohlmuth, A primal-dual active set strategy for non–linear multibody contact problems, *Comput. Methods Appl. Mech. Engrg.*, 194:3147–3166, 2005.
17. H. Khenous, P. Laborde, and Y. Renard, Comparison of two approaches for the discretization of elastodynamic contact problems, *C. R. Acad. Sci. Paris*, 342:791–796, 2006.
18. R. Kornhuber and R. Krause, Adaptive multigrid methods for Signorini's problem in linear elasticity, *Comput. Vis. Sci.*, 4:9–20, 2001.
19. T. A. Laursen and V. Chawla, Design of energy conserving algorithms for frictionless dynamic contact problems, *Internat. J. Numer. Methods Engrg.*, 40:836–886, 1997.
20. T. W. McDevitt and T. A. Laursen, A mortar-finite element formulation for frictional contact problems, *Internat. J. Numer. Methods Engrg.*, 48:1525–1547, 2000.
21. J. Simo and N. Tarnow, The discrete energy-momentum method. Conserving algorithms for nonlinear elastodynamics, *Z. Angew. Math. Phys.*, 43:757–792, 1992.
22. B. I. Wohlmuth, A mortar finite element method using dual spaces for the Lagrange multiplier, *SIAM J. Numer. Anal.*, 38:989–1012, 2000.
23. B. Yang, T. A. Laursen, and X. Meng, Two dimensional mortar contact methods for large deformation frictional sliding, *Internat. J. Numer. Methods Engrg.*, 62:1183–1225, 2005.

Some Advances in Mathematical Programming Method for Numerical Simulation of Contact Problems

H.W. Zhang, A.H. Liao, Z.Q. Xie, B.S. Chen and H. Wang

Department of Engineering Mechanics, State Key Laboratory of Structural Analysis for Industrial Equipment, Dalian University of Technology, Dalian 116024, P.R. China
E-mail: zhanghw@dlut.edu.cn

Abstract. This paper reviews some advances and applications in mathematical programming method for numerical modeling of elastic-plastic contact problems. Emphases are on the parametric variational principle and quadratic programming method used for analysis of elastic-plastic contact problems with isotropic/orthotropic friction law. The contact problem with friction between two elastic-plastic Cosserat bodies is treated in the same way as that in the conventional plastic analysis. There is no available rule for choosing a reasonable value of the penalty factors for simulation of the contact problems of Cosserat materials, and they are therefore cancelled through a special technique so that the numerical results can be of high accuracy. Two numerical examples are presented to show the efficiency of the model and algorithm presented.

Key words: Computational contact mechanics, mathematical programming method, parametric variational principle.

1 Introduction

Contact problems are very important in many fields such as mechanical and civil engineering, which have the characteristics of geometric and material discontinuity at the contact interface rather than the usual continuity property in solid mechanics [27]. The strong nonlinearity is difficult to be solved by the analytical method for problems of complex boundaries [23], and hence numerical methods are needed in the solution of engineering problems.

Conry and Seireg [4] appear to be the first to treat contact as a quadratic programming problem. Haug et al. [15] formulated the frictionless contact problem for multiple elastic bodies using quadratic programming method and solved numerical examples (wheel and beam in contact with an elastic foundation) to demonstrate its

Peter Wriggers and Udo Nackenhorst (eds.), IUTAM Symposium on Computational Contact Mechanics, 33–55.

usage. Then, over the last two decades, great efforts have been devoted to the numerical study of contact problems by mathematical programming method. Zhong and Mackerle [41] reviewed some of the main developments in finite element techniques for solution of static contact problems. Wriggers [27] and Laursen [19] gave systematically description on the methods of computational contact mechanics.

In mathematical programming methods, contact problems are considered as a special case of constrained minimization of either total or complementary potential energy such as that used in plasticity theory [22]. The minimization is formulated as a mathematical programming problem and the solution can be obtained by using either linear programming (LP) or quadratic programming (QP) techniques [20].

Alberti et al. [1] proposed the treatment of contact and friction forces for LP-based rigid-plastic/rigid-viscoplastic FEM, and analyzed some simple plane-strain and axisymmetrical forming processes. Refaat and Meguid [24] presented an algorithm to obtain the numerical solution of variational inequalities corresponding to contact problems. The algorithm was developed based on mathematical programming techniques (Sequential Linear Programming/Sequential Quadratic Programming) and avoids the use of penalty or regularization methods. Li et al. [21] presented the fundamental theory of the discontinuous deformation mechanics model for a multi-body system. A linear programming algorithm was employed to solve the resulting discrete controlling equations. Baggio and Trovalusci [2] proposed linear programming approaches for the limit analysis of rigid block structures involving friction. Ferris and Tin-Loi [10] computed the collapse load of discrete rigid block systems with frictional contact interfaces as a special constrained optimization problem (the so-called mathematical problem under equilibrium constraints, MPEC). Gilbert et al. [12] presented a computational limit analysis procedure for rigid block assemblages comprising non-associative frictional interfaces. The procedure involves solving a series of LP problems with successively modified failure surfaces (rather than working directly with the full Mixed Complementarity Problem (MCP) as others have done). Drosopoulos et al. [9] analyzed a single span masonry bridge with unilateral frictional contact interfaces by using an implementation of the classical collapse mechanism method based on linear programming. Donev et al. [6] presented a randomized linear programming algorithm for finding unjamming motions within the approximation of small displacements, focusing on the fundamental and rigorous case of packings with ideal interparticle contacts (i.e., no interparticle gaps).

Quadratic programming, unlike linear programming, is an NP-hard problem in general. It is characteristic that the objective function is quadratic function of independent variables. Givoli and Doukhovni [13] presented a finite element-quadratic programming approach for problems involving frictionless contact between an elastic body and a rigid obstacle in a general setting, and extended it to the case of large deformation contact problems. Gawecki et al. [11] presented a mathematical framework for the unilateral contact problem between an elastic-perfectly plastic beam and an elastic-perfectly plastic foundation using the quadratic programming formulations. Hu [17] presented a sensitivity-based updating algorithm incorporating the quadratic programming technique to solve the fictitious forces in contacting area. In the method, an iterative updating process based on the first-ordersensitivity analysis and the quadratic programming technique was proposed to compute the fictitious forces in contacting areas. Kontoleon and Baniotopoulos [18] applied a finite element-quadratic programming approach for the numerical treatment of problems involving frictional contact between elastic-plastic bodies and/or a rigid obstacle. The method was developed based on the theoretical results of nonsmooth mechanics and intended to effectively simulate the frictional unilateral contact problem arising on a steel base plate connection. Dostál et al. [8] presented a domain decomposition algorithm for the solution of coercive and semicoercive frictionless contact problems. The resulting problem was solved by an augmented Lagrangian type algorithm with an outer loop for the Lagrange multipliers for the equality constraints and an inner loop for the solution of the bound constrained quadratic programming problems. Dostál et al. [7] developed algorithms for the solution of special quadratic programming problems and reduced a discretized model problem by the duality theory of convex optimization to the quadratic programming problem with bound and equality constraints on the solution of 3D problems with large displacements and Coulomb friction. Su et al. [25] proposed a QP method for the crack problems involving the crack surfaces in frictional contact. This method can avoid the tiresome iteration procedure used in other work, and if there is no unloading, only one incremental step is sufficient for getting the solution. González and Abascal [14] solved the transient rolling contact of two-dimensional elastic bodies based on the use of boundary element method and mathematical programming techniques. BEM was used to compute the constant influence coefficients based on true cylinder geometry, not on the half-space approximation, while MPT was used to model the contact conditions and to solve the highly non-linear problem associated with transient rolling at each time step. Xuan and Lee [26] formulated the frictional contact problem to a QP model and constructed an iterative numerical method by solving a sequence of standard QP subproblems. The friction orientation constraints were introduced to the QP model, which can guarantee the convergence of the iterative algorithm when the larger value of friction coefficient is involved, while in the case of a smaller friction coefficient the friction orientation constraints would be inactive. Holanda and Goncalves [16] formulated unilateral contact problems as a mathematical programming problem with inequality constraints from which a linear complementarity problem was derived and solved by the Lemke algorithm. Czekanski and Meguid [5] solved the large deformation problem of dynamic frictional contact in elastic-plastic structures by use of mathematical programming and Lagrange multipliers in order to identify the candidate contact surface and contact stresses. This approach guarantees the accurate imposition of the active kinematic contact constraints and avoids the use of special contact elements.

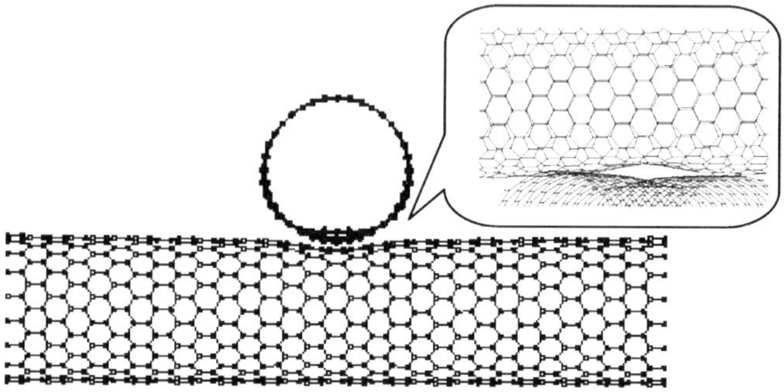

Fig. 1. Free equilibrium configuration with the initial gap 0.33 nm of two (18, 0) crossed single walled carbon nanotubes [35].

Among the different methods used for numerical analysis of elastic-plastic contact problems, the parametric variational principle (PVP) and the corresponding parametric quadratic programming (PQP method were also used successfully in the past years. Zhong et al. [42, 43] developed PVP for the analyses of contact problems and elastic-plastic structures. In Zhong's method, both contact and elastic-plastic problems can be formulated as the same form of parametric programming problems by means of finite element method. Then they are reduced to linear complementarity problems. Zhang and his co-workers further enhanced and applied the method in several areas, such as the combination of parametric programming method and iteration algorithm for the analysis of elastic or elastic-plastic contact problems [38, 39], the parametric variation principle used for coupled analysis of liquid and solid interaction occurred in saturated and partially saturated porous media [28,33], the improved quadratic programming method adopted for numerical simulation of mechanical behaviors of carbon nanotubes [35], see Figure 1.

In the work by Zhang and Schrefler [33], dynamic strain localization in saturated and partially saturated porous media was investigated with a one-dimensional model, and a gradient-dependent plasticity model was introduced to describe the plastic behavior of the solid skeleton. For the numerical analysis, a parametric variational principle and the corresponding mathematical programming algorithm were proposed to overcome the difficulties caused by the incorporation of the gradient effect, thus obtaining the numerical model used for the computation. The method was further extended to the analysis of the generally dynamic elastic-plastic strain hardening/softening problems [37], where the gradient dependent model based on the non-local theory was adopted to overcome the mesh dependent problem in the analysis of in strain localization. A parametric quadratic programming algorithm combined with both the Newmark and the precise integration methods in time domain was derived and the problem is changed into a linear complementary model. For the local and multi-scale analysis of periodic composite materials, Zhang et

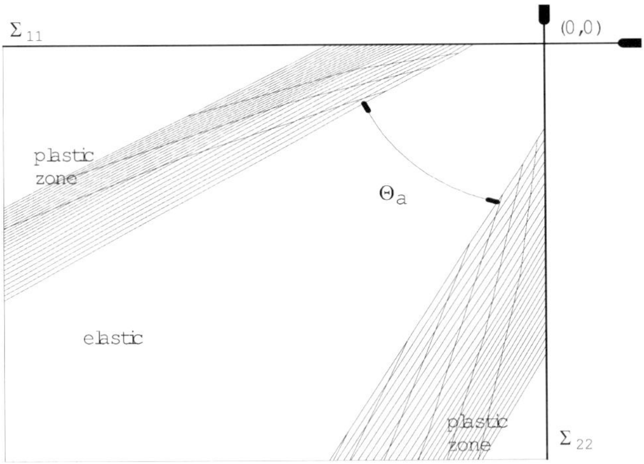

Fig. 2. Local numerical constitutive law based on micromechanical analysis [34].

al. [29] adopted and improved PVP for the micromechanical analysis of the multielastic contact body problem. The modeling approach in [29] differs somewhat from other studies because the homogenized constitutive relation for the global behaviour of periodic assemblies in elastic contact was derived with taking into account stick-slip behaviour along the contact surfaces. The body was assumed deformable with small strains. The representative volume element was treated as a multi-contact-body whose overall response reflects the continuum character of a point experiencing uniform deformation. Based on micromechanical analysis, a numerical homogenized constitutive relation for the global behaviour was defined and used for overall structural analysis at the macro-level, see Figure 2. The procedure allows a considerable reduction of the computational effort needed for finite element analyses of real composite structures and closely follows the mechanical behavior of the structure. Moreover it is applicable both to associative and non-associative "plasticity", and allowed sliding and separation between different bodies in the microstructures to be taken into account. Furthermore, a homogenized constitutive relation for assemblies of periodic inelastic bodies in contact was developed in [34] by means of a micromechanical analysis. Once the interpolation points are obtained on the macroscopic stress space, the straightforward elastic-plastic algorithm yields the response at the macroscopic level. The local stress history can be obtained at selected points of interest by a stress-recovery procedure. The procedure results in a considerable reduction of the computational effort needed for elastic-plastc finite-element analyses of real composite structures and closely followed the nonlinear mechanical behavior at the macroscopic level.

In practical analysis of contact problems, it is necessary to investigate the solution methods for three-dimensional contact problems. The number of unknowns of the LCP will greatly increase as the three-dimensional contact model being linearized, and consequently the difficulty of the solution will be increased dramatically. Therefore, it is of great significance to develop a high efficient algorithm for large-scale LCP. Zhang et al. [30] presented a non-interior smoothing algorithm based on NCP-function for numerical solution of the three-dimensional contact problems. The proposed algorithm is based on the non-differentiable NCP function and the solution of the LCP for the contact problem reduces to solving the nonlinear system of equations. The advantages of the algorithm are not only easily implemented, but also have some characters of high-efficiency and fast convergence. It is known that the friction force is generally opposite to the slip direction for isotropic friction. But in many cases, isotropic friction can not give practical solutions because the contact surfaces or the mechanical properties of materials might be anisotropic. Orthotropic friction is a special case of anisotropic friction, in which the principal friction directions are mutually orthogonal. The elastic-plastic contact problem with orthotropic friction law was studied in [32]. With the orthotropic friction law being approximated by a polyhedral law of friction, the problem is reduced to a standard linear complementarity problem. For three-dimensional contact problems, the determination of the slip direction is very difficult and needs a lot of computational efforts. Zhang et al. [32] extended the combined parametric quadratic programming and iterative method [39] to the analysis of 3D elastic-plastic contact problems with orthotropic friction. A smoothing algorithm based on NCP-function was developed for solving resulted LCP, which has advantages of easy to be understood and implemented in computer program. After the 3D elastic frictional contact problems are formulated as linear complementarity problems, the key is to find a stable and efficient method to solve the derived LCP. In the work by Zhang et al. [31], two aggregate-function-based algorithms, called respectively as self-adjusting interior point algorithm and aggregate function smoothing algorithm, were proposed for the solution of the LCP derived from the contact problems.

Recently, a finite element model for multi-body contact analysis of Cosserat materials was developed in [36]. Based on parametric virtual work principle, a quadratic programming method was proposed for the finite element analysis of frictional contact problems of Cosserat materials. The contact problem with friction was treated in the same way as in plastic analysis. The penalty factors introduced in the algorithm for contact analysis of Cosserat bodies have a direct influence on accuracy of solution. To overcome this difficulty, a special technique was introduced to cancel the penalty factors, so that the solutions can be obtained directly without tedious iterative procedures such as general incremental iterative method and therefore the method greatly simplifies the implementation of the algorithm for contact analysis of Cosserat bodies. It is found from the numerical examples that the results obtained with a Cosserat model are different from those obtained with conventional model.

2 Constitutive Equations and Orthotropic Friction Law

Constitutive equations of plasticity can be expressed as

$$
\begin{aligned}
&d\sigma_{ij} = D_{ijkl}(d\varepsilon_{kl} - d\varepsilon^p_{kl}),\\
&dd\varepsilon^p_{kl} = \lambda(\partial g/\partial\sigma_{kl}),\\
&f(\sigma_{ij}, \varepsilon^p_{kl}, \kappa, \nabla^2\kappa) \le 0,\\
&\lambda \ge 0, \lambda^T f = 0,
\end{aligned} \tag{1}
$$

where the notations of all the variables and parameters are used as in the usual way. A gradient dependent model is used to show the potential of the algorithm developed.

Without loss of generality, it is assumed that $d\kappa = h\lambda$, $\bar{c} = h\partial f/\partial\nabla^2\kappa$, and h is the hardening/softening modulus. Applying Taylor expansion to Equation (1c), we obtain the consistent equation

$$
f^0 + \mathbf{W}d\boldsymbol{\varepsilon} - M\lambda + \bar{c}\nabla^2\lambda \le 0, \tag{2}
$$

where f^0 is the initial value of the yield function, and

$$
\mathbf{W} = (\partial f/\partial\boldsymbol{\sigma})^T\mathbf{D}, \quad M = \mathbf{W}(\partial g/\partial\boldsymbol{\sigma})^T - (\partial f/\partial\boldsymbol{\varepsilon}^p)^T(\partial g/\partial\boldsymbol{\sigma})^T + (\partial f/\partial\kappa)h^T. \tag{3}
$$

Consider two candidate contact bodies, as shown in Figure 3, the common elastic-plastic contact surface is denoted by S_c. The principal friction orthotropic axes on the tangential plane at a point on S_c are denoted by τ_1 and τ_2. p_{τ_1} and p_{τ_2} are the components of tangential contact traction p_τ, Δu_{τ_1} and Δu_{τ_2} are the components of tangential contact relative displacement Δu_τ along axes τ_1 and τ_2. The corresponding coefficients of friction are denoted by μ_1 and μ_2. $p_n \le 0$ denotes the normal contact force. The form of the static friction law is the static friction law and can be expressed as: If $[(p_{\tau_1}/\mu_1)^2 + (p_{\tau_2}/\mu_2)^2]^{1/2} < -p_n$, then $\Delta u_\tau = 0$. If $[(p_{\tau_1}/\mu_1)^2 + (p_{\tau_2}/\mu_2)^2]^{1/2} = -p_n$, then there exists $\gamma \ge 0$ such that $\Delta\mu_{\tau_1} = -\gamma p_{\tau_1}/\mu_1^2$, $\Delta\mu_{\tau_2} = -\gamma p_{\tau_2}/\mu_2^2$. So the following equality holds

$$
\frac{\Delta\mu_{\tau_1}}{\Delta\mu_{\tau_2}} = \left(\frac{\mu_1^2}{\mu_2^2}\right)\left(\frac{p_{\tau_2}}{p_{\tau_2}}\right). \tag{4}
$$

Let θ_u and θ_f denote the inclination angles of the tangential relative displacement Δu_τ and tangential contact force p_τ with respect to the τ_1-axis, respectively, then Equation (4) becomes

$$
\tan\theta_u - \left(\frac{\mu_1}{\mu_2}\right)^2 \tan\theta_f, \tag{5}
$$

which indicates that the direction of tangential contact force can be different from the direction of slip for the orthotropic friction law. The following closed set is then introduced:

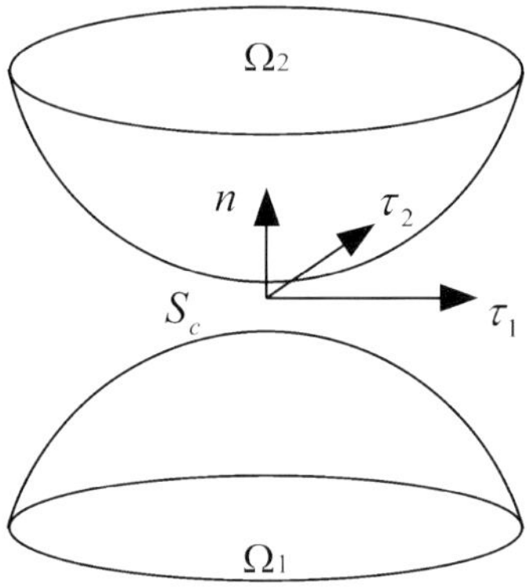

Fig. 3. Two contact bodies.

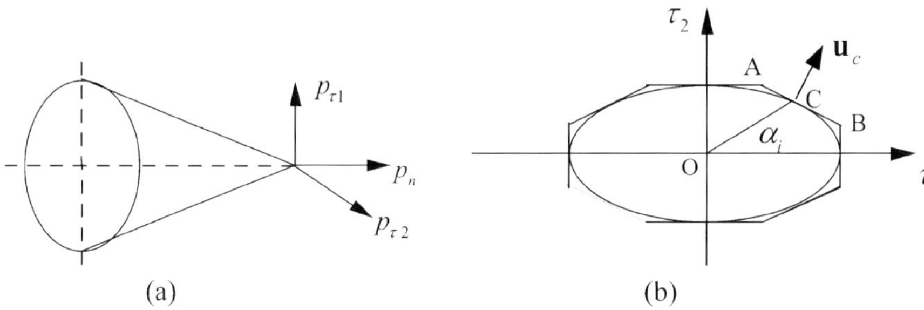

Fig. 4. Elliptic law and polyhedral form.

$$C_e = \{(p_{\tau_1}, p_{\tau_2}, p_n) : \tilde{f} = [(p_{\tau_1}/\mu_1)^2 + (p_{\tau_2}/\mu_2)^2]^{1/2} + p_n \le 0\}, \quad (6)$$

where C_e describes an elliptic cone, see Figure 4a.

For numerical simulation, a piecewise linearization approximation of the elliptic cone C_e is used in the finite element model. Figure 4b shows the cross-sectional shape of elliptic cone. The linearized approximation of elliptic cone C_e can be written as

$$\begin{aligned} C_s &= \{P_\tau : \tilde{f}_i(p_\tau, p_n) \le 0, i = 1, 2, \ldots, s\}, \\ \tilde{f}_i &= [\cos\alpha_i, \beta^2 \sin\alpha_i, \gamma_i\mu_i][p_{\tau_1}, p_{\tau_2}, p_n]^T, \end{aligned} \quad (7)$$

where $\gamma_i = (\cos^2\alpha_i + \beta^2 \sin^2\alpha_i)^{1/2}$, $\beta = \mu_1/\mu_2$, s is the number of the faces of the polyhedral cone.

The contact relative displacements are decomposed into two parts, one is elastic part (describing micro-slip) and another is plastic part (describing macro-slip). We have $\Delta\mathbf{u}_c = \Delta\mathbf{u}_c^e + \Delta\mathbf{u}_c^p$, then the relationship between the contact relative displacements and contact forces can be expressed as

$$\mathbf{p}_c = \mathbf{D}_c(\Delta\mathbf{u}_c - \Delta\mathbf{u}_c^p), \tag{8}$$

where $\Delta\mathbf{u}_c = [\Delta u_{\tau_1}, \Delta u_{\tau_2}, \varepsilon_n]^T$, $\mathbf{p}_c = [p_{\tau_1}, p_{\tau_2}, p_n]^T$, $D_{cij} = \delta_{ij}E_\tau = \delta_{ij}E_n = \delta_{ij}E$, E and E_n ($E_\tau = E_n$) are the penalty factors, $\varepsilon_n = du_n^{(1)} - du_n^{(2)} + \delta^* \geq 0$, where δ^* is the initial normal gap between contact surfaces.

In analogy with the expression of elastic-perfectly plastic constitutive model, we introduce here the slip function $\tilde{f}_i$ and slip potential function $\tilde{g}_i$ for the contact description, which are given as follows:

$$\begin{aligned}
\tilde{f}_i &= p_{\tau_1}\cos\alpha_i + \beta^2 p_{\tau_2}\sin\alpha_i + \gamma_i\mu_1 p_n \leq 0,\\
\tilde{g}_i &= p_{\tau_1}\cos\alpha_i + \beta^2 p_{\tau_2}\sin\alpha_i - c,\\
\tilde{f}_{s+1} &= p_n \geq 0, \quad \tilde{g}_{s+1} = p_n, \quad i = 1, 2, \ldots, s,
\end{aligned} \tag{9}$$

where the terms with "~" are the variables for the contact description and c is a constant. The contact plastic (slip or separation) incremental displacement vector, therefore, can be obtained by

$$\Delta\mathbf{u}_c^p = \tilde{\lambda}_k\left(\frac{\partial\tilde{g}_k}{\partial\mathbf{p}_c}\right) = \left(\frac{\partial\tilde{g}}{\partial\mathbf{p}_c}\right)^T\tilde{\boldsymbol{\lambda}}, \quad \tilde{\boldsymbol{\lambda}} \geq 0. \tag{10}$$

It is thus that the contact boundary conditions can be changed into

$$\tilde{f}_k^0 + \tilde{\mathbf{W}}_k\Delta\mathbf{u}_c - \tilde{\mathbf{M}}_k\tilde{\boldsymbol{\lambda}} \leq 0, \quad k = 1, 2, \ldots, s+1, \tag{11}$$

where

$$\tilde{\mathbf{W}}_k = \left(\frac{\partial\tilde{f}_k}{\partial\mathbf{p}_c}\right)^T\mathbf{D}_c, \quad \tilde{\mathbf{M}}_k = \left(\frac{\partial\tilde{f}_k}{\partial\mathbf{p}_c}\right)^T\mathbf{D}_c\left(\frac{\partial\tilde{\mathbf{g}}}{\partial\mathbf{p}_c}\right)^T.$$

3 Governing Equations of 3D Elastic-Plastic Contact Problems

The governing equations for 3D contact analysis can be summarized as:

Equilibrium equations

$$d\sigma_{ij,j} + db_i = 0. \tag{12}$$

Continuity equations

$$d\varepsilon_{ij} = \frac{1}{2}(du_{i,j} + du_{j,i}). \tag{13}$$

Boundary conditions

$$n_i d\sigma_{ij} = d\bar{p}_j, \quad \text{on } S_p,$$

$$du_i = d\bar{u}_i, \quad \text{on } S_u, \tag{14}$$

$$\varepsilon_n = du_n^{(1)} - du_n^{(2)} + \delta^* \geq 0, \quad \text{on } S_c, \tag{15}$$

$$\text{if } \varepsilon_n > 0 \Rightarrow p_n = p_{\tau_1} = p_{\tau_2} = 0, \tag{16}$$

$$\text{if } \varepsilon_n = 0 \Rightarrow p_n \leq 0 \text{ and } [p_{\tau_1}^2 + \beta^2 p_{\tau_2}^2]^{1/2} + \mu_1 p_n \leq 0, \tag{17}$$

$$\text{if } [p_{\tau_1}^2 + \beta^2 p_{\tau_2}^2]^{1/2} + \mu_1 p_n < 0 \Rightarrow |\Delta u_\tau| = 0,$$

$$\text{if } [p_{\tau_1}^2 + \beta^2 p_{\tau_2}^2]^{1/2} + \mu_1 p_n = 0 \Rightarrow |\Delta u_\tau| \geq 0, \tag{18}$$

where the notations of all the variables and parameters are used as in the usual way.

4 PQP Solution

Based on the description in the previous sections, the PVP of 3D elastic-plastic contact problems can be stated as follows: Among the admissible displacement fields, which satisfy Equations (13) and (14b), the actual solution makes the potential energy

$$\Pi[\lambda, \tilde{\lambda}] = \int_\Omega [du_{i,j} D_{ijkl} du_{k,l}/2 - \lambda_\alpha R_{kl\alpha} du_{k,l} - db_i du_i] d\Omega$$
$$+ \int_{S_c} (d\varepsilon_{ci} D_{cij} d\varepsilon_{ij}/2 - \tilde{\lambda}_\alpha \tilde{R}_{k\alpha} d\varepsilon_{ck}) ds - \int_{S_p} d\bar{p}_i du_i ds \tag{19}$$

minimum under the control of Equations (2) and (11), where

$$R_{kl\alpha} = (\partial g_\alpha / \partial \sigma_{ij}) D_{ijkl}, \quad \tilde{R}_{k\alpha} = (\partial \tilde{g}_\alpha / \partial p_{ci}) D_{cik},$$

$$\varepsilon_c = \Delta u_c, \quad D_{cij} = \delta_{ij} E_\tau = \delta_{ij} E_n = \delta_{ij} E. \tag{20}$$

After introducing the interpolation shape function matrix **N** and assembling the total energy of the contact system, Equation (19) can be expressed as

$$\Pi[\lambda'] = \frac{1}{2} d\mathbf{u}^T \mathbf{K} d\mathbf{u} - d\mathbf{u}^T (\Phi \boldsymbol{\lambda}' + \mathbf{p}_0 - \mathbf{p}_\delta), \tag{21}$$

where

$$\mathbf{K} = \int_\Omega \mathbf{B}^T \mathbf{D} \mathbf{B} d\Omega + \int_{S_c} \mathbf{N}^T \mathbf{D}_c \mathbf{N} ds,$$

$$\mathbf{p}_0 = \int_\Omega \mathbf{N}^T d\mathbf{b} d\Omega + \int_{S_p} \mathbf{N}^T d\bar{\mathbf{p}} ds, \quad \mathbf{p}_\delta = \int_{S_c} \mathbf{N}^T \mathbf{D}_c \mathbf{N} ds \boldsymbol{\delta}_c^*,$$

$$\Phi = \int_\Omega \mathbf{B}^T \mathbf{R}^T d\Omega + \int_{S_c} \mathbf{N}^T \tilde{\mathbf{R}}^T ds, \quad \boldsymbol{\lambda}' = [\boldsymbol{\lambda}^T, \tilde{\boldsymbol{\lambda}}^T]^T. \tag{22}$$

The control Equations (2) and (11) can be rewritten as

$$\mathbf{C}d\mathbf{u} - \mathbf{U}\boldsymbol{\lambda}' - (\mathbf{t}_0 + \mathbf{t}_\delta) + \mathbf{v}' = 0,$$
$$\mathbf{v}'^T\boldsymbol{\lambda}' = 0, \quad \mathbf{v}', \quad \boldsymbol{\lambda}' \geq 0, \tag{23}$$

where

$$\mathbf{C} = \int_\Omega \mathbf{W}\mathbf{B}d\Omega + \int_{S_c} \tilde{\mathbf{W}}\mathbf{N}ds, \quad \mathbf{U} = \int_\Omega \mathbf{M}d\Omega + \int_{S_c} \tilde{\mathbf{M}}ds,$$
$$\mathbf{t}_\delta = -\int_{S_c} \tilde{\mathbf{W}}_e ds\boldsymbol{\delta}_c^*, \quad \mathbf{t}_0 = -\int_\Omega \mathbf{f}_0 d\Omega - \int_{S_c} \tilde{\mathbf{f}}_0 ds. \tag{24}$$

From Equations (21) and (23), the following equations can be derived:

$$\mathbf{v}' - (\mathbf{U} - \mathbf{C}\mathbf{K}^{-1}\Phi)\boldsymbol{\lambda}' = -\mathbf{C}\mathbf{K}^{-1}\mathbf{p}_0 + \mathbf{t}_0 - (\mathbf{U} - \mathbf{C}\mathbf{K}^{-1}\Phi)\boldsymbol{\delta}_c^*,$$
$$\mathbf{v}'^T\boldsymbol{\lambda}' = 0, \quad \mathbf{v}', \quad \boldsymbol{\lambda}' \geq 0, \tag{25}$$

where we expand $\boldsymbol{\delta}_c^*$ to the space of $\boldsymbol{\lambda}'$.

We can see that Equation (25) is a standard linear complementarity problem. Because the penalty term exists in this equation, it is worthwhile to eliminate the penalty factor E in order to get a more accurate solution. The detailed description can be found in [43]. The matrices in Equation (25) have the forms

$$\mathbf{K} = \begin{bmatrix} \mathbf{K}_{11} & \mathbf{K}_{12} \\ \mathbf{K}_{21} & \mathbf{K}_{22} \times \mathbf{K}'_{22} \end{bmatrix}, \quad \mathbf{p}_0 = \begin{bmatrix} \mathbf{p}_{0i} \\ \mathbf{p}_{0o} \end{bmatrix}, \mathbf{t}_0 = \begin{bmatrix} \mathbf{t}_{0i} \\ \mathbf{t}_{0o} \end{bmatrix},$$
$$\mathbf{C} = \begin{bmatrix} \mathbf{C}_{11} & \mathbf{C}_{12} \\ 0 & E\mathbf{C}'_{22} \end{bmatrix}, \quad \boldsymbol{\Phi} = \begin{bmatrix} \boldsymbol{\Phi}_{11} & 0 \\ \boldsymbol{\Phi}_{21} & E\boldsymbol{\Phi}'_{22} \end{bmatrix}, \quad \mathbf{U} = \begin{bmatrix} \mathbf{U}_{11} & 0 \\ 0 & E\mathbf{U}'_{22} \end{bmatrix}. \tag{26}$$

Through a series of derivations, Equation (25) can be divided into

$$\mathbf{v} - (\mathbf{U}_{11} - \mathbf{C}_{11}\mathbf{K}_{11}^{-1}\boldsymbol{\Phi}_{11})\boldsymbol{\lambda} - (\mathbf{C}_{11}\mathbf{K}_{11}^{-1}\mathbf{K}_{12} - \mathbf{C}_{12})\Phi'_{22}\tilde{\boldsymbol{\lambda}}'$$
$$= -\mathbf{C}_{11}\mathbf{K}_{11}^{-1}\mathbf{p}_{0i} + \mathbf{t}_{0i} - (\mathbf{C}_{11}\mathbf{K}_{11}^{-1}\mathbf{K}_{12} - \mathbf{C}_{12})\boldsymbol{\delta}_c^*$$
$$\tilde{\mathbf{v}} - \mathbf{C}'_{22}(\mathbf{K}_{21}\mathbf{K}_{11}^{-1}\boldsymbol{\Phi}_{11} - \boldsymbol{\Phi}_{21})\boldsymbol{\lambda} - \mathbf{C}'_{22}\mathbf{K}^*_{22}\boldsymbol{\Phi}'_{22}\tilde{\boldsymbol{\lambda}}$$
$$= -\mathbf{C}'_{22}\mathbf{p}_0^* + \mathbf{t}_{0o} - \mathbf{C}'_{22}\mathbf{K}^*_{22}\boldsymbol{\delta}_c^*$$
$$\mathbf{v}^T\boldsymbol{\lambda} = 0, \quad \tilde{\mathbf{v}}^T\tilde{\boldsymbol{\lambda}} = 0, \quad \mathbf{v}, \boldsymbol{\lambda}, \tilde{\mathbf{v}}, \tilde{\boldsymbol{\lambda}} \geq 0, \tag{27}$$

where

$$\mathbf{K}^*_{22} = \mathbf{K}_{22} - \mathbf{K}_{21}\mathbf{K}_{11}^{-1}\mathbf{K}_{12}, \quad \mathbf{p}_0^* = \mathbf{p}_{0o} - \mathbf{K}_{21}\mathbf{K}_{11}^{-1}\mathbf{p}_{0i}.$$

Obviously, the expressions of the stiffness matrix and the load vectors in above equations are the same as those in the substructuring analysis, so the substructure technique can be used to treat the contact problems and the algorithm presented here will be easily implemented in a code for substructuring analysis. The design of contact elements description can be found in [40].

5 Constitutive Equations and Frictional Contact Algorithm of Cosserat Bodies

In the Cosserat theory, an independent rotation vector $d\omega_i$ is introduced besides the translational displacement vector du_i. The curvature tensor $d\chi_{ij}$ and the couple stress tensor dm_{ij} are introduced. So the strain increments in the elastic-plastic analysis of Cosserat bodies can be expressed as

$$d\varepsilon_{ij} = d\varepsilon_{ij}^e + d\varepsilon_{ij}^p, \tag{28}$$

$$d\chi_{ij} = d\chi_{ij}^e + d\chi_{ij}^p. \tag{29}$$

The plastic components can be deduced from the flow rules

$$d\varepsilon_{ij}^p = \lambda_\alpha \frac{\partial g_\alpha}{\partial \sigma_{ij}}, \quad d\chi_{ij}^p = \lambda_\alpha \frac{\partial g_\alpha}{\partial m_{ij}}. \tag{30}$$

In the elastic-plastic Cosserat theory, m_{ij} and χ_{ij}^p are considered in the construction of the yield function. As a result, the yield function for the Cosserat continuum can be defined as follows:

$$f_\alpha(\sigma_{ij}, \varepsilon_{ij}^p, m_{ij}, \chi_{ij}^p, k) \le 0, \quad \lambda_\alpha \begin{cases} \ge 0, & f_\alpha = 0, \\ = 0, & f_\alpha < 0. \end{cases} \tag{31}$$

The yield function $f_\alpha(\sigma_{ij}, \varepsilon_{ij}^p, m_{ij}, \chi_{ij}^p, k)$ can be rewritten approximately in the linear terms of Taylor's series expansion at the current state as

$$f_\alpha = f_\alpha^0 + \mathbf{W}^u d\varepsilon + \mathbf{W}^\omega d\chi - \mathbf{M}\lambda, \tag{32}$$

where

$$\begin{aligned} \mathbf{W}^u &= \left(\frac{\partial f_\alpha}{\partial \boldsymbol{\sigma}}\right)^T \mathbf{D}, \quad \mathbf{W}^\omega = \left(\frac{\partial f_\alpha}{\partial \mathbf{m}}\right)^T \mathbf{G}, \\ \mathbf{M} &= \mathbf{W}^u \left(\frac{\partial g}{\partial \boldsymbol{\sigma}}\right)^T + \mathbf{W}^\omega \left(\frac{\partial g}{\partial \mathbf{m}}\right)^T - \left(\frac{\partial f_\alpha}{\partial \boldsymbol{\varepsilon}^p}\right)^T \left(\frac{\partial g}{\partial \boldsymbol{\sigma}}\right)^T \\ &\quad - \left(\frac{\partial f_\alpha}{\partial \mathbf{m}^p}\right)^T \left(\frac{\partial g}{\partial \boldsymbol{\sigma}}\right)^T - \left(\frac{\partial f_\alpha}{\partial k}\right) h^T. \end{aligned} \tag{33}$$

For the contact problem, to consider the effect of couple stress on the contact surfaces, it is possible to assume that the following conditions hold on the contact surface between two Cosserat bodies

$$\bar{M}_{cj} = -\bar{M}(p_n, p_t, du, d\omega), \quad j = 3 \text{ on } S_c, \tag{34}$$

where $\bar{M}(p_n, p_t, du, d\omega)$ is the function of the contact pressure forces and relative displacements. To simplify the computations, in our numerical examples, $\bar{M}_{c3}$ is assumed to be zero. However, it is possible to assume that $\bar{M}(p_n, p_t, du, d\omega)$ should be obtained by laboratory test. Without loss of generality, it is assumed that the following constitutive relationship in incremental form holds

$$d\bar{M}_{cj} = K(p_n)d\omega_{cj}, \quad j = 3 \text{ on } S_c, \tag{35}$$

where $K(p_n)$ is a contact state dependent "stiffness" and is defined as

$$K(p_n) = C_0\left(-\frac{p_n}{H}\right)^{\alpha}, \tag{36}$$

in which C_0, H, α are material parameters.

Adopting the similar method as that proposed above, the complementary equations for contact problem of Cosserat bodies are obtained

$$\left.\begin{aligned} &\tilde{f}_k^0 + \tilde{\mathbf{W}}_k d\mathbf{u}_c - \tilde{\mathbf{M}}_k\tilde{\boldsymbol{\lambda}} + \tilde{\upsilon}_k = 0 \\ &\tilde{\lambda}_k\tilde{\upsilon}_k = 0, \quad \tilde{\lambda}_k, \tilde{\upsilon}_k \geq 0, \quad k = 1,2,3 \end{aligned}\right\} \quad \text{on } S_c, \tag{37}$$

where $\tilde{\upsilon}_k$ is the slack vector, $\tilde{f}_k^0$ the initial value of the yield function at time t_0, and

$$\tilde{W}_k = \left(\frac{\tilde{\partial} f_k}{\partial \mathbf{p}_c}\right)^T \mathbf{D}_c, \quad \tilde{M}_k = \left(\frac{\tilde{\partial} f_k}{\partial \mathbf{p}_c}\right)^T \mathbf{D}_c\left(\frac{\partial \tilde{\mathbf{g}}}{\partial \mathbf{p}_c}\right)^T.$$

6 Governing Equations for Contact Problems of Cosserat Bodies

The governing equations of the elastic-plastic contact problems of Cosserat bodies are given as follows:

Equilibrium equations

$$d\sigma_{ij,j} + db_i = 0, \tag{38}$$

$$dM_{ij,i} + e_{jik}d\sigma_{ik} + d\psi_j = 0, \quad j = 3. \tag{39}$$

Continuity equations

$$d\varepsilon_{ij} = du_{j,i} + e_{jik}d\omega_k, \tag{40}$$

$$d\chi_{ji} = d\omega_{j,i}, \quad j = 3, i = 1,2. \tag{41}$$

Constitutive equations

$$\begin{aligned} d\sigma_{ij} &= D_{ijkl}(d\varepsilon_{kl} - d\varepsilon_{kl}^p), \\ dm_{ij} &= G_{ijkl}(d\chi_{ij} + d\chi_{ij}^p), \quad k = 3, \end{aligned} \tag{42}$$

$$dM_{ij}n_i = dM_{cj} = K(p_n)d\omega_{cj}, \quad j = 3 \text{ on } S_c, \tag{43a}$$

$$f_\alpha(\sigma_{ij}, \varepsilon_{ij}^p, m_{ij}, \chi_{ij}^p, k) \le 0,$$
$$\lambda_\alpha \begin{cases} \ge 0, & f_\alpha = 0, \\ = 0, & f_\alpha < 0. \end{cases} \tag{43b}$$

Boundary conditions:

$$du_i = d\bar{u}_i \quad \text{on } S_u, \tag{44a}$$

$$n_j d\sigma_{ij} = d\bar{p}_i \quad \text{on } S_p, \tag{44b}$$

$$d\omega_i = d\bar{\omega}_i \quad \text{on } S_\omega, \tag{44c}$$

$$dM_{ij}n_i = d\bar{\phi}_j, \quad j = 3 \text{ on } S_M, \tag{44d}$$

$$\left.\begin{array}{l} \tilde{f}_k^0 + \tilde{W}_k d\mathbf{u}_c - \tilde{M}_k\tilde{\boldsymbol{\lambda}} + \tilde{\upsilon}_k = 0 \\ \tilde{\lambda}_k\tilde{\upsilon}_k = 0, \quad \tilde{\lambda}_k, \tilde{\upsilon}_k \ge 0, \quad k = 1, 2, 3 \end{array}\right\} \text{ on } S_c. \tag{44e}$$

7 PQP Solution for Contact Problems of Cosserat Bodies

According to the parametric variational principle, among admissible solutions which satisfy the strain-displacement relations (40) and (41) and the boundary conditions (44a) and (44c), the exact solution under the control of system (43b) and (44e) yields the following virtual work formulation

$$\begin{aligned} \delta\Pi_e^p = &\int_\Omega d\sigma_{ij}\delta d\varepsilon_{ij}d\Omega + \int_\Omega dM_{ij}\delta d\chi_{ij}d\Omega - \int_\Omega \lambda_\alpha R_{kl\alpha}\delta d\varepsilon_{kl}d\Omega \\ &- \int_\Omega \lambda_\alpha R'_{kl\alpha}\delta d\chi_{kl}d\Omega - \int_\Omega db_i\delta du_i d\Omega - \int_\Omega d\Psi_i\delta d\omega_i d\Omega \\ &- \int_{S_P} d\bar{T}_i\delta du_i dS - \int_{S_M} d\bar{\psi}_i\delta d\omega_i dS + \int_{S_c} D_{cij}du_{ci}\delta du_{cj}dS \\ &- \int_{S_c} \tilde{\lambda}_\alpha\tilde{R}_{k\alpha}\delta du_{ck}dS + \int_{S_c} d\omega_{c3}K(p_n)\delta d\omega_{c3}dS, \end{aligned} \tag{45}$$

where

$$\begin{aligned} &R_{kl\alpha} = \left(\frac{\partial g_\alpha}{\partial\sigma_{ij}}\right)D_{ijkl}, \quad R'_{kl\alpha} = \left(\frac{\partial g_\alpha}{\partial m_{ij}}\right)G_{ijkl}, \quad \tilde{R}_{k\alpha} = \left(\frac{\partial\tilde{g}_\alpha}{\partial p_{ci}}\right)D_{cik}, \\ &D_{c11} = E_t, \quad D_{c22} = E_n, \quad D_{c12} = D_{c21} = 0. \end{aligned} \tag{46}$$

After introducing the interpolation shape function matrices $\mathbf{N}^u$ and $\mathbf{N}^\omega$ into Equation (45), the finite element equations of the problem can be expressed as

$$\begin{bmatrix} \mathbf{K}^{\omega\omega} + \tilde{\mathbf{K}}^{\omega\omega} + \mathbf{K}^{\omega_c\omega_c} & \mathbf{K}^{\omega u} \\ \mathbf{K}^{u\omega} & \mathbf{K}^{uu} + \mathbf{K}^{u_c u_c} \end{bmatrix} \cdot \begin{bmatrix} d\boldsymbol{\omega}' \\ d\mathbf{u}' \end{bmatrix}$$

$$- \begin{bmatrix} {}^1\boldsymbol{\Phi}^\omega + {}^2\boldsymbol{\Phi}^\omega & 0 \\ \boldsymbol{\Phi}^u & \mathbf{K}^{u_c\tilde{\lambda}} \end{bmatrix} \cdot \begin{bmatrix} \boldsymbol{\lambda} \\ \tilde{\boldsymbol{\lambda}} \end{bmatrix} - \begin{bmatrix} \mathbf{P}^\omega + \bar{\mathbf{P}}^\omega \\ \mathbf{P}^u + \bar{\mathbf{P}}^u - \mathbf{P}^{uc} \end{bmatrix} = 0, \tag{47}$$

where

$$\mathbf{K}^{uu} = \int_\Omega \mathbf{B}^{u^T} \mathbf{D} \mathbf{B}^u d\Omega, \quad \mathbf{K}^{u\omega} = \int_\Omega \mathbf{B}^{u^T} \mathbf{D} \mathbf{e} \mathbf{N}^\omega d\Omega,$$

$$\mathbf{K}^{\omega u} = \int_\Omega \mathbf{N}^{\omega^T} \mathbf{e}^T \mathbf{D} \mathbf{B}^u d\Omega, \quad \tilde{\mathbf{K}}^{\omega\omega} = \int_\Omega \mathbf{N}^{\omega^T} \mathbf{e}^T \mathbf{D} \mathbf{e} \mathbf{N}^\omega d\Omega,$$

$$\mathbf{K}^{\omega\omega} = \int_\Omega \mathbf{B}^{\omega^T} \mathbf{G} \mathbf{B}^\omega d\Omega, \quad \boldsymbol{\Phi}^u = \int_\Omega \mathbf{B}^{u^T} \mathbf{R}^T d\Omega,$$

$${}^1\boldsymbol{\Phi}^\omega = \int_\Omega \mathbf{N}^{\omega^T} \mathbf{R}^T \mathbf{e} d\Omega, \quad {}^2\boldsymbol{\Phi}^\omega = \int_\Omega \mathbf{B}^{\omega^T} \mathbf{R}'^T d\Omega,$$

$$\mathbf{P}^u = \int_\Omega \mathbf{N}^{u^T} d\mathbf{b} d\Omega, \quad \mathbf{P}^\Omega = \int_\Omega \mathbf{N}^{\omega^T} d\boldsymbol{\Psi} d\Omega,$$

$$\tilde{\mathbf{P}}^u = \int_{S_P} \mathbf{N}^{u^T} d\bar{\mathbf{T}} dS, \quad \tilde{\mathbf{P}}^\omega = \int_{S_M} \mathbf{N}^{\omega^T} d\bar{\boldsymbol{\Psi}} S,$$

$$\mathbf{K}^{u_c u_c} = \int_{S_c} \mathbf{N}^{u^T} \mathbf{D}_c \mathbf{N}^u dS, \quad \mathbf{P}^{uc} = \int_{S_c} \mathbf{N}^{u^T} \mathbf{D}_c \mathbf{N}^u dS \boldsymbol{\delta}_c^*,$$

$$\mathbf{K}^{u_c\tilde{\lambda}} = \int_{S_c} \mathbf{N}^{u^T} \tilde{\mathbf{R}}^T dS, \quad \mathbf{K}^{\omega_c\omega_c} = \int_{S_c} \mathbf{N}^{\omega^T} K(p_n) \mathbf{N}^\omega dS.$$

Equation (47) can be rewritten as

$$\mathbf{K} d\mathbf{u} - \boldsymbol{\Phi}\boldsymbol{\lambda}' - \mathbf{p}_0 + \mathbf{p}_\delta = 0, \tag{48}$$

where

$$\mathbf{K} = \begin{bmatrix} \mathbf{K}^{\omega\omega} + \tilde{\mathbf{K}}^{\omega\omega} + \mathbf{K}^{\omega_c\omega_c} & \mathbf{K}^{\omega u} \\ \mathbf{K}^{u\omega} & \mathbf{K}^{uu} + \mathbf{K}^{u_c u_c} \end{bmatrix},$$

$$\boldsymbol{\Phi} = \begin{bmatrix} {}^1\boldsymbol{\Phi}^\omega + {}^2\boldsymbol{\Phi}^\omega & 0 \\ \boldsymbol{\Phi}^u + \mathbf{K}^{u_c\tilde{\lambda}} & \end{bmatrix}, \quad \mathbf{P}_0 = \begin{bmatrix} \mathbf{P}^\omega + \tilde{\mathbf{P}}^\omega \\ \mathbf{P}^u + \tilde{\mathbf{P}}^u \end{bmatrix},$$

$$\mathbf{P}_\delta = \begin{bmatrix} 0 \\ \mathbf{P}^{uc} \end{bmatrix}, \quad \boldsymbol{\lambda}' = [\boldsymbol{\lambda}^T, \tilde{\boldsymbol{\lambda}}^T]^T, \quad d\mathbf{u} = [d\boldsymbol{\omega}'^T, d\mathbf{u}'^T]^T$$

and its governing Equations (43b) and (44e) can be discretized as

$$\begin{aligned} &\mathbf{C}d\mathbf{u} - \mathbf{U}\boldsymbol{\lambda}' - (\mathbf{t}_0 + \mathbf{t}_\delta) + \mathbf{v}' = 0, \\ &\mathbf{v}'^T\boldsymbol{\lambda}' = 0, \quad \mathbf{v}', \boldsymbol{\lambda}' \geq 0, \end{aligned} \tag{49}$$

where

$$\mathbf{C} = [\mathbf{C}^2 \ \mathbf{C}^1],$$

$$\mathbf{C}^1 = \int_\Omega \mathbf{W}^u\mathbf{B}^u d\Omega + \int_{S_c} \tilde{\mathbf{W}}\mathbf{N}^u dS, \quad \mathbf{C}^2 = \int_\Omega \mathbf{W}^u\mathbf{e}\mathbf{N}^\omega d\Omega + \int_\Omega \mathbf{W}^u\mathbf{B}^\omega d\Omega,$$

$$\mathbf{U} = \int_\Omega \mathbf{M}d\Omega + \int_{S_c} \tilde{\mathbf{M}}dS, \quad \mathbf{t}_0 = -\int_\Omega \mathbf{f}^0_\alpha d\Omega - \int_{S_c} \tilde{\mathbf{f}}^0_k dS, \quad \mathbf{t}_\delta = -\int_{S_c} \tilde{\mathbf{W}}_k dS\boldsymbol{\delta}^*_c.$$

From Equations (48) and (49) and by a series of derivatives, we finally obtain

$$\begin{aligned} &\mathbf{v}' - (\mathbf{U} - \mathbf{C}\mathbf{K}^{-1}\Phi)\boldsymbol{\lambda}' = -\mathbf{C}\mathbf{K}^{-1}\mathbf{p}_0 + \mathbf{t}_0 - (\mathbf{U} - \mathbf{C}\mathbf{K}^{-1}\Phi)\boldsymbol{\delta}^*_c, \\ &\mathbf{v}'^T\boldsymbol{\lambda}' = 0, \quad \mathbf{v}', \boldsymbol{\lambda}' \geq 0. \end{aligned} \tag{50}$$

We can see that Equation (50) is quite similar to Equation (23). The only difference is that the coefficient matrices in Equation (50) include the terms corresponding to the rotational freedom $d\boldsymbol{\omega}'$. So all the derived procedure developed for the conventional contact model can be used for the elastic-plastic contact analysis of Cosserat bodies. The final discretized equations will be the same as those given in Equations (27).

All numerical algorithms and models mentioned above are implemented in the finite element analysis and design optimization software system JIGFEX.

8 Numerical Examples

8.1 Computation of Turbine

This example is used to show the capability of the computer program for the contact analysis of conventional engineering problems. Figure 5 shows a turbocharger compressor used in diesel locomotive. The compressor impeller is made of identical segments in the circumferential direction and hence constitutes a rotationally periodic structure. The impeller with a shaft hole 34 mm in diameter is mounted onto the sleeve and the compressor shaft via interference fit. The material of the sleeve and shaft is steel and the material of the impeller is aluminum. The three parts are assembled together slowly using a press (normally with oil applied at the intersection to act as a lubricant). Figure 6 shows the finite element mesh of the impeller. The effects of temperature and pressure loading are not considered as they are small in comparison with the centrifugal force due to the high speed of rotation. The main

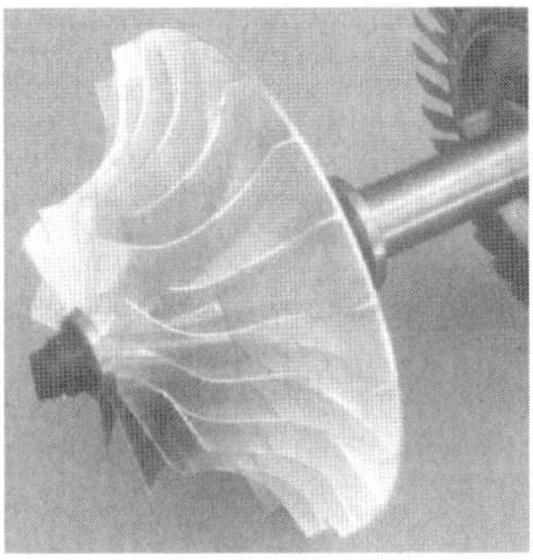

Fig. 5. The turbocharger compressor.

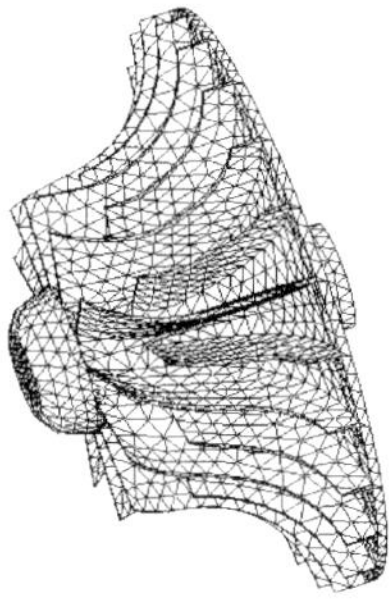

Fig. 6. Computation model.

loads applied to the compressor are centrifugal forces caused by high-speed rotation when operating. The load cases are computed for various rotational speeds n (0, 24000, 25000, 27000, 29000 rpm).

In Figures 7 and 8, solid curves represent the contact stress distribution on the external surface of shaft sleeve (surface between impeller and shaft sleeve), dashed curves represent the contact stress distribution on the internal surface of shaft sleeve (surface between shaft sleeve and shaft). Contact stresses distribution of both the external and internal surface of shaft sleeve along the axial direction are illustrated in Figure 7 with four rotational speeds. Along the axial direction, normal displacement of contact point pair of between impeller and shaft sleeve increases with the increase of the radial dimension of impeller (Figure 8). Computed results show there is obvious relationship between the mass distribution of impeller and normal displacement of contact points of impeller and shaft sleeve within the certain rotational speed.

8.2 Contact Analysis of Cosserat Bodies

This is a multibody contact problem, shown in Figure 9. The geometric parameters of the structure are: $a = 0.3$ cm, $R = 0.4$ cm, $b = 0.6$ cm. Materials parameters used

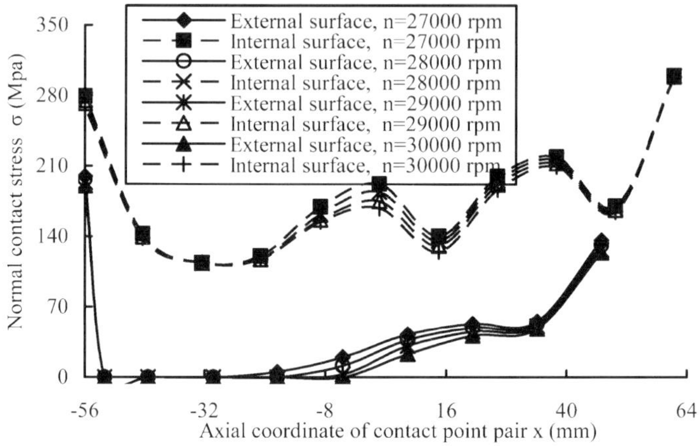

Fig. 7. Normal contact stress distributions along the axial direction under different rotational speeds.

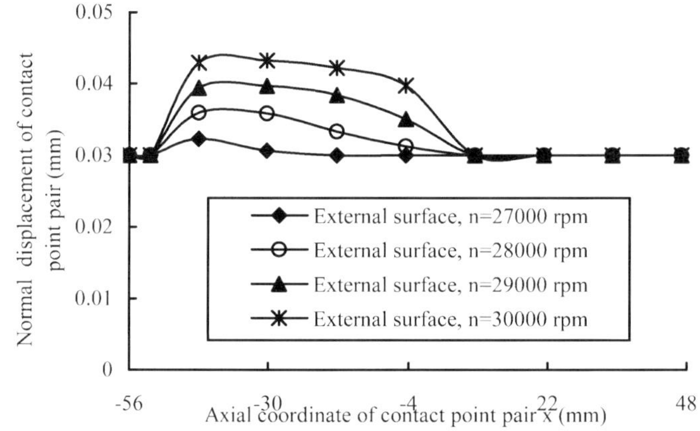

Fig. 8. Variation of normal displacement under different rotational speeds.

are: for material 1, $E = 50$ GPa, $\nu = 0.3$, $\mu_c = 10$ GPa, $l_c = 0.1$ cm; for material 2, $E = 3$ GPa, $\nu = 0.3$, $\mu_c = 1$ GPa, $l_c = 0.1$ cm; initial yield stress $\bar{\sigma} = 0.6$ GPa, hardening modulus $h = 3$ GPa; for material 3, $E = 40$ GPa, $\nu = 0.3$, $\mu_c = 10$ GPa, $l_c = 0.1$ cm. The load P puts on the top surface of the structure is 1.0 kN/cm. It is assumed that materials 1 and 2 hold on the elastic state.

The finite element mesh used in the computation is shown in Figure 10. Here, we only discuss the contact stress distribution along the upper contact surface, as shown in Figures 11 and 12. Figure 13 shows the variation of shear stresses on the contact surface in elastic-plastic state with the change of μ_c. Form the results, the differ-

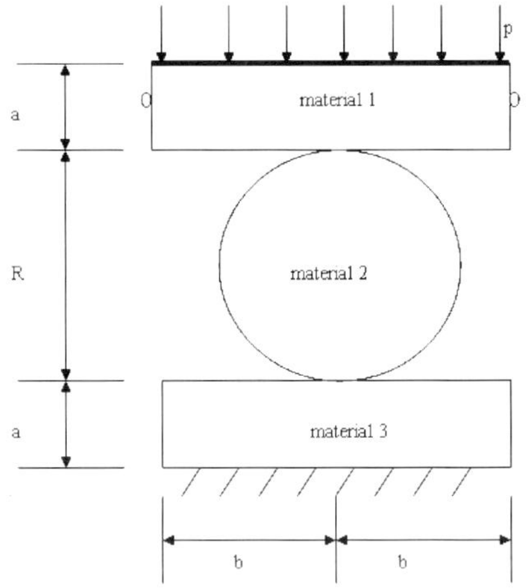

Fig. 9. Structure computed in numercal example 8.2.

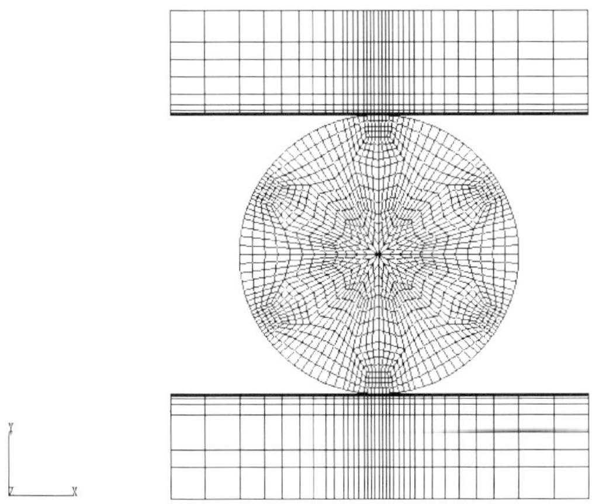

Fig. 10. Finite element model.

ences between the elastic and the elastic-plastic solutions are clearly observed. It can be found that the material parameters of Cosserat material have important effects on the mechanical response of the contact bodies. With a certain value of material parameters, the shear stress distribution along the contact surface will vanish.

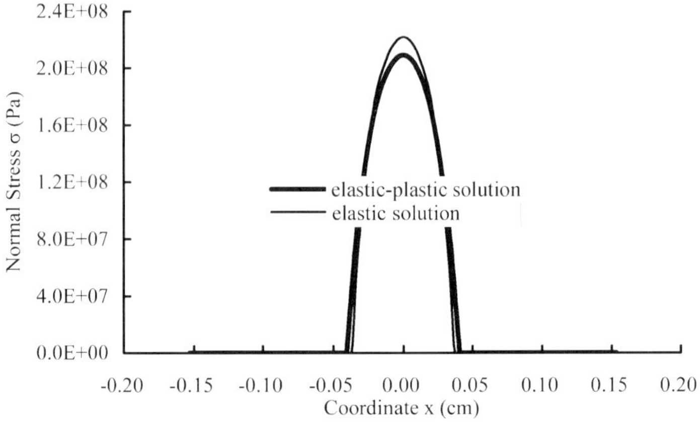

Fig. 11. Normal contact stress distribution along the upper contact surface.

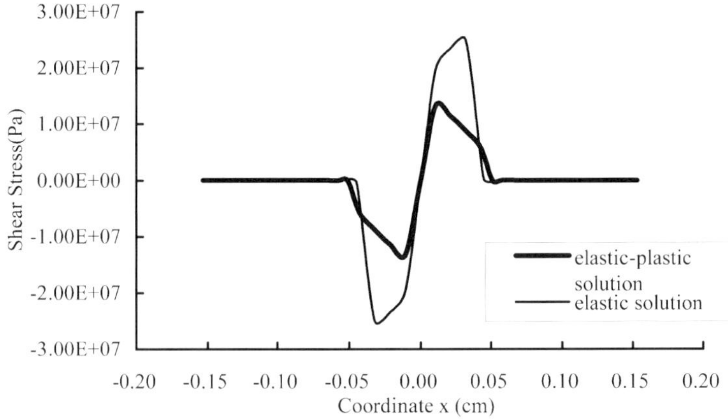

Fig. 12. Shear contact stress distribution along the upper contact surface.

9 Conclusions

Some advances and applications in mathematical programming method for numerical modeling of elastic-plastic contact problems are reviewed in this paper. With the orthotropic friction law being approximated by a polyhedral law of friction, the problem is reduced to a standard linear complementarity problem. The contact problem with friction between two Cosserat bodies is treated in the same way as that in plastic analysis. The numerical algorithm is implemented in the finite element analysis and design optimization software system JIFEX. Numerical results are presented to show the effectiveness of the model and algorithm proposed.

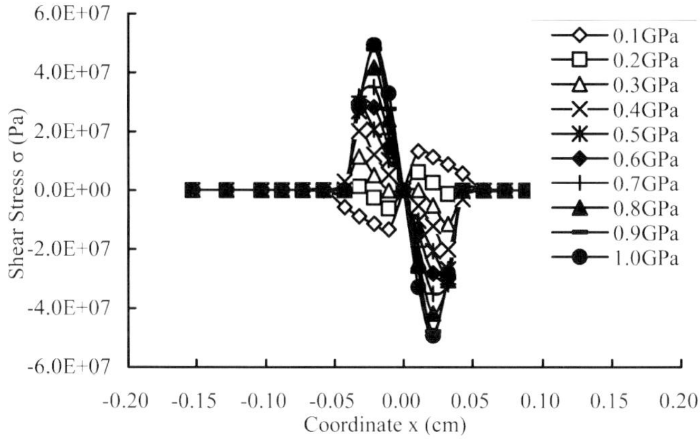

Fig. 13. Shear stress distribution along the upper contact surface with different μ_c used for the middle disk in the elastic-plastic state.

Acknowledgements

The supports of the National Natural Science Foundation of China (50679013, 10225212, 10421202, 10640420176, 10511120288), the Program for Changjiang Scholars and Innovative Research Team in University of China (PCSIRT) and the National Key Basic Research Special Foundation of China (2005CB321704) are gratefully acknowledged.

References

1. Alberti N., Cannizzaro L. and Riccobono R. (1987) A new numerical method for axiymmetrical forming process. *CIRP Annals – Manufacturing Technology*, 36(1):131-133.
2. Baggio C. and Trovalusci P. (1998) Limit analysis for no-tension and frictional three dimensional discrete systems. *Mechanics of Structures and Machines*, 26:287–304.
3. Christensen P.W. (2002) A semi-smooth Newton method for elasto-plastic contact problems. *International Journal of Solids and Structures*, 39:2323–2341.
4. Conry T.F. and Seireg A. (1971) A mathematical prograrnming method for design of elastic bodies in contact. *Journal of Applied Mechanics*, 38:387–392.
5. Czekanski A. and Meguid S.A. (2006) On the use of variational inequalities to model impact problems of elasto-plastic media. *International Journal of Impact Engineering*, 32:1485–1511.
6. Donev A., Torquato S., Stillinger F.H. and Connelly R. (2004) A linear programming algorithm to test for jamming in hard-sphere packings. *Journal of Computational Physics*, 197:139–166.
7. Dostál Z., David H. and Radek K. (2005) FETI based algorithms for contact problems: Scalability, large displacements and 3D Coulomb friction. *Computer Methods in Applied Mechanics and Engineering*, 194:395–409.

8. Dostál Z., Francisco A.M., Gomes N. and Sandra A.S. (2000) Solution of contact problems by FETI domain decomposition with natural coarse space projections. *Computer Methods in Applied Mechanics and Engineering*, 190:1611–1627.
9. Drosopoulos G.A., Stavroulakis G.E. and Massalas C.V. (2006) Limit analysis of a single span masonry bridge with unilateral frictional contact interfaces. *Engineering Structures*, 28:1864–1873.
10. Ferris M.C. and Tin-Loi F. (2001) Limit analysis of frictional block assemblies as a mathematical program with complementarity constraints. *International Journal of Mechanical Sciences*, 43:209–224.
11. Gawecki A., Kuczma M.S. and Krüger P. (1998) Analysis of slackened skeletal structures by mathematical programming. *Computers & Structures*, 69:639–654.
12. Gilbert M., Casapulla C. and Ahmed H.M. (2006) Limit analysis of masonry block structures with non-associative frictional joints using linear programming. *Computers & Structures*, 84:873–887.
13. Givoli D. and Doukhovni I. (1996) Finite element-quadratic programming approach for contact problems with geometrical nonlinearity. *Computers & Structures*, 61(1):31–41.
14. González J.A. and Abascal R. (2002) Solving 2D transient rolling contact problems using the BEM and mathematical programming techniques. *International Journal for Numerical Methods in Engineering*, 53:843–874.
15. Haug E., Chand R. and Pan K. (1977) Multibody elastic contact analysis by quadratic programming. *Journal of Optimization Theory and Applications*, 21:189–198.
16. Holanda A.S. de and Goncalves P.B. (2003) Postbuckling analysis of plates resting on a tensionless elastic foundation. *Journal of Engineering Mechanics*, 4:438–448.
17. Hu N. (1999) Buckling analysis of delaminated laminates with consideration of contact in buckling mode. *International Journal for Numerical Methods in Engineering*, 44:1457–1479.
18. Kontoleon M.J. and Baniotopoulos C.C. (2000) Computational aspects on the frictional unilateral contact problem arising on steel base plate connections. *Computers & Structures*, 78:303–309.
19. Laursen T.A. (2002) *Computational Contact and Impact Mechanics: Fundamentals of Modeling Interfacial Phenomena in Nonlinear Finite Element Analysis*. Springer-Verlag, Heidelberg.
20. Lee S.S. (1994) A computational method for frictional contact problem using finite element method. *International Journal for Numerical Methods in Engineering*, 37:217–228.
21. Li Y., Luan M. and Feng X. (2000) Fundamentals and formulations of discontinuous deformation mechanics model and its linear programming algorithm. *J. Dalian University of Technology*, 40(3):351–357.
22. Maier G. (1971) Incremental plastic analysis in the presence of large displacements and physical instabilizing effects. *International Journal of Solids and Structures*, 7:345–372.
23. Mijar A.R. and Arora J.S. (2000) Review of formulations for elastostatic fricional contact problems. *Structural and Multidisciplinary Optimization*, 20:167–189.
24. Refaat M.H. and Meguid S.A. (1998) A new strategy for the solution of frictional contact problems. *International Journal for Numerical Methods in Engineering*, 43:1053–1068.
25. Su R.K.L., Zhu Y. and Leung A.Y.T. (2002) Parametric quadratic programming method for elastic contact fracture analysis. *International Journal of Fracture*, 117:143–157.
26. Xuan Z.C. and Lee K.H. (2003) Interior point surrogate dual algorithm for unilateral problems. *Acta Mechanica*, 166:149–167.

27. Wriggers P. (2002) *Computational Contact Mechanics*. Wiley Press, New York.
28. Zhang H.W. (1995) Parametric Variational Principle for elastic-plastic consolidation analysis of saturated porous media. *International Journal for Numerical and Analytical Methods in Geomechanics*, 19:851–867.
29. Zhang H.W., Galvanetto U. and Schrefler B.A. (1999) Local analysis and global nonlinear behaviour of periodic assemblies of bodies in elastic contact. *Computational Mechanics*, 24(4):217–229.
30. Zhang H.W., He S.Y. and Li X.S. (2004) Non-interior smoothing algorithm for frictional contact problems. *Applied Mathematics and Mechanics*, 25(1):47–58.
31. Zhang H.W., He S.Y. and Li X.S. (2005) Two aggregate-function-based algorithms for analysis of 3D frictional contact by linear complementarity problem formulation. *Computer Methods in Applied Mechanics and Engineering*, 194(50–52):5139–5158.
32. Zhang H.W., He S.Y., Li X.S. and Wriggers P. (2004) A new algorithm for numerical solution of 3D elastoplastic contact problems with orthotropic friction law. *Computational Mechanics*, 34(1):1–14.
33. Zhang H.W. and Schrefler B.A. (2000) Gradient-dependent plasticity model and dynamic strain localisation analysis of saturated and partially saturated porous media: One dimensional model. *European Journal of Solid Mechanics, A/Solids*, 19(3):503–524.
34. Zhang H.W. and Schrefler B.A. (2000) Global constitutive behaviour of periodic assemblies of inelastic bodies in contact. *Mechanics of Composite Materials and Structures*, 7(4):355–382.
35. Zhang H.W., Wang J.B., Ye H.F. and Wang L. (2006) Parametric variational principle and quadratic programming algorithm for van der Waals force simulation of parallel and cross nanotubes. *International Journal of Solids and Structures*, in press.
36. Zhang H.W., Wang H., Wriggers P. and Schrefler B.A. (2005) A finite element model for contact analysis of multiple Cosserat bodies. *Computational Mechanics*, 36(6):444–458.
37. Zhang H.W., Zhang X.W. and Chen J.S. (2003) A new algorithm for numerical solution of dynamic elastic-plastic hardening and softening problems. *Computers & Structures*, 81(17):1739–1749.
38. Zhang H.W., Zhong W.X. and Gu Y.X. (1995) A combined programming and iteration algorithm for finite element analysis of three-dimensional contact problems. *Acta Mechanic Sinica*, 11(4):318–326.
39. Zhang H.W., Zhong W.X. and Gu Y.X. (1998) A combined parametric quadratic programming and iteration method for 3D elastic-plastic frictional contact problem analysis. *Computer Methods in Applied Mechanics and Engineering*, 155:307–324.
40. Zhang H.W., Zhong X.W., Wu C.H. and Liao A.H. (2006) Some advances and applications in quadratic programming method for numerical modeling of elastoplastic contact problems. *International Journal of Mechanical Science*, 2(48):176–189.
41. Zhong Z.H. and Mackerle J. (1992) Static contact problems – A review. *Engineering computations*, 9:3–37.
42. Zhong W.X. and Sun S.M. (1988) A finite element method for elasto-plastic structure and contact problem by parametric quadratic programming. *International Journal for Numerical Methods in Engineering*, 26(6):2723–2738.
43. Zhong W.X., Zhang H.W. and Wu C.W. (1997) *Parametric Variational Principle and Its Applications in Engineering*. Science Press, Beijing.

Finite Element Analysis of Nonsmooth Frictional Contact

Anna Pandolfi[1] and Michael Ortiz[2]

[1] *Dipartimento di Ingegneria Strutturale, Politecnico, 20133 Milano, Italy*
E-mail: pandolfi@stru.polimi.it
[2] *Engineering and Applied Science Division, California Institute of Technology, Pasadena, CA 91125, U.S.A.*
E-mail: ortiz@aero.caltech.edu

Abstract. A nonsmooth contact class of algorithms were introduced by Kane et al. [1] and extended to the case of friction by Pandolfi et. al [2]. The formulation specifically addresses geometries for which neither normals nor gap functions can be properly defined, e.g. bodies with corners. The formulation provides the incremental displacements in variational form, following from a minimum principle. Selected numerical examples of application of the algorithm are presented here.

Key words: Frictional contact, non-smooth contact, finite elements, variational formulations.

1 Introduction

The nonsmooth contact algorithm of Kane et al. [1] was extended to the case of friction in [2]. The nonsmooth contact approach specifically addresses multi-body nonsmooth geometries, for which neither normals nor gap functions can be defined. Examples of such situations are ballistic impacts of brittle targets, that often result in the formation of large numbers of fragments which undergo complex collision sequences before eventually scattering [3]. During fragmentation, the corners of many angular fragments may come together at a point, which precludes the definition of a gap function as a means of detecting – and constraining – their interpenetration. Additionally, the nonsmooth character of the fragments has the consequence that, in the contact region, normals are not defined in the usual sense. We refer to contact processes involving the simultaneous interaction between many angular bodies as nonsmooth contact.

Most contact algorithms proposed to date envision two smooth bodies in contact and use a gap function to constrain or penalize interpenetration (see, e.g., [4–6], and references in [1, 2]). Such approaches, although effective and widely used, do not

Peter Wriggers and Udo Nackenhorst (eds.), IUTAM Symposium on Computational Contact Mechanics, 57–76.

appear to be readily applicable to the analysis of nonsmooth contact. In fact, most of them can be viewed as a regularization of the well known Signorini condition, which relates the contact forces to a convenient measure of the distance between the contacting bodies. Frictionless nonsmooth contact algorithms, as discussed in [1], use nonsmooth analysis [7] to characterize the contact forces and to provide the analytical tools for treating time-discretized approximations. Goals and techniques described in [1] are quite different from the ones used in previous approaches based on nonsmooth analysis [8,9]. This approach does not require any assumption of smoothness of the boundary and enables the analysis of contact between sharp features such as edges and vertices. As noted in [10], nonsmooth algorithms bear a resemblance to those which are suggested by the mathematical theory of plasticity [11, 12], especially as regards the use of closest-point projections, although the admissible sets which arise in contact problems are generally nonconvex (which precludes the direct application of convex analysis).

The complex phenomenon of frictional contact has been addressed widely in the literature (see, e.g., [13–17], and other references in [2]). Frictional effects are generally accounted for by the introduction of a friction law which relates the sliding velocity to the tangential contact forces. Frictional tractions can be exerted without sliding, i.e., under stick conditions; when sliding occurs, the frictional tractions always oppose the sliding velocity and are, therefore, dissipative. In [2], the variational structure of the nonsmooth contact theory [1] was extended in a satisfactory way to the case of frictional contact. The treatment of friction fits within the general variational framework for dynamic and dissipative problems [18–20]. The approach relies on time-discretization and leads to the formulation of minimum principles characterizing the solutions of the incremental problem. The function to be minimized contains both conservative and dissipative terms, and may be regarded as an incremental potential energy. The minimization of the incremental potential energy delivers a solution consistent with the equations of motion of solids in frictional contact, accounting for stick-slip conditions.

A further qualifying feature of the proposed algorithms is that they do not require the definition "a priori" of master and slave surfaces. This allows for the analysis of problems involving unknown or variable contact surfaces or self-contact.

2 Frictionless Contact

The algorithm here considered, based on the tools of nonsmooth analysis [7], specifically applies to the finite-element solution of contact problems.

We consider a deformable body occupying a domain $B_0 \subset \mathbb{R}^3$ in its reference configuration and a finite element discretization. We restrict our attention to elastic bodies, so that $\phi(\boldsymbol{\varphi})$ represents the strain energy density. The motions of the body in the time interval $[0, T]$ are described by deformation mappings $\boldsymbol{\varphi} : B_0 \times [0, T] \to \mathbb{R}^3$. For a fixed $t \in [0, T]$, the deformation mappings $\boldsymbol{\varphi}(\cdot, t)$

define a finite-dimensional space X. We are concerned with deformation mappings that, in presence of impenetrability constraints, do not entail separation or interpenetration of matter. Such deformation mappings define the admissible set C (one-to-one mappings in X) [1, 2, 7].

In presence of impenetrability constraints, the action functional $I[\boldsymbol{\varphi}]$ of the system can be written as:

$$I[\boldsymbol{\varphi}] = \int_0^T \left[\frac{1}{2} \dot{\boldsymbol{\varphi}}^T \mathbf{M} \dot{\boldsymbol{\varphi}} - \phi(\boldsymbol{\varphi}) - I_C(\boldsymbol{\varphi}) + \mathbf{f}^{\text{ext}} \cdot \boldsymbol{\varphi} \right] dt \tag{1}$$

where $\mathbf{M}$ is the mass matrix of the solid, $\mathbf{f}^{\text{ext}}(t)$ are the external forces, and the term $I_C(\boldsymbol{\varphi})$ added to the energy of the solid represents the indicator function of the admissible set C:

$$I_C(\boldsymbol{\varphi}) = \begin{cases} 0, & \text{if } \boldsymbol{\varphi} \in C; \\ \infty, & \text{otherwise.} \end{cases} \tag{2}$$

As usually done in the so-called barrier methods, the term $I_C(\boldsymbol{\varphi})$ in the energy effectively bars the trajectories from violating the interpenetration constraints.

As shown in [1], the absolutely continuous trajectories $\boldsymbol{\varphi}(t)$ that render stationary the action functional are the solutions of the equation:

$$\mathbf{0} \in \mathbf{M}\ddot{\boldsymbol{\varphi}} + \mathbf{f}^{\text{int}}(\boldsymbol{\varphi}) + \partial I_C(\boldsymbol{\varphi}) - \mathbf{f}^{\text{ext}}, \tag{3}$$

where $\mathbf{f}^{\text{int}}(\boldsymbol{\varphi}) = \partial\phi(\boldsymbol{\varphi})$ are the internal forces and $\partial I_C(\boldsymbol{\varphi}) = \mathbf{f}^{\text{con}}(\boldsymbol{\varphi})$ is the generalized gradient of the indicator function, consisting of all the contact force fields corresponding to a contact configuration. In the nonsmooth case, the set of contact reactions may include cones centered at the corners of the surfaces in contact [7]. The generalized gradient reduces to the ordinary derivative at points where the function is continuously differentiable. Thus, if the surfaces in contact are smooth, $\partial I_C(\boldsymbol{\varphi})$ consists of all normal and compressive force fields over the area of contact. The problem is completed by the inclusion of suitable initial conditions on positions and velocities.

Characterizations of the admissible set C in case of polyhedra can be found in [1, 2]. Polyhedra result from a triangulation of the domain of the solid by simplices. By enforcing the absence of intersection between a given pair of boundary simplices (triangles in three dimensions), the admissible set has the form of a collection of algebraic inequality constraints on the nodal displacements:

$$g_\alpha(\boldsymbol{\varphi}) \geq 0, \quad \alpha = 1, \ldots, N, \tag{4}$$

N being the total number of possible contact constraints. Thus, the indicator function can be expressed as:

$$C = \{\boldsymbol{\varphi} : B_{0h} \times [0, T] \rightarrow \mathbb{R}^3 \mid g_\alpha(\boldsymbol{\varphi}) \geq 0, \quad \alpha = 1, \ldots, N\}, \tag{5}$$

where B_{0h} denotes the discretized body. As consequence, the system of normal contact forces $\mathbf{N}_\alpha$ corresponding to constraint α may be written in the form:

$$\mathbf{N}_\alpha = \lambda_\alpha \nabla g_\alpha(\boldsymbol{\varphi}_\alpha), \tag{6}$$

where λ_α is a scalar multiplier which is subject to unilateral constraints, and $\boldsymbol{\varphi}_\alpha$ is the collection of position vectors of the nodes attached to the pair of simplices involved in constraint α. The global contact system $\mathbf{f}^{\mathrm{con}}(\boldsymbol{\varphi})$ is obtained by the assembly (symbolically indicated with $\mathcal{A}$) of all the local normal force systems in the sense of the finite elements:

$$\mathbf{f}^{\mathrm{con}}(\boldsymbol{\varphi}) = \mathcal{A}(\{\mathbf{N}_\alpha,\ \alpha = 1, \ldots, N\}). \tag{7}$$

The invariance properties of C render I_C (and the action I) invariant under translations and rotations, thus preserve linear and angular momentum [21]. Additionally, the time independence of the Lagrangian provides global energy conservation [1,2].

3 Frictional Contact

In the presence of friction, the force system in Equation (3) has to include a frictional force field:

$$\mathbf{0} \in \mathbf{M}\ddot{\boldsymbol{\varphi}} + \mathbf{f}^{\mathrm{int}}(\boldsymbol{\varphi}) + \partial I_C(\boldsymbol{\varphi}) + \mathbf{f}^{\mathrm{fri}} - \mathbf{f}^{\mathrm{ext}}. \tag{8}$$

The frictional forces $\mathbf{f}^{\mathrm{fri}}$ are required to be self-equilibrated (they do not work under rigid-body motions) and tangential to the surfaces in contact (they do not work under conditions of normal separation of the bodies). Additionally, we assume that the magnitude of the frictional forces depends on the normal pressure through a suitable friction law. Such restrictions render the formulation of frictional force systems a difficult task, especially in a finite deformation framework.

The proper definition of frictional forces in the context of the discretized contact problem at the hand requires the definition of the associated sliding velocity field, which must be derived carefully considering one by one all the involved impenetrability constraints.

3.1 Sliding Velocity Field

We denote by $\dot{\boldsymbol{\varphi}}_\alpha$ the collection of velocities of the nodes contained in the pair of simplices involved in constraint α. The local velocity field $\dot{\boldsymbol{\varphi}}_\alpha$ may include rigid-body components and normal opening or closure modes, so that it can be decomposed into:

$$\dot{\boldsymbol{\varphi}}_\alpha = \mathbf{c}_\alpha + \boldsymbol{\beta}_\alpha \times \boldsymbol{\varphi}_\alpha + \lambda_\alpha \nabla g_\alpha + \dot{\boldsymbol{\varphi}}_\alpha^{\mathrm{sli}}, \tag{9}$$

where $\mathbf{c}_\alpha + \boldsymbol{\beta}_\alpha \times \boldsymbol{\varphi}_\alpha$ is a general rigid velocity field and $\lambda_\alpha \nabla g_\alpha$ represents a normal separation velocity field between the two simplices in contact. The constant vector

$\mathbf{c}_\alpha$ describes a rigid translation, the vector $\boldsymbol{\beta}_\alpha$ an angular velocity, and the scalar λ_α a normal separation velocity. Note that the sliding velocity field has to vanish identically under rigid-body motions (i.e., it must have zero total linear and angular momentum; thus it may be regarded as a deformation mode), and does not produce work under the normal contact force.

One way to obtain the sliding velocity field is to remove the rigid and normal components from the full local velocity field, by solving the following linear problem in the unknowns $\dot{\boldsymbol{\varphi}}_\alpha^{\mathrm{sli}}$, $\mathbf{c}_\alpha$, $\boldsymbol{\beta}_\alpha$ and λ_α:

$$\dot{\boldsymbol{\varphi}}_\alpha^{\mathrm{sli}} = \dot{\boldsymbol{\varphi}}_\alpha - \mathbf{c}_\alpha - \boldsymbol{\beta}_\alpha \times \boldsymbol{\varphi}_\alpha - \lambda_\alpha \nabla g_\alpha, \tag{10}$$

$$\mathcal{R}(\mathbf{M}_\alpha \dot{\boldsymbol{\varphi}}_\alpha^{\mathrm{sli}}) = \mathbf{0}, \tag{11}$$

$$\mathcal{M}(\mathbf{M}_\alpha \dot{\boldsymbol{\varphi}}_\alpha^{\mathrm{sli}}; \boldsymbol{\varphi}_\alpha) = \mathbf{0}, \tag{12}$$

$$\mathbf{N}_\alpha \cdot \dot{\boldsymbol{\varphi}}_\alpha^{\mathrm{sli}} = \nabla g_\alpha(\boldsymbol{\varphi}_\alpha) \cdot \dot{\boldsymbol{\varphi}}_\alpha^{\mathrm{sli}} = 0, \tag{13}$$

where $\mathbf{M}_\alpha$ is the local stiffness matrix and the operators $\mathcal{R}$ and $\mathcal{M}$ return the resultant and resultant moment of the local vector field which they are applied to, respectively. Equation (13), that imposes the orthogonality conditions between $\dot{\boldsymbol{\varphi}}_\alpha^{\mathrm{sli}}$ and the normal contact forces $\mathbf{N}_\alpha$, shows that the constraint function g_α is not affected by the sliding velocity field. Therefore, the vector fields $\dot{\boldsymbol{\varphi}}_\alpha^{\mathrm{sli}}$ and $\dot{\boldsymbol{\varphi}}_\alpha$ are related linearly through the projection operator $\mathbf{P}_\alpha(\boldsymbol{\varphi}_\alpha)$

$$\dot{\boldsymbol{\varphi}}_\alpha^{\mathrm{sli}} = \mathbf{P}_\alpha(\boldsymbol{\varphi}_\alpha)\dot{\boldsymbol{\varphi}}_\alpha. \tag{14}$$

Details of the derivation of (14) can be found in [2].

3.2 Frictional Forces

We now postulate the existence of a frictional dissipation potential $\psi_\alpha^*(\dot{\boldsymbol{\varphi}}_\alpha; \boldsymbol{\varphi}_\alpha)$ such that the local frictional force system $\mathbf{R}_\alpha$ derives as:

$$\mathbf{R}_\alpha = -\partial_{\dot{\boldsymbol{\varphi}}_\alpha} \psi_\alpha^*(\dot{\boldsymbol{\varphi}}_\alpha; \boldsymbol{\varphi}_\alpha). \tag{15}$$

The dependence of ψ_α^* on $\boldsymbol{\varphi}_\alpha$ is regarded as parametric. In order to represent a true frictional dissipation function, $\psi_\alpha^*(\dot{\boldsymbol{\varphi}}_\alpha; \boldsymbol{\varphi}_\alpha)$ must depend on $\dot{\boldsymbol{\varphi}}_\alpha$ only through the sliding velocity field $\dot{\boldsymbol{\varphi}}_\alpha^{\mathrm{sli}}$:

$$\psi_\alpha^* = \psi_\alpha^*(\dot{\boldsymbol{\varphi}}_\alpha^{\mathrm{sli}}; \boldsymbol{\varphi}_\alpha). \tag{16}$$

It is interesting to note that the frictional forces $\mathbf{R}_\alpha$ do work on the sliding velocities only. In particular, in analogy with the relationship between strain and stresses, the sliding local velocity field $\dot{\boldsymbol{\varphi}}_\alpha^{\mathrm{sli}}$, which vanishes for local rigid-body motions, plays the role of a deformation rate, and the local frictional forces $\mathbf{R}_\alpha$ play the role of the conjugate stresses.

The global frictional force array $\mathbf{f}^{\text{fri}}$ is obtained by assembling all the local frictional forces arrays $\mathbf{R}_\alpha$ pertaining to each constraint:

$$\mathbf{f}^{\text{fri}} = \mathcal{A}(\{\mathbf{R}_\alpha,\ \alpha = 1, \ldots, N\}). \tag{17}$$

Since all the local velocity arrays derive from compatible global velocity fields, the resulting frictional force field $\mathbf{f}^{\text{fri}}$ satisfies the relation:

$$\sum_{\alpha=1}^{N} \mathbf{R}_\alpha \cdot \dot{\boldsymbol{\varphi}}_\alpha = \mathbf{f}^{\text{fri}} \cdot \dot{\boldsymbol{\varphi}}, \tag{18}$$

which implies that $\mathbf{f}^{\text{fri}}$ inherits the potential structure from the local fields:

$$\mathbf{f}^{\text{fri}} = -\partial_{\dot{\boldsymbol{\varphi}}} \psi^*(\dot{\boldsymbol{\varphi}}; \boldsymbol{\varphi}), \tag{19}$$

where

$$\psi^*(\dot{\boldsymbol{\varphi}}; \boldsymbol{\varphi}) = \sum_{\alpha=1}^{N} \psi_\alpha^*(\dot{\boldsymbol{\varphi}}_\alpha; \boldsymbol{\varphi}_\alpha) \tag{20}$$

is referred to as the global frictional dissipation function.

In the particular case of Coulomb's law, the frictional dissipation potential is:

$$\psi_\alpha^*(\dot{\boldsymbol{\varphi}}_\alpha; \boldsymbol{\varphi}_\alpha) = \sum_{\text{nodes}} \mu |\mathbf{N}_\alpha(\boldsymbol{\varphi}_\alpha)| |\dot{\boldsymbol{\varphi}}_\alpha^{\text{sli}}|, \tag{21}$$

where μ is the coefficient of friction. The sum extends over the nodes of the pair of boundary simplices involved in the contact constraint α. The frictional forces follow Equation (15) in the form

$$\begin{cases} \mathbf{R}_\alpha = \mu\, |\mathbf{N}_\alpha(\varphi_\alpha)|\, \mathbf{P}_\alpha^T\, \dot{\boldsymbol{\varphi}}_\alpha^{\text{sli}} / |\dot{\boldsymbol{\varphi}}_\alpha^{\text{sli}}| & \text{if } \dot{\boldsymbol{\varphi}}_\alpha^{\text{sli}} \neq \mathbf{0}, \\ \mathbf{R}_\alpha \in \{\mathbf{F}_\alpha \text{ s. t. } |\mathbf{F}_\alpha| \leq \mu\, |\mathbf{N}_\alpha(\varphi_\alpha)| \text{ for all local nodes}\}, & \text{if } \dot{\boldsymbol{\varphi}}_\alpha^{\text{sli}} = \mathbf{0}. \end{cases} \tag{22}$$

These relations apply node by node. The first case corresponds to slip conditions, whereas the second case corresponds to stick. In the stick case, the frictional forces are not determined uniquely by Coulomb's law and follow from equilibrium.

4 Time-Discretization

Time-discretization furnishes the key device enabling the variational formulation of dynamical and frictional dissipation problems in terms of a minimum principle. To derive the algorithm, we consider an incremental solution procedure where $\boldsymbol{\varphi}$ is approximated at discrete times $t_n = n\Delta t$, and specialize Equation (3) to time t_{n+1}, with the result:

$$\mathbf{0} \in \mathbf{M}\ddot{\boldsymbol{\varphi}}_{n+1} + \mathbf{f}^{\text{int}}(\boldsymbol{\varphi}_{n+1}) + \mathbf{f}^{\text{con}}(\boldsymbol{\varphi}_{n+1}) - \mathbf{f}^{\text{ext}}_{n+1}, \tag{23}$$

where internal and contact forces are given by:

$$\mathbf{f}^{\text{int}}(\boldsymbol{\varphi}_{n+1}) = \partial\phi(\boldsymbol{\varphi}_{n+1}), \qquad \mathbf{f}^{\text{con}}(\boldsymbol{\varphi}_{n+1}) = \partial I_C(\boldsymbol{\varphi}_{n+1}). \tag{24}$$

The Newmark family of algorithms provides a general class of implicit/explicit time stepping procedures:

$$\begin{aligned} \boldsymbol{\varphi}_{n+1} &= \boldsymbol{\varphi}_n + \triangle t\dot{\boldsymbol{\varphi}}_n + \triangle t^2[(1/2-\beta)\,\ddot{\boldsymbol{\varphi}}_n + \beta\,\ddot{\boldsymbol{\varphi}}_{n+1}], \\ \dot{\boldsymbol{\varphi}}_{n+1} &= \dot{\boldsymbol{\varphi}}_n + \triangle t\,[(1-\gamma)\,\ddot{\boldsymbol{\varphi}}_n + \gamma\,\ddot{\boldsymbol{\varphi}}_{n+1}]. \end{aligned} \tag{25}$$

The algorithm is explicit when $\beta = 0$. In order to derive a robust contact algorithm, it is necessary to use a fully implicit treatment for the contact force system $\partial I_C(\boldsymbol{\varphi}_{n+1})$, while the remainder of the terms in Equation (23) may be treated either implicitly or explicitly. To distinguish the two contributions, we start by splitting the accelerations in Equation (23) into the sum of a term due to the internal forces and a term due to contact forces:

$$\ddot{\boldsymbol{\varphi}}_{n+1} = \ddot{\boldsymbol{\varphi}}^{\text{int}}_{n+1} + \ddot{\boldsymbol{\varphi}}^{\text{con}}_{n+1}, \tag{26}$$

where:

$$\ddot{\boldsymbol{\varphi}}^{\text{con}}_{n+1} = -\mathbf{M}^{-1}\mathbf{f}^{\text{con}}(\boldsymbol{\varphi}_{n+1}), \qquad \ddot{\boldsymbol{\varphi}}^{\text{int}}_{n+1} = \mathbf{M}^{-1}[\mathbf{f}^{\text{ext}} - \mathbf{f}^{\text{int}}(\boldsymbol{\varphi}_{n+1})]. \tag{27}$$

The Newmark equations are thus written in the form:

$$\begin{aligned} \boldsymbol{\varphi}_{n+1} &= \boldsymbol{\varphi}^{\text{pre}}_{n+1} + \beta\triangle t^2\ddot{\boldsymbol{\varphi}}^{\text{int}}_{n+1} + \frac{\triangle t^2}{2}\,\ddot{\boldsymbol{\varphi}}^{\text{con}}_{n+1}, \\ \dot{\boldsymbol{\varphi}}_{n+1} &= \dot{\boldsymbol{\varphi}}^{\text{pre}}_{n+1} + \gamma\,\triangle t\,\ddot{\boldsymbol{\varphi}}^{\text{int}}_{n+1} + \triangle t\,\ddot{\boldsymbol{\varphi}}^{\text{con}}_{n+1}, \end{aligned} \tag{28}$$

where we introduce the predictors:

$$\begin{aligned} \boldsymbol{\varphi}^{\text{pre}}_{n+1} &= \boldsymbol{\varphi}_n + \triangle t\,\dot{\boldsymbol{\varphi}}_n + (\frac{1}{2} - \beta)\triangle t^2\ddot{\boldsymbol{\varphi}}^{\text{int}}_n, \\ \dot{\boldsymbol{\varphi}}^{\text{pre}}_{n+1} &= \dot{\boldsymbol{\varphi}}_n + (1-\gamma)\,\triangle t\,\ddot{\boldsymbol{\varphi}}^{\text{int}}_n. \end{aligned} \tag{29}$$

By using Equations (26–29), Equation (23) may be recast as a system of nonlinear algebraic equations to be solved for $\boldsymbol{\varphi}_{n+1}$:

$$\mathbf{0} \in \frac{2}{\triangle t^2}\mathbf{M}(\boldsymbol{\varphi}_{n+1} - \boldsymbol{\varphi}^{\text{pre}}_{n+1}) + 2\beta[\mathbf{f}^{\text{int}}(\boldsymbol{\varphi}_{n+1}) - \mathbf{f}^{\text{ext}}_{n+1}] + \mathbf{f}^{\text{con}}(\boldsymbol{\varphi}_{n+1}), \tag{30}$$

that in the following will be called "contact equation". Once $\boldsymbol{\varphi}_{n+1}$ is computed, the accelerations follow from Equation (27), with the result:

$$\ddot{\boldsymbol{\varphi}}^{\text{con}}_{n+1} = \frac{2}{\triangle t^2}(\boldsymbol{\varphi}_{n+1} - \boldsymbol{\varphi}^{\text{pre}}_{n+1}) - 2\beta\ddot{\boldsymbol{\varphi}}^{\text{int}}_{n+1}, \tag{31}$$

while velocities and displacements are computed from Equations (28).

The solution $\boldsymbol{\varphi}_{n+1}$ of the contact equation (30) is the crucial part of the algorithm: it can be obtained exactly, by a variational characterization, or in an approximate way.

4.1 Exact Solution: Variational Characterization

Consider the following functional:

$$f(\varphi_{n+1}) = \|\varphi_{n+1} - \varphi_{n+1}^{\text{pre}}\|_K^2 + 2\beta[\phi(\varphi_{n+1}) - \mathbf{f}_{n+1}^{\text{ext}} \cdot \varphi_{n+1}], \tag{32}$$

where

$$\|\mathbf{u}\|_K \equiv \frac{1}{\Delta t}\sqrt{\mathbf{u}^T\mathbf{M}\mathbf{u}} \tag{33}$$

may be interpreted as a kinetic-energy norm. We observe that the contact equation (30) may be expressed in the form:

$$\mathbf{0} \in \partial\big(f(\varphi_{n+1}) + I_C(\varphi_{n+1})\big). \tag{34}$$

Thus, the stable solutions of Equation (34) may now be identified with the solutions of the minimization problem:

$$\min_{\varphi_{n+1}\in X} f(\varphi_{n+1}) + I_C(\varphi_{n+1}), \tag{35}$$

or, equivalently, with the solutions of the constrained minimization problem:

$$\min_{\varphi_{n+1}\in C} f(\varphi_{n+1}). \tag{36}$$

This is a standard nonlinear optimization problem, which may be solved by a variety of methods [22–24]. An essential part of the solution of (36) is the determination of the active constraints, which has the effect of resolving the precise sequence of the collisions which take place in many-body problems.

The algorithms possess a geometrical structure similar to that of the closest-point return mapping algorithms of plasticity [11, 12], where the closest-point projection here has to be interpreted in energetic terms [1]. In the special case of explicit integration, φ_{n+1} is indeed the closest point to $\varphi_{n+1}^{\text{pre}}$ within C in the sense of the kinetic norm (37). It bears emphasis, however, that the closest-point projection onto a nonconvex set is set-valued in general and that, in contrast to closest-point return mapping algorithms of plasticity, the solution deformation mapping φ_{n+1} may be nonunique.

In the explicit case $\beta = 0$ and the objective function $f(\varphi_{n+1})$ reduces to:

$$f(\varphi_{n+1}) = \|\varphi_{n+1} - \varphi_{n+1}^{\text{pre}}\|_K^2. \tag{37}$$

i.e., a quadratic programming problem. Provided that the mass matrix is diagonal, in the explicit case the global optimization problem decomposes into uncoupled local problems, each involving a small number of degrees of freedom. The local problems are set up by first detecting all intersections between faces. The intersecting faces are then grouped in accordance to their respective adjacencies, with every disjoint group defining a local problem. Therefore, each local problem may be solved independently.

4.1.1 Frictional Dissipation

It has been demonstrated that time-discretization represents a powerful device for the definition of minimum principles for dissipative problems [18–20, 25]. We thus extend the variational principles (35–36) (that remain in force) to the frictional case by adding to the objective function an incremental (or time-discretized) frictional dissipation, obtaining:

$$f(\boldsymbol{\varphi}_{n+1}) = \| \boldsymbol{\varphi}_{n+1} - \boldsymbol{\varphi}_{n+1}^{\text{pre}} \|_K^2 + 2\beta\, [\phi(\boldsymbol{\varphi}_{n+1}) - \mathbf{f}_{n+1}^{\text{ext}} \cdot \boldsymbol{\varphi}_{n+1}] \\ + \Delta t\, \psi^* \,(\Delta\dot{\boldsymbol{\varphi}}; \boldsymbol{\varphi}_{n+1})\,. \tag{38}$$

The incremental frictional dissipation in (38) is computed on the basis of a simple finite-difference approximation of the velocity, independent of the Newmark relations:

$$\Delta\dot{\boldsymbol{\varphi}} = \frac{\boldsymbol{\varphi}_{n+1} - \boldsymbol{\varphi}_n}{\Delta t}. \tag{39}$$

Such discretization of the frictional dissipation retains its meaning in the static case. The variational principle in presence of friction is necessarily incremental, even in the static case, as required by the dissipative nature of friction.

In the case of friction, the contact equation (30) becomes:

$$\mathbf{0} \in \frac{2}{\Delta t^2}\mathbf{M}(\boldsymbol{\varphi}_{n+1} - \boldsymbol{\varphi}_{n+1}^{\text{pre}}) + 2\beta[\mathbf{f}^{\text{int}}(\boldsymbol{\varphi}_{n+1}) - \mathbf{f}_{n+1}^{\text{ext}}] + \partial I_C(\boldsymbol{\varphi}_{n+1}) + \mathbf{f}_{n+1}^{\text{fri}}. \tag{40}$$

Observe that in this case the full contact forces are given by:

$$\mathbf{f}_{n+1}^{\text{con}} = \partial I_C(\boldsymbol{\varphi}_{n+1}) + \mathbf{f}_{n+1}^{\text{fri}}, \tag{41}$$

and the frictional forces follow as:

$$\mathbf{f}_{n+1}^{\text{fri}} = -\frac{\partial}{\partial \boldsymbol{\varphi}_{n+1}}\left[\Delta t\, \psi^*\left(\frac{\boldsymbol{\varphi}_{n+1} - \boldsymbol{\varphi}_n}{\Delta t}; \boldsymbol{\varphi}_{n+1}\right)\right] \\ = -\delta_1 \psi^*\left(\frac{\boldsymbol{\varphi}_{n+1} - \boldsymbol{\varphi}_n}{\Delta t}; \boldsymbol{\varphi}_{n+1}\right) - \Delta t\, \delta_2 \psi^*\left(\frac{\boldsymbol{\varphi}_{n+1} - \boldsymbol{\varphi}_n}{\Delta t}; \boldsymbol{\varphi}_{n+1}\right). \tag{42}$$

Up to admissible truncation errors, they are consistent with the frictional law in Equation (19). Thus, Equation (40) itself furnishes a consistent time discretization of the full equations of motion (8). It should be noted that the spurious second order term in (42) is of order $O(\Delta t)$, and can therefore be regarded as an admissible contribution to the truncation error of the algorithm. In this manner Coulomb's law of friction can be formulated in variational form as a minimum principle, despite its "nonassociated" character.

4.2 Approximate Solution: Penalty Approach

When numerous constraints are involved, as in the case of impact on a brittle target, the nonsmooth contact algorithm above described slows down considerably, requiring prohibitive computational times. This difficulty can be overcome by adopting an approximate procedure based on a penalty approach [27]. In the spirit of barrier methods, we approximate the indicator function as the sum of quadratic terms, one for each violated constraint:

$$I_C(\boldsymbol{\varphi}) \approx \frac{1}{2} p \sum_{\alpha=1}^{N} g_\alpha^2(\boldsymbol{\varphi}), \tag{43}$$

where p is a penalty parameter. The contact forces in this case derive as:

$$\partial I_C(\boldsymbol{\varphi}_{n+1}) \approx p \sum_{\alpha=1}^{N} g_\alpha(\boldsymbol{\varphi}) \nabla g_\alpha(\boldsymbol{\varphi}). \tag{44}$$

Within the time-discretized algorithm above described, the contact forces are approximated as:

$$\mathbf{f}^{\text{con}}(\boldsymbol{\varphi}_{n+1}^{\text{pre}}) = p \sum_{\alpha=1}^{N} g_\alpha(\boldsymbol{\varphi}_{n+1}^{\text{pre}}) \nabla g_\alpha(\boldsymbol{\varphi}_{n+1}^{\text{pre}}), \tag{45}$$

and the contact equation (40) is written as:

$$\frac{2}{\Delta t^2}\mathbf{M}(\boldsymbol{\varphi}_{n+1} - \boldsymbol{\varphi}_{n+1}^{\text{pre}}) + 2\beta[\mathbf{f}^{\text{int}}(\boldsymbol{\varphi}_{n+1}) - \mathbf{f}_{n+1}^{\text{ext}}] + \mathbf{f}^{\text{con}}(\boldsymbol{\varphi}_{n+1}^{\text{pre}}) + \mathbf{f}_{n+1}^{\text{fri}} = \mathbf{0}. \tag{46}$$

The frictionless explicit case is especially of computational interest, because it allows to compute directly the (approximate) solutions as:

$$\boldsymbol{\varphi}_{n+1} = \boldsymbol{\varphi}_{n+1}^{\text{pre}} + \frac{\Delta t^2}{2} \mathbf{M}^{-1} \mathbf{f}^{\text{con}}(\boldsymbol{\varphi}_{n+1}^{\text{pre}}). \tag{47}$$

Expression (47) can be extended in an approximate way to the frictional case, by accounting for the contribution of frictional forces:

$$\boldsymbol{\varphi}_{n+1} = \boldsymbol{\varphi}_{n+1}^{\text{pre}} + \frac{\Delta t^2}{2} \mathbf{M}^{-1} \left[\mathbf{f}^{\text{con}}(\boldsymbol{\varphi}_{n+1}^{\text{pre}}) + \mathbf{f}^{\text{fri}}(\boldsymbol{\varphi}_{n+1}^{\text{pre}})\right]. \tag{48}$$

The frictional forces are explicitly computed from the predictor displacement and velocity fields defined in Equation (29). In particular, for each impenetrability constraint, the sliding velocities are approximated as:

$$\dot{\boldsymbol{\varphi}}_\alpha^{\text{sli}} \approx \mathbf{P}_\alpha(\boldsymbol{\varphi}_\alpha^{\text{pre}})\, \dot{\boldsymbol{\varphi}}_\alpha^{\text{pre}}. \tag{49}$$

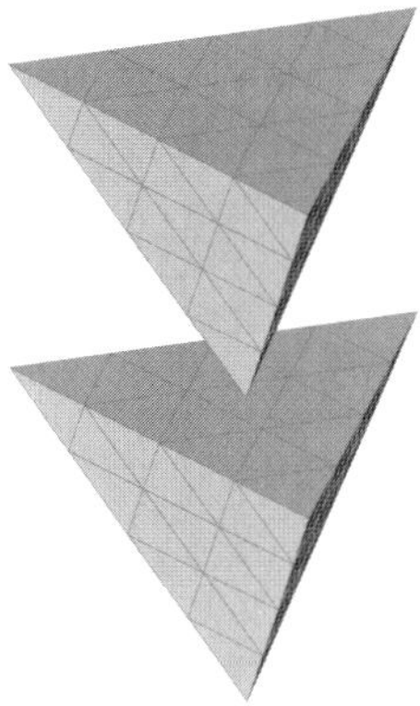

Fig. 1. Vertex on face collision of two tetrahedra. Computational mesh and initial configuration.

5 Numerical Applications

We present a selection of numerical applications of the nonsmooth contact algorithm in dynamics and in statics. All the subsequent calculations are carried out using the explicit algorithm with $\beta = 0$ and $\gamma = 0.5$. The time step is chosen as a fraction of the Levy–Courant stable time step for explicit integration.

5.1 Collisions between Tetrahedra

The nonsmooth characteristics of the algorithm are demonstrated by the collision corner–to–face between two unconstrained deformable tetrahedra. The constrained minimization problem (36) is solved using Spellucci's implementation of sequential quadratic programming [22].

The material of the bodies is assumed to obey a hyperelastic Neo-Hookean constitutive law extended to the compressible range. The strain energy density has the form:

$$W(\mathbf{F}) = \frac{\lambda}{2}(\log J)^2 - \mu \log J + \frac{\mu}{2}\mathrm{tr}(\mathbf{C}), \tag{50}$$

where $\mathbf{F} = \nabla_0 \boldsymbol{\varphi}$ is the deformation gradient; $\mathbf{C} = \mathbf{F}^T\mathbf{F}$ is the right Cauchy–Green deformation tensor; $J = \det(F)$ is the Jacobian of the deformation; and tr denotes the trace. The material constants used in calculations are $\lambda = 115.4$ GPa, $\mu = 79.6$ GPa, and a referential mass density $\rho_0 = 7,800$ kg/m^3. The friction coefficient μ is given values ranging from 0 to 1. The two tetrahedra have the same volume and mass.

The mesh – comprising 70 nodes and 16 elements – and the configuration at the point of impact are depicted in Figure 1. The initial velocity of the top tetrahedron has the components $v_x = 0.21$ mm/μs, $v_y = 0$ and $v_z = -0.21$ mm/μs; the

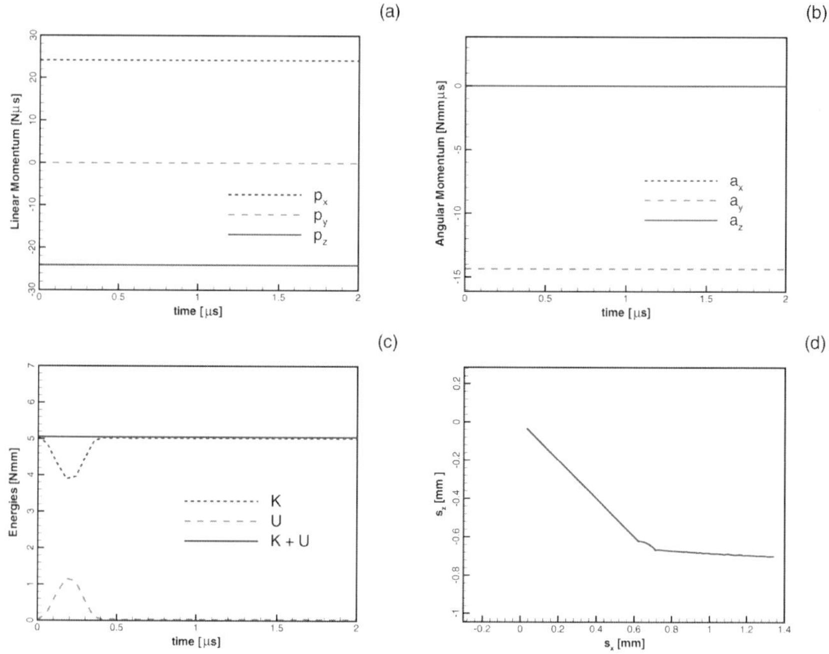

Fig. 2. Nonsmooth frictionless vertex-on-face collision of two unconstrained tetrahedra: (a) Linear momentum history. (b) Angular momentum history. (c) Evolution of elastic, kinetic and total energy. (d) Trajectory of a node close to the centroid of the top tetrahedron [2].

second tetrahedron is at rest. The time history of the frictionless collision, $\mu = 0$, is described in Figure 2 through (a) the linear momentum history; (b) the angular momentum history; (c) the kinetic (K), elastic (U) and the total $(K + U)$ mechanical energy history; and (d) the trajectory of the node closest to the centroid of the top tetrahedron. The first three figures demonstrate the good energy-momentum conservation properties of the algorithm. Note, for example, the exchange between kinetic and elastic energy during the contact phase in Figure 2c; both energies recover their initial value after the collision phase. Also, note the deflection of the the trajectory of the top tetrahedron as a result of the collision in Figure 2d: the two tetrahedra exchange the vertical component of the linear momentum, and conserve at the same time the horizontal or tangential component.

Figure 3 shows the results for the frictional cases. We observe that the total mechanical energy decays monotonically in time, Figure 3a, and that the dissipation rate increases with the coefficient of friction. In the presence of friction, the trajectories of the centroid of the top tetrahedron, Figure 3b, are in marked contrast to the frictionless trajectories. Thus, the top tetrahedron catches at its tip and transfers tangential

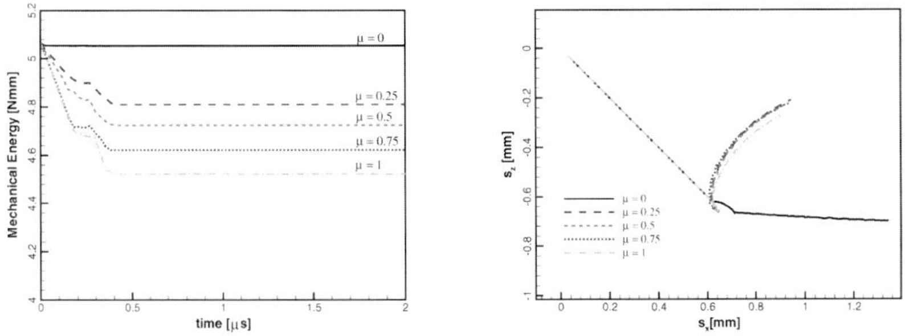

Fig. 3. Nonsmooth frictional vertex-to-face collision of two unconstrained tetrahedra: (a) Total mechanical energy histories for different values of the friction coefficient. (b) Trajectories of a node close to the centroid of the top tetrahedron for different values of the friction coefficient [2].

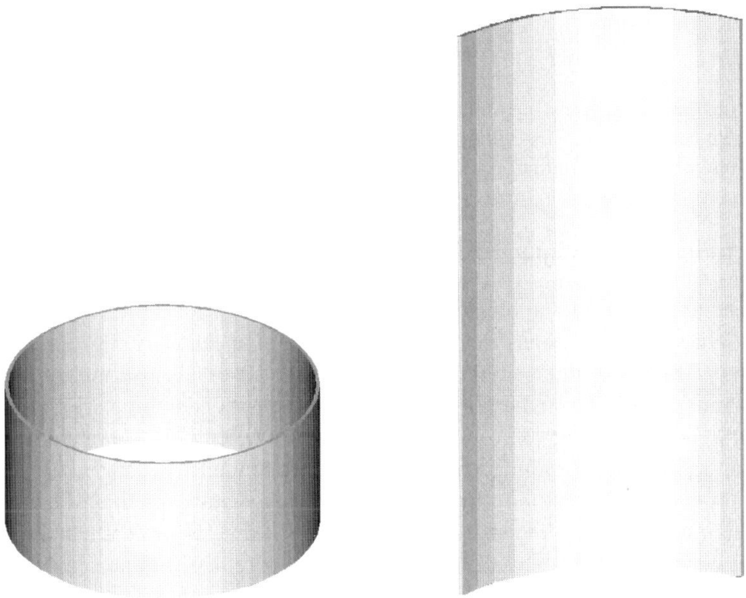

Fig. 4. Crushing of clamped aluminum shells, initial configurations. The meshes consists of: (left) 8121 nodes and 3881 10-node tetrahedral elements; (right) 2906 nodes and 726 10-node tetrahedral elements.

momentum to the bottom tetrahedron. As a consequence of this frictional interaction, both tetrahedra acquire angular momentum and start to rotate.

Fig. 5. Slow-speed compression of the clamped aluminum can. Top left: Initial deviation from the straight configuration. Top right: Intermediate stage with self-contact of the surfaces of the body. Bottom: Final configuration.

5.2 Crushing of Aluminum Shells

In a second example of application we evaluate the performance of the dynamic contact algorithm – without friction – in the case of inelasticity, by simulating the crushing of thin aluminum shells. We consider two cases: (i) a short circular cylinder (20 mm radius, 20 mm height) undergoing a slow-speed compression; (ii) a long circular cylinder (20 mm radius, 60 mm height; modelled only 1/4 of the solid) undergoing an impulsive loading. Both the shells are clamped at the top and bottom; the loading is transmitted in the form of (i) a progressive displacement of the top nodes, or (ii) an initial velocity imposed for a short lapse of time, to simulate an impact.

The geometry of the two shells is shown in Figure 4. Aluminum is assumed to obey J_2-plasticity with power-law hardening and rate dependency, and linear thermal softening. The temperature field is calculated locally assuming adiabatic conditions. The state variables are updated by recourse to the method of extension of Cuitiño and Ortiz [26]; the details of the constitutive model and of the numerical algorithm may be found elsewhere [26]. The properties of aluminum assumed in the calculations are collected in Table 1.

Results of the explicit dynamic calculations are shown in Figures 5–7. In Figure 5 three snapshots of the progressive crushing are drawn. In the first image it is possible to observe the configuration change due to buckling. The second image shows the be-

Fig. 6. Final configuration of the impacted aluminum shell. Initial velocity imposed to the top clamped nodes: 100 m/s.

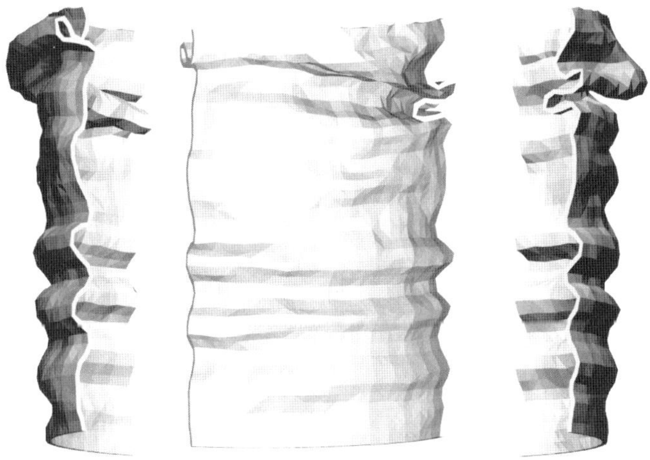

Fig. 7. Final configuration of the impacted aluminum shell. Initial velocity imposed to the top nodes: 1000 m/s.

ginning of the contact process, as a consequence of the folding of the thin walls. The third image shows the crushed configuration, where multiple self-contact situations can be observed. Figure 6 shows the two lateral views and the font view of the final configuration of the shell impacted at an end. The impact is simulated by applying for a short lapse of time a constant velocity $v = 100$ m/s to the nodes of the im-

Table 1. Mechanical parameters for aluminum adopted in the numerical simulations.

Material Property			
Young Modulus	E	GPa	69
Poisson's ratio	ν		0.33
Mass density	ρ	kg/m^3	2719
Initial yield stress	σ_0	MPa	90
Reference plastic strain	ε_0^p		10^{-3}
Reference plastic strain rate	$\dot{\varepsilon}_0^p$	s^{-1}	10^4
Rate sensitivity exponent	m		100
Hardening exponent	n		25
Heat capacity	c_v	J/kg K	904
Melting temperature	T_m	°C	700
Thermal exponent	α		1
Taylor–Quinney coefficient	β		0.9

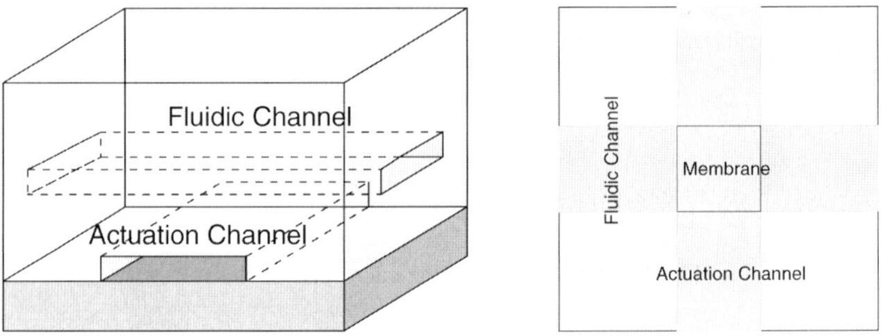

Fig. 8. Architecture of a standard MSL microvalve. The size of the channels ranges 10–100 μm in width and 10–50 μm in height. The thickness of the separation membrane ranges 3–15 μm.

pacted end. It is possible to observe that the impacted side of the shell folds causing the self-contact of the internal surface of the shell. Figure 7 shows the corresponding result for a velocity $v = 1000$ m/s. For testing purposes, in the latter calculation the penalty version of the algorithm, Equation (47), was used. It can be observed that the enforcement of the impenetrability constraints is achieved reasonably also with the approximate algorithm.

5.3 Closure of Microvalves

As last example of application of nonsmooth contact algorithms, we present the closure of a microvalve under the pressure of an actuation fluid. Microvalves, fabricated by means of Multilayer-Soft-Lithography (MSL), are widely used in biology and in

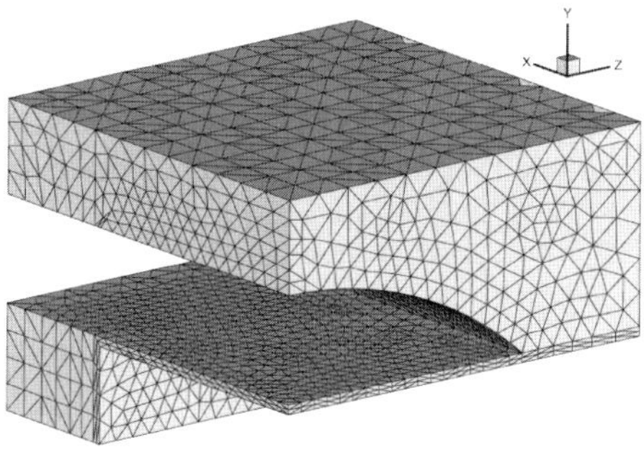

Fig. 9. Finite element mesh of the microvalve. It consists of 3898 nodes and 2021 10-node tetrahedral elements.

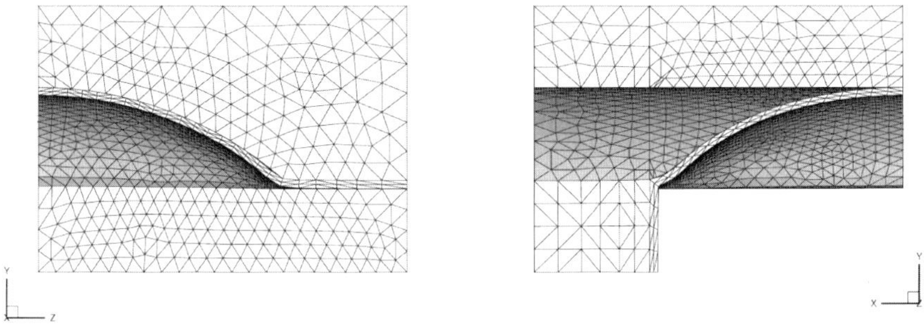

Fig. 10. Configuration of the microvalve at closure.

biochemistry [28]. The architecture of a standard microvalve is shown in Figure 8. The small size of the microvalve allows for high frequency open-closure cycles (50–100 Hz). Although fast, the mechanical activation of the valve can be considered quasistatic, since the magnitude of the inertia forces is very small. To analyze the behavior of the microvalve, we used an explicit procedure derived from standard time-stepping algorithms developed for dynamics. We adopt the algorithmic form of the adaptive dynamic relaxation procedure described in [29]. The contact algorithm is activated in the penalty version, Equation (47).

The finite element mesh adopted is depicted in Figure 9. Because of the symmetries, only 1/4 of the microvalve is discretized. The size of the whole valve is $300 \times 300 \times 200$ μm. The material obeys a nearly-incompressible hyperleastic neo-Hookean model described by the following strain energy density:

$$W(\mathbf{F}) = \frac{\mu}{2}\left[tr(\mathbf{C}) - 3 - \log J\right], \tag{51}$$

where the variables have the same meaning as in Equation (50). The value assumed for the shear modulus is $\mu = 0.6$ MPa. Figure 10 shows the closure of the valve under 9 kPa pressure applied to the bottom (actuation) channel. The contact involves a wide region, and although the approximated version of the algorithm has been used, there is no violation of the constraints. As the example demonstrates, the nonsmooth algorithm originally developed for dynamic contact can be extended successfully to quasi-static problems.

6 Conclusions

The contact formulation here discussed takes nonsmooth analysis [7] as a basis for the development of contact algorithms capable of dealing with complex contact situations, involving multiple bodies with corners [1]. A remarkable feature of the formulation is that it accords all bodies an equal role without differentiating between master and slave bodies. The incremental displacements, that pertain to the time-discretized incremental problem, follow from a minimum principle, which is radically different from the variational principles of classical mechanics [18]. The extension of the algorithm to friction retains the variational structure of the formulation [2]. Friction is accounted for simply by adding to the objective function a term which measures the incremental work of friction. As a consequence, friction, although in analogy to so-called nonassociative plasticity rules, is amenable to a variational characterization. Numerical results are consistent with an algorithmic form of conservation of linear and angular momenta proper of the Newmark algorithm [25].

Numerical applications of the algorithm may require the use of approximated tools, e.g. expressing the indicator function I_C in terms of a penalty coefficient. In spite of the fact that in the explicit dynamic case some of the impenetrability constraints can be violated, the penalty approach is successful when applied to the analysis of contact in quasi-static cases, in combination with the dynamic relaxation algorithms.

Acknowledgment

AP acknowledges the support of the Italian MIUR PRIN-2005 programme, Protocol No. 2005085973.

References

1. Kane C, Repetto EA, Ortiz M, Marsden JE (1999) Finite element analysis of nonsmooth contact, *Computer Methods in Applied Mechanics and Engineering*, 180(1):1–26.
2. Pandolfi A, Kane C, Ortiz M, Marsden JE (2002) Time-discretized variational formulation of non-smooth frictional contact, *International Journal for Numerical Methods in Engineering*, 53(8):1801–1830.

3. Camacho GT, Ortiz M (1996) Computational modelling of impact damage in brittle materials, *International Journal of Solids and Structures*, 33:2899–2938.
4. Stein E, Wriggers P (1982) Calculation of impact contact problems of thin elastic shells taking into account geometrical nonlinearities within the contact region, *Computer Methods in Applied Mechanics and Engineering* 34(1–3):861–880.
5. Wriggers P, Simo JC (1985) A note on tangent stiffness for fully nonlinear contact problems, *Communications in Applied Numerical Methods*, 1(5):199–203.
6. Zavarise G, Wriggers P, Stein E, Schrefler BA (1992) Real contact mechanisms and finite-element formulation. A coupled thermomechanical approach, *International Journal for Numerical Methods in Engineering*, 19(3):173–182.
7. Clarke FH (1983) *Optimization and Nonsmooth Analysis*. John Wiley & Sons, New York.
8. Panagiotopoulos PD (1985) *Inequality Problems in Mechanics and Applications*. Birkhauser, Boston, MA.
9. Kikuchi N, Oden JT (1988) *Contact Problems in Elasticity: A Study of Variational Inequalities and Finite Element Methods*. SIAM, Philadelphia, PA.
10. Laursen TA, Govindjee S (1994) A note on the treatment of frictionless contact between nonsmooth surfaces in fully nonlinear problems, *Communications in Numerical Methods in Engineering*, 10(11):869–878.
11. Ortiz M, Pinsky PM, Taylor RL (1983) Operator split methods for the numerical solution of the elastoplastic dynamic problem, *Computer Methods in Applied Mechanics and Engineering*, 39(2):137–157.
12. Pinsky PM, Ortiz M, Taylor RL (1983) Operator split methods for the numerical solution of the finite-deformation elastoplastic dynamic problem, *Computers & Structures*, 17(3):345–359.
13. Wriggers P, Van TV, Stein E (1990) Finite-element formulation of large deformation impact-contact problems with friction, *Computers & Structures*, 37(3):319–331.
14. Moreau JJ (1988) *Unilateral Contact and Dry Friction in Finite Freedom*. CISM, Udine.
15. Simo JC, Laursen TA (1992) An augmented Lagrangian treatment of contact problems involving friction, *Computer & Structures*, 42:97–116.
16. Laursen TA (1999) On the development of thermodynamically consistent algorithms for thermomechanical frictional contact, *Computer Methods in Applied Mechanics and Engineering*, 177:273–287.
17. Pfeiffer F (1999) Unilateral problems of dynamics, *Archive of Applied Mechanics*, 69:503–527.
18. Radovitzky R, Ortiz M (1999) Error estimation and adaptive meshing in strongly nonlinear dynamic problems, *Computer Methods in Applied Mechanics and Engineering*, 172:203–240.
19. Ortiz M, Stainier L (1999) The variational formulation of viscoplastic constitutive updates, *Computer Methods in Applied Mechanics and Engineering*, 171:419–444.
20. Ortiz M, Repetto EA (1999) Nonconvex energy minimization and dislocation structures in ductile single crystals, *Journal of the Mechanics and Physics of Solids*, 47(2):397–462.
21. Marsden JE, Ratiu TS (1983) *Mechanics and Symmetry*. Texts in Applied Mathematics, Vol. 17. Springer-Verlag, Berlin.
22. Spellucci P (1993) Technical Report 1651. Technical University at Darmstadt.
23. Goldfarb D, Idnani A (1983) A numerically stable dual method for solving strictly quadratic programs, *Mathematical Programming*, 27:1–33.
24. Boggs PT, Tolle JW (1995) Sequential quadratic programming, *Acta Numerica*, 4:1–51.

25. Kane C, Marsden JE, Ortiz M, West M (2000) Variational integrators and the Newmark algorithm for conservative and dissipative mechanical systems, *International Journal for Numerical Methods in Engineering*, 49(10):1295–1325.
26. Cuitiño AM, Ortiz M (1992) A material-independent method for extending stress update algorithms from small-strain plasticity to finite plasticity with multiplicative cinematics, *Engineering Computations*, 9:437–451.
27. Mota A, Klug WS, Pandolfi A, Ortiz M (2002) Finite-element simulation of firearm injury to the human cranium, *Computational Mechanics*, 31:115–121.
28. Studer V, Hang G, Pandolfi A, Ortiz M, Anderson WF, Quake SR (2004) Scaling properties of a low-actuation pressure microfluidic valve, *Journal of Applied Physics*, 95(1):393–398.
29. Oakley DR, Knight NFJ (1995) Adaptive dynamic relaxation algorithm for non-linear hyperelastic structures. Part I: Formulation, *Computer Methods in Applied Mechanics and Engineering*, 126:67–89.

Energy-Momentum Conserving Schemes for Frictionless Dynamic Contact Problems. Part I: NTS Method

Peter Betsch and Christian Hesch

Chair of Computational Mechanics, Department of Mechanical Engineering, University of Siegen, 57068 Siegen, Germany
E-mail: {betsch, hesch}@imr.mb.uni-siegen.de

Abstract. Dynamic contact problems in elasticity are treated within a finite element framework by employing the well-established node-to-segment method. A new formulation of the algorithmic forces of contact is proposed which makes possible the design of energy-momentum conserving integrators. The numerical example presented herein indicates that the present approach provides enhanced numerical stability.

Key words: Large deformation contact problems, conserving time integration, node-to-segment method.

1 Introduction

The present work deals with large deformation contact problems within a nonlinear finite element framework. For a survey of previous developments in this field we refer to the books by Laursen [11] and Wriggers [20]. We restrict our attention to the frictionless dynamic contact of elastic bodies. The corresponding semi-discrete system can be classified as Hamiltonian system with symmetry. Consequently, the momentum maps associated with specific symmetries as well as the total energy are conserved quantities of the underlying finite-dimensional system.

Energy-momentum conserving schemes (and energy decaying variants thereof) have previously been developed in the framework of nonlinear elastodynamics in order to meet the numerical stability requirements of finite-deformation problems, see, for example, Simo and Tarnow [18] and Betsch and Steinmann [4].

Finite deformation contact problems put even higher demands on the numerical stability properties of time-stepping schemes. It is thus not surprising that recently published works aim at the extension of energy-momentum schemes to the realm of contact/impact problems. To this end Laursen and Chawla [12] enforce the discrete

Peter Wriggers and Udo Nackenhorst (eds.), IUTAM Symposium on Computational Contact Mechanics, 77–96.

gap rate rather than the constraint of impenetrability. Similarly, Armero and Petöcz [2] modify the contact constraint to achieve the desired conservation properties. Consequently, in both works the impenetrability condition is violated in general. Alternatively, Laursen and Love [13] enforce the constraint of impenetrability and achieve algorithmic energy conservation by introducing a so-called discrete contact velocity. However, this approach requires the solution of quadratic equations which turn out to be unsolvable in some events. It is further worth noting that the aforementioned developments have been carried out within the framework of the node-to-segment (NTS) formulation of the contact problem.

In the present work we aim at the design of energy-momentum schemes for contact problems in the framework of the NTS formulation. From the outset we regard the semi-discrete contact problem as finite-dimensional Hamiltonian system subject to (holonomic) contact constraints. Accordingly, the equations of motion assume the form of differential-algebraic equations (DAEs). Energy-momentum schemes emanating from the direct discretization of the DAEs have been recently developed, see Gonzalez [8] and Betsch and Steinmann [5]. Based on these developments our approach to the design of energy-momentum schemes makes use of the invariance properties of the discrete contact constraints by exploring the representation theorem due to Cauchy.

An outline of the rest of the paper is as follows. Section 2 deals with the Hamiltonian formulation of semi-discrete elastodynamics. In this connection, the incorporation of algebraic constraints is treated and the relevant conservation properties are outlined. In Section 3 the energy-momentum conserving discretization of the underlying DAEs is dealt with. In particular, the notion of a discrete gradient is introduced in conjunction with Cauchy's representation theorem. This approach is then particularized to the NTS description of the contact problem. After the treatment of a representative numerical example in Section 4, conclusions are drawn in Section 5.

2 Hamiltonian Formulation of Semi-Discrete Elastodynamics

We start with the space finite element discretization of nonlinear elastodynamics. In particular, we aim at the Hamiltonian formulation of the resulting semi-discrete problem. Further details of the present discretization approach may be found in Simo [16, chapter IV] and the works cited therein.

2.1 The Free Semi-Discrete Elastic Body

We first focus on the space discretization of the free elastic body (i.e. pure Neumann boundary conditions). Let $\mathcal{B}$ be a regular region in n_{dim}-dimensional Euclidean space ($n_{\text{dim}} \leq 3$) occupied by the reference configuration of the elastic body. Furthermore, let $\mathbb{I} = [0, T]$ denote the time interval of interest. From a kinematic point of view

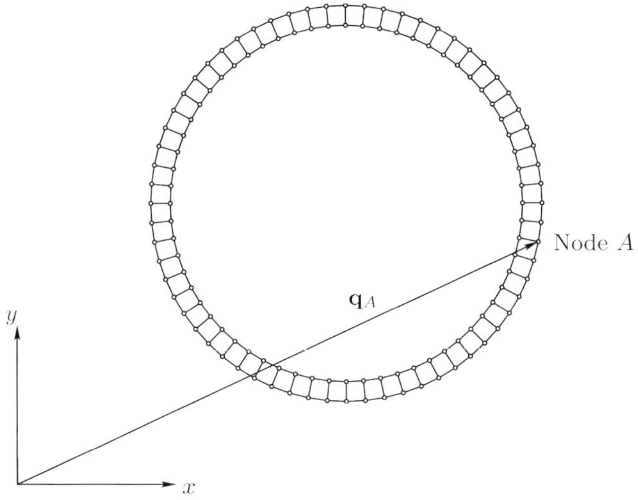

Fig. 1. Planar ($n_{\text{dim}} = 2$) sketch of a free semi-discrete body.

the standard displacement-based finite element approach employs an approximation of the deformation field $\boldsymbol{\varphi}$: $\mathcal{B} \times \mathbb{I} \to \mathbb{R}^{n_{\text{dim}}}$ of the form

$$\boldsymbol{\varphi}(\mathbf{X}, t) = \sum_{A=1}^{n_{\text{node}}} N_A(\mathbf{X})\mathbf{q}_A(t). \tag{1}$$

Within the material (or Lagrangian) description of motion $\boldsymbol{\varphi}(\mathbf{X}, t)$ describes the position of material point $\mathbf{X}$ of body $\mathcal{B}$ at time t. Moreover, $N_A : \mathcal{B} \to \mathbb{R}$ are global shape functions associated with the nodes $A = 1, \ldots, n_{\text{node}}$ and $\mathbf{q}_A : \mathbb{I} \to \mathbb{R}^{n_{\text{dim}}}$ denotes the position vector at time $t \in \mathbb{I}$ of the nodal point A (Figure 1). Accordingly, possible configurations of the semi-discrete dynamical system at hand are characterized by

$$\mathbf{q} = (\mathbf{q}_1, \ldots, \mathbf{q}_{n_{\text{node}}}) \in \mathbb{R}^{n_{\text{dof}}}, \tag{2}$$

where $n_{\text{dof}} = n_{\text{dim}} \cdot n_{\text{node}}$. The material velocity is defined by $\mathbf{v} = \partial\boldsymbol{\varphi}/\partial t = \dot{\boldsymbol{\varphi}}$ such that the finite element approximation implies

$$\mathbf{v}(\mathbf{X}, t) = \sum_{A=1}^{n_{\text{node}}} N_A(\mathbf{X})\mathbf{v}_A(t) \tag{3}$$

with $\mathbf{v}_A = \dot{\mathbf{q}}_A$. Moreover, the finite element approximation (1) gives rise to the discrete deformation gradient

$$\mathbf{F} = \frac{\partial\boldsymbol{\varphi}}{\partial\mathbf{X}} = \sum_{A=1}^{n_{\text{node}}} \mathbf{q}_A \otimes \nabla N_A(\mathbf{X}). \tag{4}$$

Then the discrete version of the deformation tensor (or right Cauchy–Green tensor) $\mathbf{C} = \mathbf{F}^T\mathbf{F}$ can be written as

$$\mathbf{C} = \sum_{A,B=1}^{n_{\text{node}}} \mathbf{q}_A \cdot \mathbf{q}_B \nabla N_A \otimes \nabla N_B. \tag{5}$$

Hyperelastic material behavior is modeled by means of a scalar-valued strain energy density function W($\mathbf{C}$) such that the second Piola-Kirchhoff stress tensor can be calculated via

$$\mathbf{S} = 2DW(\mathbf{C}), \tag{6}$$

where $DW(\mathbf{C}) = \partial W/\partial \mathbf{C}$. Then the discrete strain energy function is given by

$$V^{\text{int}}(\mathbf{q}) = \int_{\mathcal{B}} W(\mathbf{C}) dV. \tag{7}$$

For simplicity we assume that the external forces acting on the body can be derived from a potential function

$$V^{\text{ext}} = -\int_{\mathcal{B}} \varrho_R \mathbf{b} \cdot \boldsymbol{\varphi}\, dV - \int_{\partial\mathcal{B}_\sigma} \bar{\mathbf{t}} \cdot \boldsymbol{\varphi}\, dA, \tag{8}$$

where $\varrho_R : \mathcal{B} \to \mathbb{R}_+$ denotes the reference mass density, $\mathbf{b} : \mathcal{B} \times \mathbb{I} \to \mathbb{R}^{n_{\text{dim}}}$ is the applied body force and $\bar{\mathbf{t}}$ is the prescribed traction boundary condition on $\partial\mathcal{B}_\sigma \times \mathbb{I}$. In view of (1) one obtains

$$V^{\text{ext}}(\mathbf{q}) = -\sum_{A=1}^{n_{\text{node}}} \mathbf{q}_A \cdot \mathbf{F}_A^{\text{ext}} \tag{9}$$

with prescribed external nodal forces

$$\mathbf{F}_A^{\text{ext}} = \int_{\mathcal{B}} N_A \varrho_R \mathbf{b}\, dV + \int_{\partial\mathcal{B}_\sigma} N_A \bar{\mathbf{t}}\, dA. \tag{10}$$

The kinetic energy of the body at time t is given by

$$T = \frac{1}{2} \int_{\mathcal{B}} \varrho_R \mathbf{v} \cdot \mathbf{v}\, dV \tag{11}$$

such that substitution from (3) into (11) leads to

$$T(\mathbf{v}) = \frac{1}{2} \sum_{A,B=1}^{n_{\text{node}}} M_{AB} \mathbf{v}_A \cdot \mathbf{v}_B = \frac{1}{2} \mathbf{v} \cdot \mathbf{M}\mathbf{v}, \tag{12}$$

where

$$M_{AB} = \int_{\mathcal{B}} \varrho_R N_A N_B \, dV \tag{13}$$

are the coefficients of the consistent mass matrix. Note that $\mathbf{M}$ consists of diagonal sub-matrices $\mathbf{M}_{AB} = M_{AB}\mathbf{I}_{n_{\mathrm{dim}}}$ with $A, B = 1, \ldots, n_{\mathrm{node}}$. The Lagrangian of the finite-dimensional dynamical system under consideration is given by $L(\mathbf{q}, \mathbf{v}) = T(\mathbf{v}) - (V^{\mathrm{int}}(\mathbf{q}) + V^{\mathrm{ext}}(\mathbf{q}))$. To perform the transition to the Hamiltonian formulation we introduce the conjugate momenta

$$\mathbf{p} = \frac{\partial L}{\partial \mathbf{v}} = \mathbf{M}\mathbf{v}. \tag{14}$$

Then the Hamiltonian function follows from the Legendre transformation of $L(\mathbf{q}, \mathbf{v})$ with respect to $\mathbf{v}$ as $H(\mathbf{q}, \mathbf{p}) = \mathbf{p} \cdot \mathbf{v} - L(\mathbf{q}, \mathbf{v})$, with the velocities $\mathbf{v}$ being replaced by the momenta in (14). Accordingly, the Hamiltonian of the free semi-discrete elastic body can be written in the form

$$H(\mathbf{q}, \mathbf{p}) = \frac{1}{2}\mathbf{p} \cdot \mathbf{M}^{-1}\mathbf{p} + V^{\mathrm{int}}(\mathbf{q}) + V^{\mathrm{ext}}(\mathbf{q}). \tag{15}$$

Consequently, the equations of motion can be written in canonical Hamiltonian form

$$\begin{aligned} \dot{\mathbf{q}} &= \frac{\partial H}{\partial \mathbf{p}} = \mathbf{M}^{-1}\mathbf{p}, \\ \dot{\mathbf{p}} &= -\frac{\partial H}{\partial \mathbf{q}} = \mathbf{F}^{\mathrm{ext}} - \mathbf{F}^{\mathrm{int}}(\mathbf{q}). \end{aligned} \tag{16}$$

In this connection, the internal forces are given by

$$\mathbf{F}^{\mathrm{int}}(\mathbf{q}) = \nabla V^{\mathrm{int}}(\mathbf{q}). \tag{17}$$

A more compact description of the equations of motion can be achieved by introducing the vector of phase space coordinates

$$\mathbf{z} = (\mathbf{q}, \mathbf{p}) \in \mathbb{R}^{2n_{\mathrm{dof}}}. \tag{18}$$

Then the equations of motion pertaining to the semi-discrete free elastic body can alternatively be written as

$$\dot{\mathbf{z}} = \mathbb{J}\nabla H(\mathbf{z}). \tag{19}$$

In the last equation $\mathbb{J} \in \mathbb{R}^{2n_{\mathrm{dof}} \times 2n_{\mathrm{dof}}}$ is the canonical symplectic matrix

$$\mathbb{J} = \begin{bmatrix} \mathbf{0} & \mathbf{I} \\ -\mathbf{I} & \mathbf{0} \end{bmatrix} \tag{20}$$

where $\mathbf{I}$ and $\mathbf{0}$ are the $n_{\mathrm{dof}} \times n_{\mathrm{dof}}$ identity and zero matrices. Note that $\mathbb{J}^T = \mathbb{J}^{-1} = -\mathbb{J}$ and $\mathbb{J}^2 = -\mathbb{I}$, where $\mathbb{I}$ denotes the $(2n_{\mathrm{dof}} \times 2n_{\mathrm{dof}})$ identity matrix.

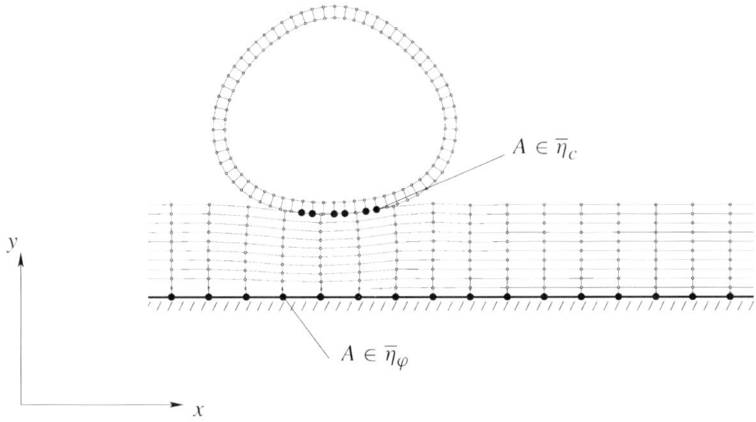

Fig. 2. The planar two-body contact problem.

2.2 Constrained Semi-Discrete Elastic Bodies

We next focus on specific boundary conditions which restrict the motion of the semi-discrete elastic body. These restrictions can be characterized by geometric constraints acting on the boundary nodes of the discrete system at hand. In particular, we distinguish between Dirichlet-type boundary conditions and constraints due to contact. For the present purposes it suffices to consider the planar two-body contact problem (Figure 2).

Assume that n_{node} denotes the total number of nodes due to the space discretization of the two elastic bodies, so that $\eta = \{1, \ldots, n_{\text{node}}\}$ is the set of node numbers associated with the discrete two-body system. Further let $\overline{\eta} \subset \eta$ be the set of node numbers lying on the boundaries of the two-body system. The relevant boundary conditions can be characterized by algebraic constraints of the form

$$\mathbf{\Phi}(\mathbf{q}) = \mathbf{0}. \tag{21}$$

In the case of Dirichlet-type boundary conditions we have

$$\mathbf{q}_A(t) = \overline{\boldsymbol{\varphi}}(\mathbf{X}_A, t), \quad \text{for } A \in \overline{\eta}_\varphi, \tag{22}$$

where $\overline{\boldsymbol{\varphi}}_t$ is prescribed and $\overline{\eta}_\varphi \subset \overline{\eta}$ is the set of node numbers belonging to the Dirichlet boundary. Similarly, if the two bodies are in contact and provided that 'active' nodes $A \in \overline{\eta}_c \subset \overline{\eta} - \overline{\eta}_\varphi$ lying on the contact surface have been detected, additional constraints of the form (21) arise (see Section 3.5 for further details). Due to the presence of the constraints (21), the equations of motion can now be written in the form

$$\begin{aligned}\dot{\mathbf{q}} &= \frac{\partial H}{\partial \mathbf{p}},\\ \dot{\mathbf{p}} &= -\frac{\partial H}{\partial \mathbf{q}} - D\boldsymbol{\Phi}(\mathbf{q})^T\boldsymbol{\lambda},\\ \mathbf{0} &= \boldsymbol{\Phi}(\mathbf{q}),\end{aligned} \tag{23}$$

where $\boldsymbol{\Phi}(\mathbf{q}) \in \mathbb{R}^m$ are the relevant constraint functions, $D\boldsymbol{\Phi}(\mathbf{q})$ is the corresponding constraint Jacobian and $\boldsymbol{\lambda} \in \mathbb{R}^m$ are Lagrange multipliers which determine the size of the constraint forces in $(23)_2$. Similar to (19), the set of differential-algebraic equations (DAEs) in (23) can be rewritten in compact form by introducing the augmented Hamiltonian

$$\mathcal{H}_\lambda(\mathbf{z}) = \frac{1}{2}\mathbf{p}\cdot\mathbf{M}^{-1}\mathbf{p} + \mathcal{V}_\lambda(\mathbf{q}), \tag{24}$$

where

$$\mathcal{V}_\lambda(\mathbf{q}) = V^{\text{int}}(\mathbf{q}) + V^{\text{ext}}(\mathbf{q}) + \boldsymbol{\lambda}\cdot\boldsymbol{\Phi}(\mathbf{q}) \tag{25}$$

is an augmented potential function. Now the differential part of the DAEs can be written as

$$\dot{\mathbf{z}} = \mathbb{J}\nabla\mathcal{H}_\lambda(\mathbf{z}), \tag{26}$$

which, of course, has to be supplemented with the algebraic constraints $(23)_3$.

2.3 Conservation Properties of the Semi-Discrete System

We next focus on the conservation properties relevant for the constrained finite-dimensional mechanical system under consideration. Our goal is the subsequent design of time-stepping schemes which inherit these conservation properties from the semi-discrete system. Since algorithmic conservation of total linear momentum is rather trivial we focus on the conservation properties associated with the total angular momentum and the total energy.

2.3.1 Conservation of the Total Angular Momentum

In this section we elaborate on the connection between specific symmetry properties of the augmented Hamiltonian (24) and associated momentum maps (see Marsden and Ratiu [14] for more background information on these issues). In particular, we focus on the rotational invariance of the augmented Hamiltonian which implies conservation of the total angular momentum. Accordingly, assume that the augmented Hamiltonian is invariant under rotations such that

$$\mathcal{H}_\lambda(\mathbf{Q}\circ\mathbf{z}) := \mathcal{H}_\lambda(\mathbf{Q}\mathbf{q}_1,\ldots,\mathbf{Q}\mathbf{q}_N,\mathbf{Q}\mathbf{p}_1,\ldots,\mathbf{Q}\mathbf{p}_N) \equiv \mathcal{H}_\lambda(\mathbf{z}) \tag{27}$$

for all proper orthogonal matrices $\mathbf{Q} \in SO(3)$, where $SO(3)$ denotes the rotation group. A one-parameter group of rotation matrices can be written in the form $\mathbf{Q}_\varepsilon =$

$\exp(\varepsilon\hat{\boldsymbol{\xi}}) \in SO(3)$ where $\hat{\boldsymbol{\xi}} \in so(3)$ is a skew-symmetric matrix. In the present case a closed-form expression of $\exp(\varepsilon\hat{\boldsymbol{\xi}})$ is given by the Rodrigues formula (see, for example, Marsden and Ratiu [14, chapter 9]). The invariance property (27) implies

$$0 = \left.\frac{d}{d\varepsilon}\right|_{\varepsilon=0} \mathcal{H}_\lambda(\exp(\varepsilon\hat{\boldsymbol{\xi}}) \circ \mathbf{z}) = \nabla \mathcal{H}_\lambda(\mathbf{z}) \cdot \left.\frac{d}{d\varepsilon}\right|_{\varepsilon=0} \exp(\varepsilon\hat{\boldsymbol{\xi}}) \circ \mathbf{z} = \nabla \mathcal{H}_\lambda(\mathbf{z}) \cdot \boldsymbol{\xi}_P(\mathbf{z}), \tag{28}$$

where $\boldsymbol{\xi}_P(\mathbf{z})$ denotes the infinitesimal generator. The associated momentum map is defined by

$$\boldsymbol{\xi}_P(\mathbf{z}) = \mathbb{J}\nabla J_\xi(\mathbf{z}) \tag{29}$$

with

$$J_\xi(\mathbf{z}) = \mathbf{J}(\mathbf{z}) \cdot \boldsymbol{\xi} \tag{30}$$

Here, $\boldsymbol{\xi} \in \mathbb{R}^3$ is the axial vector of $\hat{\boldsymbol{\xi}}$ (i.e. $\hat{\boldsymbol{\xi}}\mathbf{a} = \boldsymbol{\xi} \times \mathbf{a}$ for all $\mathbf{a} \in \mathbb{R}^3$) and $\mathbf{J} \in \mathbb{R}^3$ is the total angular momentum given by

$$\boxed{\mathbf{J}(\mathbf{z}) = \sum_{A=1}^{n_{\text{node}}} \mathbf{q}_A \times \mathbf{p}_A} \tag{31}$$

Now a straightforward calculation yields

$$\begin{aligned} \frac{d}{dt} J_\xi(\mathbf{z}) &= \nabla J_\xi(\mathbf{z}) \cdot \dot{\mathbf{z}} \\ &= \nabla J_\xi(\mathbf{z}) \cdot \mathbb{J}\nabla \mathcal{H}_\lambda(\mathbf{z}) \\ &= \boldsymbol{\xi}_P(\mathbf{z}) \cdot \mathbb{J}^2\nabla \mathcal{H}_\lambda(\mathbf{z}) \\ &= -\boldsymbol{\xi}_P(\mathbf{z}) \cdot \nabla \mathcal{H}_\lambda(\mathbf{z}) \\ &= 0. \end{aligned} \tag{32}$$

Thus rotational invariance of the augmented Hamiltonian implies conservation of the total angular momentum.

2.3.2 *Conservation of the Total Energy*

Due to the skew-symmetry of $\mathbb{J}$ one obtains

$$\nabla \mathcal{H}_\lambda(\mathbf{z}) \cdot \dot{\mathbf{z}} = \nabla \mathcal{H}_\lambda(\mathbf{z}) \cdot \mathbb{J}\nabla \mathcal{H}_\lambda(\mathbf{z}) = 0. \tag{33}$$

With regard to (24) and (15), the augmented Hamiltonian can be written as

$$\mathcal{H}_\lambda(\mathbf{z}) = H(\mathbf{z}) + \boldsymbol{\lambda} \cdot \boldsymbol{\Phi}(\mathbf{q}). \tag{34}$$

Accordingly, (33) can be written in the form

$$
\begin{aligned}
\nabla H(\mathbf{z}) \cdot \dot{\mathbf{z}} + \boldsymbol{\lambda} \cdot D\boldsymbol{\Phi}(\mathbf{q})\dot{\mathbf{q}} &= 0, \\
\frac{d}{dt} H(\mathbf{z}) + \boldsymbol{\lambda} \cdot \frac{d}{dt}\boldsymbol{\Phi}(\mathbf{q}) &= 0.
\end{aligned}
\tag{35}
$$

As a consequence of the geometric constraints $(23)_3$, the consistency condition $d\boldsymbol{\Phi}(\mathbf{q})/dt = \mathbf{0}$ has to be satisfied. Thus (35) yields $dH(\mathbf{z})/dt = 0$, which implies conservation of the total energy.

3 Energy-Momentum Scheme

We next outline the design of a time-stepping scheme which is able to reproduce for any step-size the crucial conservation properties summarized above.

Concerning the time discretization of the DAEs (23), we apply the Galerkin-based approach developed by Betsch and Steinmann [5]. To this end, we consider a characteristic time-step $\Delta t = t_{n+1} - t_n$ and restrict our attention to linear approximations (the so-called mG(1) method in [5]) of the form

$$
\mathbf{z}^h(\alpha) = (1-\alpha)\mathbf{z}_n + \alpha\mathbf{z}_{n+1} \quad \text{for} \quad \alpha \in [0, 1]. \tag{36}
$$

In this connection all quantities at t_n, such as $\mathbf{z}_n$, can be regarded as being given. Note that (36) leads to a globally continuous approximation of the phase space coordinates. In contrast to that, the Lagrange multipliers are assumed to be piecewise constant in each time-step, i.e.

$$
\boldsymbol{\lambda}^h = \boldsymbol{\lambda}_{n+1}. \tag{37}
$$

The mG(1) method yields

$$
\mathbf{z}_{n+1} - \mathbf{z}_n = \Delta t \mathbb{J} \int_0^1 \nabla \mathcal{H}_{\lambda^h}(\mathbf{z}^h)\, d\alpha. \tag{38}
$$

It is shown in [5] that the application of a specific quadrature formula for the evaluation of the time integral in (38) has a strong impact on the conservation properties of the resulting time-stepping scheme. In the present work we choose

$$
\int_0^1 \nabla \mathcal{H}_{\lambda^h}(\mathbf{z}^h)\, d\alpha \approx \overline{\nabla} \mathcal{H}_{\lambda_{n+1}}(\mathbf{z}_n, \mathbf{z}_{n+1}), \tag{39}
$$

where $\overline{\nabla}\mathcal{H}_\lambda(z_n, z_{n+1})$ is a discrete gradient (or derivative) in the sense of Gonzalez [7]. It is shown in [7] that the discrete gradient can be designed such that the desired conservation properties are satisfied and specific consistency and accuracy requirements are met. To achieve this goal we aim at a reparametrization of the augmented Hamiltonian which incorporates the invariance properties in a natural way. For example, assume that the rotational invariance property (27) holds and that the augmented Hamiltonian depends only on $\mathbb{S}(\mathbf{z})$, where

$$\mathbb{S}(\mathbf{z}) = \mathbb{S}(\mathbf{z}_1, \ldots, \mathbf{z}_N) = \{\mathbf{y}_A \cdot \mathbf{y}_B \,,\, 1 \le A \le B \le n_{\text{node}} \,,\, \mathbf{y}_A \in \{\mathbf{q}_A, \mathbf{p}_A\}\} \tag{40}$$

is the set of (quadratic) invariants of $\mathbf{z} \in \mathbb{R}^{2n_{\text{dof}}}$. It is worth mentioning that this approach is in accordance with Cauchy's Representation Theorem (see, for example, Truesdell and Noll [19, section 11]). Accordingly, the augmented Hamiltonian can now be written in the form

$$\mathcal{H}_\lambda(\mathbf{z}) = \widetilde{\mathcal{H}}_\lambda(\boldsymbol{\pi}(\mathbf{z})), \tag{41}$$

where the vector of relevant invariants

$$\boldsymbol{\pi}(\mathbf{z}) = \begin{bmatrix} \pi_1(\mathbf{z}) \\ \vdots \\ \pi_d(\mathbf{z}). \end{bmatrix} \tag{42}$$

has been introduced. Note that the components $\pi_i(\mathbf{z})$ depend only on $\mathbb{S}(\mathbf{z})$. In the following we make use of Gonzalez' [7] definition of the discrete gradient. Accordingly, in the present context, the discrete gradient of the augmented Hamiltonian assumes the form

$$\boxed{\overline{\nabla}\mathcal{H}_\lambda(\mathbf{z}_n, \mathbf{z}_{n+1}) = D\boldsymbol{\pi}(\mathbf{z}_{n+\frac{1}{2}})^T \overline{\overline{\nabla}}\widetilde{\mathcal{H}}_\lambda(\boldsymbol{\pi}(\mathbf{z}_n), \boldsymbol{\pi}(\mathbf{z}_{n+1}))} \tag{43}$$

with

$$\begin{aligned}\overline{\overline{\nabla}}\widetilde{\mathcal{H}}_\lambda(\boldsymbol{\pi}_n, \boldsymbol{\pi}_{n+1}) &= \nabla\widetilde{\mathcal{H}}_\lambda(\boldsymbol{\pi}_{n+\frac{1}{2}}) \\ &+ \frac{\widetilde{\mathcal{H}}_\lambda(\boldsymbol{\pi}_{n+1}) - \widetilde{\mathcal{H}}_\lambda(\boldsymbol{\pi}_n) - \nabla\widetilde{\mathcal{H}}_\lambda(\boldsymbol{\pi}_{n+\frac{1}{2}}) \cdot (\boldsymbol{\pi}_{n+1} - \boldsymbol{\pi}_n)}{\| \boldsymbol{\pi}_{n+1} - \boldsymbol{\pi}_n \|^2}(\boldsymbol{\pi}_{n+1} - \boldsymbol{\pi}_n)\end{aligned} \tag{44}$$

and

$$\begin{aligned}\mathbf{z}_{n+\frac{1}{2}} &= \tfrac{1}{2}(\mathbf{z}_n + \mathbf{z}_{n+1}), \\ \boldsymbol{\pi}_{n+\frac{1}{2}} &= \tfrac{1}{2}(\boldsymbol{\pi}_n + \boldsymbol{\pi}_{n+1}).\end{aligned} \tag{45}$$

To summarize, the mG(1) method with quadrature formula (39) yields the following time-stepping scheme:

$$(H) = \left\{ \begin{array}{l} \text{Let the initial values per time step } \mathbf{z}_n \text{ and the step-size } \Delta t \text{ be given. Find} \\ \mathbf{z}_{n+1} \text{ and } \boldsymbol{\lambda}_{n+1} \text{ as the solution of the algebraic system of equations} \\ \qquad \begin{aligned}\mathbf{z}_{n+1} &= \mathbf{z}_n + \Delta t \mathbb{J}\overline{\nabla}\mathcal{H}_{\lambda_{n+1}}(\mathbf{z}_n, \mathbf{z}_{n+1}) \\ \mathbf{0} &= \boldsymbol{\Phi}(\mathbf{q}_{n+1})\end{aligned} \end{array} \right. \tag{46}$$

Remark 1. In essence, the scheme (46) is equivalent to the method proposed by Gonzalez [8]. We further remark that in addition to the constraints on configuration level $(46)_2$, the constraints on momentum level, i.e. $d\boldsymbol{\Phi}(\mathbf{q})/dt = D\boldsymbol{\Phi}(\mathbf{q})\mathbf{M}^{-1}\mathbf{p} = \mathbf{0}$ can be enforced at the end of the time step by adjusting the GGL-type [6] technique to the present conserving framework, see [5] for further details. However, numerical tests revealed no significant improvement of the numerical performance which would justify the additional computational effort.

Remark 2. If a function $f(\mathbf{z})$ is merely quadratic it may be written as $f(\mathbf{z}) = \widetilde{f}(\boldsymbol{\pi}(\mathbf{z})) = \mathbf{a} \cdot \boldsymbol{\pi}(\mathbf{z})$, with constant $\mathbf{a} \in \mathbb{R}^d$. It can be easily verified that the corresponding discrete gradient is given by

$$\overline{\nabla} f(\mathbf{z}_n, \mathbf{z}_{n+1}) = D\boldsymbol{\pi}\left(\mathbf{z}_{n+\frac{1}{2}}\right)^T \mathbf{a} = \nabla f\left(\mathbf{z}_{n+\frac{1}{2}}\right). \tag{47}$$

Accordingly, in this case the discrete gradient coincides with the standard gradient evaluated in $\mathbf{z}_{n+\frac{1}{2}}$.

3.1 Algorithmic Conservation Properties

Similar to the continuous case dealt with before, we next verify that the scheme (46) indeed satisfies the relevant conservation laws.

3.1.1 Algorithmic Conservation of the Total Angular Momentum

The fundamental theorem of calculus gives

$$\begin{aligned}
J_\xi(\mathbf{z}_{n+1}) - J_\xi(\mathbf{z}_n) &= \int_0^1 \nabla J_\xi\left(\mathbf{z}^h(\alpha)\right) \cdot \left(\mathbf{z}^h(\alpha)\right)' \, d\alpha \\
&= \int_0^1 \boldsymbol{\xi}_P\left(\mathbf{z}^h(\alpha)\right) d\alpha \cdot \mathbb{J}(\mathbf{z}_{n+1} - \mathbf{z}_n) \\
&= \boldsymbol{\xi}_P\left(\mathbf{z}_{n+\frac{1}{2}}\right) \cdot \mathbb{J}(\mathbf{z}_{n+1} - \mathbf{z}_n) \\
&= \boldsymbol{\xi}_P\left(\mathbf{z}_{n+\frac{1}{2}}\right) \cdot \mathbb{J}^2 \Delta t \overline{\nabla} \mathcal{H}_{\lambda_{n+1}}(\mathbf{z}_n, \mathbf{z}_{n+1}) \\
&= -\Delta t \overline{\nabla} \mathcal{H}_{\lambda_{n+1}}(\mathbf{z}_n, \mathbf{z}_{n+1}) \cdot \boldsymbol{\xi}_P\left(\mathbf{z}_{n+\frac{1}{2}}\right) \\
&= -\Delta t \overline{\overline{\nabla}} \widetilde{\mathcal{H}}_{\lambda_{n+1}}\left(\boldsymbol{\pi}(\mathbf{z}_n), \boldsymbol{\pi}(\mathbf{z}_{n+1})\right) \cdot D\boldsymbol{\pi}(\mathbf{z}_{n+\frac{1}{2}})\boldsymbol{\xi}_P\left(\mathbf{z}_{n+\frac{1}{2}}\right) \\
&= 0,
\end{aligned} \tag{48}$$

where, similar to (28), use has been made of the property

$$\mathbf{0} = \left.\frac{d}{d\varepsilon}\right|_{\varepsilon=0} \boldsymbol{\pi}(\exp(\varepsilon\hat{\boldsymbol{\xi}}) \circ \mathbf{z}) = D\boldsymbol{\pi}(\mathbf{z})\boldsymbol{\xi}_P(\mathbf{z}) \tag{49}$$

which holds due to the rotational invariance of the vector-valued function $\boldsymbol{\pi}(\mathbf{z})$. Equation (48) corroborates algorithmic conservation of the total angular momentum.

3.1.2 Algorithmic Conservation of the Total Energy

Similar to (33), in the discrete setting we get

$$\frac{1}{\Delta t}\overline{\nabla}\mathcal{H}_{\lambda_{n+1}}(\mathbf{z}_n, \mathbf{z}_{n+1}) \cdot (\mathbf{z}_{n+1} - \mathbf{z}_n) = \overline{\nabla}\mathcal{H}_{\lambda_{n+1}} \cdot \mathbb{J}\overline{\nabla}\mathcal{H}_{\lambda_{n+1}} = 0. \tag{50}$$

On the other hand, with regard to the discrete gradient (43), we obtain

$$\begin{aligned}
\overline{\nabla}\mathcal{H}_{\lambda_{n+1}}(\mathbf{z}_n, \mathbf{z}_{n+1}) \cdot (\mathbf{z}_{n+1} - \mathbf{z}_n) \\
&= \overline{\overline{\nabla}}\widetilde{\mathcal{H}}_{\lambda_{n+1}}\big(\boldsymbol{\pi}(\mathbf{z}_n), \boldsymbol{\pi}(\mathbf{z}_{n+1})\big) \cdot D\boldsymbol{\pi}(\mathbf{z}_{n+\frac{1}{2}}) \cdot (\mathbf{z}_{n+1} - \mathbf{z}_n) \\
&= \overline{\overline{\nabla}}\widetilde{\mathcal{H}}_{\lambda_{n+1}}\big(\boldsymbol{\pi}(\mathbf{z}_n), \boldsymbol{\pi}(\mathbf{z}_{n+1})\big) \cdot \big(\boldsymbol{\pi}(\mathbf{z}_{n+1}) - \boldsymbol{\pi}(\mathbf{z}_n)\big) \\
&= \widetilde{\mathcal{H}}_{\lambda_{n+1}}\big(\boldsymbol{\pi}(\mathbf{z}_{n+1})\big) - \widetilde{\mathcal{H}}_{\lambda_{n+1}}\big(\boldsymbol{\pi}(\mathbf{z}_n)\big) \\
&= \mathcal{H}_{\lambda_{n+1}}(\mathbf{z}_{n+1}) - \mathcal{H}_{\lambda_{n+1}}(\mathbf{z}_n) \\
&= H(\mathbf{z}_{n+1}) - H(\mathbf{z}_n) + \boldsymbol{\lambda}_{n+1} \cdot \big(\boldsymbol{\Phi}(\mathbf{q}_{n+1}) - \boldsymbol{\Phi}(\mathbf{q}_n)\big) \\
&= H(\mathbf{z}_{n+1}) - H(\mathbf{z}_n),
\end{aligned} \tag{51}$$

where use has been made of (34), (44), (46) and the fact that the invariants $\boldsymbol{\pi}(\mathbf{z})$ are quadratic functions. Comparison of (50) and (51) yields

$$H(\mathbf{z}_{n+1}) = H(\mathbf{z}_n) \tag{52}$$

which confirms algorithmic conservation of the total energy.

3.2 Final Form of the Energy-Momentum Scheme

We next exploit the specific (separable) form of the augmented Hamiltonian (24) to recast the energy-momentum scheme (46) in an alternative form which is especially well-suited for the computer implementation. With regard to (12), the kinetic energy in (24) can be written as

$$T(\mathbf{p}) = \frac{1}{2}\sum_{A,B=1}^{n_{\text{node}}} M_{AB}^{-1}\mathbf{p}_A \cdot \mathbf{p}_B \tag{53}$$

and is thus merely a quadratic function of the nodal momenta. In this connection, the inverse of the mass matrix is composed of diagonal sub-matrices $\mathbf{M}_{AB}^{-1} = M_{AB}^{-1}\mathbf{I}_{n_{\text{dim}}}$ $(A, B = 1, \ldots, n_{\text{node}})$. Taking into account Remark 2, the discrete gradient of the augmented Hamiltonian (43) can be written in simplified form

$$\overline{\nabla}\mathcal{H}_\lambda(\mathbf{z}_n, \mathbf{z}_{n+1}) = \begin{bmatrix} \overline{\nabla}_q \mathcal{V}_\lambda(\mathbf{q}_n, \mathbf{q}_{n+1}) \\ \overline{\nabla}_p T(\mathbf{p}_n, \mathbf{p}_{n+1}) \end{bmatrix} = \begin{bmatrix} \overline{\nabla}_q \mathcal{V}_\lambda(\mathbf{q}_n, \mathbf{q}_{n+1}) \\ \mathbf{M}^{-1}\mathbf{p}_{n+\frac{1}{2}} \end{bmatrix}. \tag{54}$$

Accordingly, application of the discrete gradient is confined to the augmented potential function (25). That is, (43) boils down to

$$\boxed{\overline{\nabla}_q \mathcal{V}_\lambda(\mathbf{q}_n, \mathbf{q}_{n+1}) = D\boldsymbol{\pi}(\mathbf{q}_{n+\frac{1}{2}})^T \overline{\overline{\nabla}} \widetilde{\mathcal{V}}_\lambda\big(\boldsymbol{\pi}(\mathbf{q}_n), \boldsymbol{\pi}(\mathbf{q}_{n+1})\big)} \tag{55}$$

Now the energy-momentum scheme (46) gives rise to the following algorithmic problem:

$$(L) = \begin{cases} \text{Let the initial values per time step } (\mathbf{q}_n, \mathbf{v}_n) \text{ and the step-size } \Delta t \text{ be given.} \\ \text{Find } (\mathbf{q}_{n+1}, \mathbf{v}_{n+1}) \text{ and } \boldsymbol{\lambda}_{n+1} \text{ as the solution of the algebraic system of equations} \\ \qquad \mathbf{q}_{n+1} - \mathbf{q}_n = \dfrac{\Delta t}{2}\big(\mathbf{v}_n + \mathbf{v}_{n+1}\big) \\ \mathbf{M}(\mathbf{v}_{n+1} - \mathbf{v}_n) = -\Delta t \overline{\nabla}_q V(\mathbf{q}_n, \mathbf{q}_{n+1}) - \Delta t \sum\limits_{l=1}^{m} (\lambda_l)_{n+1} \overline{\nabla}_q \Phi_l(\mathbf{q}_n, \mathbf{q}_{n+1}) \\ \qquad \mathbf{0} = \boldsymbol{\Phi}(\mathbf{q}_{n+1}) \end{cases} \tag{56}$$

In $(56)_2$ the potential energy function is given by $V(\mathbf{q}) = V^{\text{int}}(\mathbf{q}) + V^{\text{ext}}(\mathbf{q})$. We refer to [3] for details of the implementation of the energy-momentum scheme (56).

3.3 Application to Planar Problems

The application of the scheme (56) essentially depends on specific parametrizations of the discrete strain energy function $V^{\text{int}}(\mathbf{q})$ and the constraint functions $\Phi_l(\mathbf{q})$ in terms of appropriate invariants. We shall illustrate this procedure by considering planar problems, i.e. $n_{\text{dim}} = 2$ and $\mathbf{q}_A \in \mathbb{R}^2$ $(A \in \eta)$.

As before, we focus on the case of rotational invariance. If a scalar-valued function $\gamma(\mathbf{q}_1, \ldots, \mathbf{q}_{n_{\text{node}}})$ is invariant under the proper orthogonal group, then Cauchy's representation theorem (Truesdell and Noll [19, section 11] or Antman [1, chapter 8]) implies that $\gamma(\mathbf{q})$ depends only on the set of quadratic invariants $\mathbb{I}(\mathbf{q}) = \mathbb{S}(\mathbf{q}) \bigcup \mathbb{T}(\mathbf{q})$, where

$$\begin{aligned} \mathbb{S}(\mathbf{q}) &= \{\mathbf{q}_A \cdot \mathbf{q}_B \,,\, 1 \leq A \leq B \leq n_{\text{nodes}}\}, \\ \mathbb{T}(\mathbf{q}) &= \{\det([\mathbf{q}_A, \mathbf{q}_B]) \,,\, 1 \leq A < B \leq n_{\text{nodes}}\}. \end{aligned} \tag{57}$$

We first deal with the discrete strain energy function. Thereafter, we focus on the constraint functions corresponding to a specific contact formulation.

3.4 Treatment of the Discrete Strain Energy Function

Applying numerical integration to the evaluation of the discrete strain energy function (7) yields

$$V^{\text{int}}(\mathbf{q}) = \sum_{m=1}^{n_{gp}} \widetilde{W}\big(\boldsymbol{\pi}^{(m)}(\mathbf{q})\big) w^{(m)} \tag{58}$$

with

$$\boldsymbol{\pi}^{(m)}(\mathbf{q}) = \begin{bmatrix} C_{11}^{(m)}(\mathbf{q}) \\ C_{22}^{(m)}(\mathbf{q}) \\ C_{12}^{(m)}(\mathbf{q}) \end{bmatrix} \tag{59}$$

and the components of the discrete deformation tensor (5)

$$C_{ij}^{(m)}(\mathbf{q}) = \sum_{A,B=1}^{n_{\text{node}}} \mathbf{q}_A \cdot \mathbf{q}_B \big(\mathbf{e}_i \cdot \nabla N_A(\mathbf{X}^{(m)})\big)\big(\mathbf{e}_j \cdot \nabla N_B(\mathbf{X}^{(m)})\big). \tag{60}$$

In this connection, index m refers to specific quadrature points ($m \in \{1, \ldots, n_{gp}\}$) with associated coordinates $\mathbf{X}^{(m)}$ and 'weights' $w^{(m)}$. It is obvious from (60), that the components of the discrete deformation tensor depend only on $\mathbb{S}(\mathbf{q})$ and thus qualify as invariants. With regard to (55), the discrete gradient of $V^{\text{int}}(\mathbf{q})$ can now be written as

$$\overline{\nabla}_q V^{\text{int}}(\mathbf{q}_n, \mathbf{q}_{n+1}) = \sum_{m=1}^{n_{gp}} D\boldsymbol{\pi}^{(m)}(\mathbf{q}_{n+\frac{1}{2}})^T \overline{\overline{\nabla}} \widetilde{W}\big(\boldsymbol{\pi}^{(m)}(\mathbf{q}_n), \boldsymbol{\pi}^{(m)}(\mathbf{q}_{n+1})\big) w^{(m)}. \tag{61}$$

It is worth noting that $\overline{\overline{\nabla}} \widetilde{W}$ can be linked to the algorithmic constitutive relation proposed by Simo and Gonzalez [17, section 4], see also Gonzalez [9].

3.5 Node-to-Segment Contact Element

We next focus on the use of node-to-segment (NTS) contact elements for the enforcement of the constraint of impenetrability. In particular, we aim at a representation of the NTS contact constraints in terms of appropriate invariants.

Consider a representative NTS contact element depicted in Figure 3. Suppose that the slave node X_S is in contact with the master segment $\overline{Y_1 Y_2}$ given by

$$\mathbf{y}(\xi) = \frac{1}{2}(1 - \xi)\mathbf{y}_1 + \frac{1}{2}(1 + \xi)\mathbf{y}_2 \tag{62}$$

for $\xi \in (-1, 1)$. The point of contact on the master segment $\overline{Y_1 Y_2}$, denoted by $\bar{\mathbf{y}} = \mathbf{y}(\bar{\xi})$, is determined by the minimization of $f(\xi) = \|\mathbf{x}_S - \mathbf{y}(\xi)\|$, which yields the condition

$$\big(\mathbf{x}_S - \mathbf{y}(\bar{\xi})\big) \cdot \mathbf{y}'(\bar{\xi}) = 0. \tag{63}$$

Inserting from (62) into (63) yields

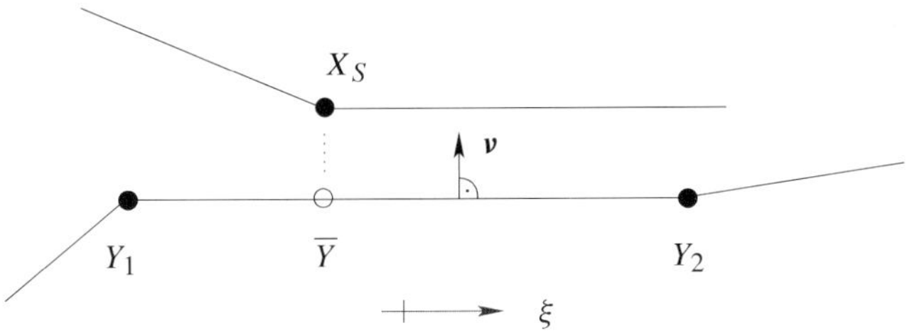

Fig. 3. Representative node-to-segment contact element

$$\bar{\xi} = \frac{\left(\mathbf{x}_S - \frac{1}{2}(\mathbf{y}_1 + \mathbf{y}_2)\right) \cdot \frac{1}{2}(\mathbf{y}_2 - \mathbf{y}_1)}{\|\frac{1}{2}(\mathbf{y}_2 - \mathbf{y}_1)\|^2} . \tag{64}$$

The condition of impenetrability leads to the definition of the constraint function

$$\phi_S(\overline{\mathbf{q}}) = \left(\mathbf{x}_S - \mathbf{y}(\bar{\xi})\right) \cdot \boldsymbol{\nu}(\bar{\xi}), \tag{65}$$

where

$$\overline{\mathbf{q}} = (\mathbf{x}_S\, , \mathbf{y}_1\, , \mathbf{y}_2) \in \mathbb{R}^6 \tag{66}$$

contains the relevant nodal position vectors and $\boldsymbol{\nu}(\bar{\xi}) \in \mathbb{R}^2$ is the unit outward normal to the master segment at $\overline{\mathbf{y}} = \mathbf{y}(\bar{\xi})$. With regard to (62), the tangent vector

$$\mathbf{y}'(\xi) = \frac{1}{2}(\mathbf{y}_2 - \mathbf{y}_1) \tag{67}$$

is constant. Accordingly, in the present case $\boldsymbol{\nu}$ does not depend on ξ and may be written as

$$\boldsymbol{\nu} = -\boldsymbol{\Lambda}(\mathbf{y}_2 - \mathbf{y}_1)/\|\mathbf{y}_2 - \mathbf{y}_1\| \tag{68}$$

with the constant matrix

$$\boldsymbol{\Lambda} = \begin{bmatrix} 0 & 1 \\ -1 & 0 \end{bmatrix} . \tag{69}$$

Note that $\boldsymbol{\Lambda}^T = \boldsymbol{\Lambda}^{-1} = -\boldsymbol{\Lambda}$ and $\boldsymbol{\Lambda}^2 = -\mathbf{I}_2$. To summarize, the constraint function (65) can be written in the form

$$\phi_S(\overline{\mathbf{q}}) = \left(\mathbf{x}_S - \frac{1}{2}(1 - \bar{\xi})\mathbf{y}_1 - \frac{1}{2}(1 + \bar{\xi})\mathbf{y}_2\right) \cdot \boldsymbol{\nu}. \tag{70}$$

Then a straightforward calculation yields the associated constraint Jacobian

$$D\phi_S(\overline{\mathbf{q}})^T = \begin{bmatrix} \boldsymbol{\nu} \\ -\frac{1}{2}(1 - \bar{\xi})\boldsymbol{\nu} \\ -\frac{1}{2}(1 + \bar{\xi})\boldsymbol{\nu} \end{bmatrix} . \tag{71}$$

3.5.1 NTS Contact Element in Terms of Invariants

Next we recast the NTS contact constraint in terms of quadratic invariants. Substituting (68) into (70) yields

$$\phi_S(\overline{\mathbf{q}}) = \frac{(\mathbf{y}_2 - \mathbf{y}_1) \cdot \mathbf{\Lambda}\big(\mathbf{x}_S - \frac{1}{2}(\mathbf{y}_1 + \mathbf{y}_2)\big)}{\|\mathbf{y}_2 - \mathbf{y}_1\|}\,, \tag{72}$$

where use has been made of the skew-symmetry of $\mathbf{\Lambda}$. Upon definition of the two quadratic invariants

$$\begin{aligned} \pi_1(\overline{\mathbf{q}}) &= (\mathbf{y}_2 - \mathbf{y}_1) \cdot \mathbf{\Lambda}\Big(\mathbf{x}_S - \tfrac{1}{2}(\mathbf{y}_1 + \mathbf{y}_2)\Big), \\ \pi_2(\overline{\mathbf{q}}) &= (\mathbf{y}_2 - \mathbf{y}_1) \cdot (\mathbf{y}_2 - \mathbf{y}_1) \end{aligned} \tag{73}$$

the constraint function (72) can be recast in the form

$$\boxed{\phi_S(\overline{\mathbf{q}}) = \tilde{\phi}_S\big(\boldsymbol{\pi}(\overline{\mathbf{q}})\big) = \frac{\pi_1(\overline{\mathbf{q}})}{\sqrt{\pi_2(\overline{\mathbf{q}})}}} \tag{74}$$

With regard to (55), the discrete gradient of $\phi_S(\overline{\mathbf{q}})$ can now be written as

$$\overline{\nabla}_{\overline{q}}\phi_S(\overline{\mathbf{q}}_n, \overline{\mathbf{q}}_{n+1}) = D\boldsymbol{\pi}(\overline{\mathbf{q}}_{n+\frac{1}{2}})^T \overline{\overline{\nabla}}\tilde{\phi}_S\big(\boldsymbol{\pi}(\overline{\mathbf{q}}_n), \boldsymbol{\pi}(\overline{\mathbf{q}}_{n+1})\big) \tag{75}$$

with

$$D\boldsymbol{\pi}(\overline{\mathbf{q}}) = \begin{bmatrix} (\mathbf{y}_2^T - \mathbf{y}_1^T)\mathbf{\Lambda} & (\mathbf{x}_S^T - \mathbf{y}_2^T)\mathbf{\Lambda} & (\mathbf{y}_1^T - \mathbf{x}_S^T)\mathbf{\Lambda} \\ \mathbf{0}^T & 2(\mathbf{y}_1^T - \mathbf{y}_2^T) & 2(\mathbf{y}_2^T - \mathbf{y}_1^T) \end{bmatrix}. \tag{76}$$

Note that $\pi_2(\overline{\mathbf{q}})$ depends only on $\mathbb{S}(\mathbf{q})$. Moreover, $\pi_1(\overline{\mathbf{q}})$ may also be written in the alternative form

$$\pi_1(\overline{\mathbf{q}}) = \mathbf{y}_2 \cdot \mathbf{\Lambda}\mathbf{x}_S + \mathbf{y}_1 \cdot \mathbf{\Lambda}\mathbf{y}_2 + \mathbf{x}_S \cdot \mathbf{\Lambda}\mathbf{y}_1\,. \tag{77}$$

Since $\mathbf{a} \cdot \mathbf{\Lambda}\mathbf{b} = \det([\mathbf{a}, \mathbf{b}])$ for any $\mathbf{a}\,, \mathbf{b} \in \mathbb{R}^2$, $\pi_1(\overline{\mathbf{q}})$ depends only on $\mathbb{T}(\mathbf{q})$.

Remark 3. So far the emphasis has been put on rotational invariance. However, the invariants (73) and (60) are as well invariant under translations. For example, concerning (73), it can be easily verified that

$$\pi_\alpha(\mathbf{x}_S + \mathbf{c}\,, \mathbf{y}_1 + \mathbf{c}\,, \mathbf{y}_2 + \mathbf{c}) = \pi_\alpha(\mathbf{x}_S\,, \mathbf{y}_1\,, \mathbf{y}_2) \tag{78}$$

for $\alpha = 1, 2$ and any $\mathbf{c} \in \mathbb{R}^2$. Translational invariance of the augmented Hamiltonian is associated with conservation of the total linear momentum.

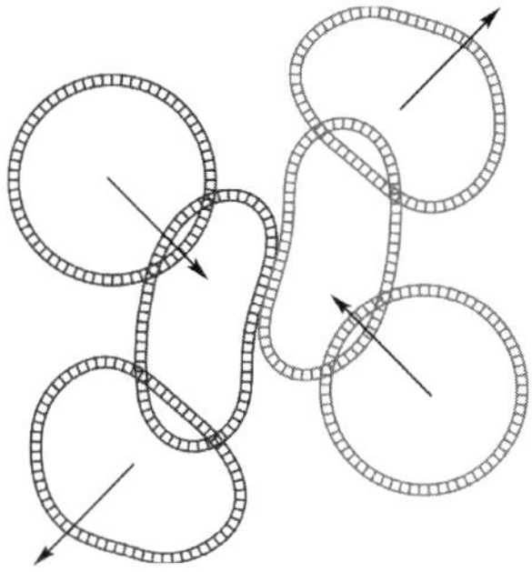

Fig. 4. Snapshots of the motion.

4 Numerical Example

The present numerical example deals with the impact of two elastic rings. Similar examples have been previously considered by Wriggers et al. [21] and Laursen and Love [13]. This example is especially well-suited to check the algorithmic conservation properties.

Sixty-four isoparametric displacement-based bi-linear finite elements have been used to discretize each initially circular ring (cf. Figure 1). The material behavior of both rings is assumed to be governed by the St. Venant–Kirchhoff material model with Young's modulus $E = 100$ and Poisson's ratio $\nu = 0.1$. The mass density of both rings is $\varrho_R = 0.001$. The two rings move towards each other with an initial velocity of $v_0 = 10$. In the simulations documented below a time-step of $\Delta t = 0.01$ has been used.

To illustrate the simulated motion snapshots of the two rings at successive points in time are depicted in Figure 4. After the initial free-flight phase contact takes place within the time interval of approximately [6, 16]. Since no external forces/torques act on the present two-body system the total linear momentum as well as the total angular momentum are conserved quantities. These momenta are indeed conserved by the proposed algorithm, see Figure 5 for a plot of the total angular momentum. Furthermore, algorithmic conservation of the total energy follows from Figure 6. To demonstrate the robustness of the proposed approach the calculations have also been carried out with a time-step of $\Delta t = 1.0$. In fact, the algorithmic conservation properties still hold. It can be observed from Figure 7 that despite the large time-steps the present method is able capture main characteristics of the impact problem under consideration.

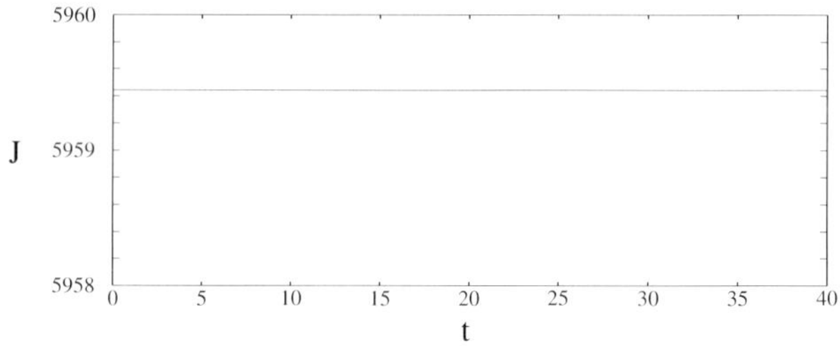

Fig. 5. Total angular momentum versus time

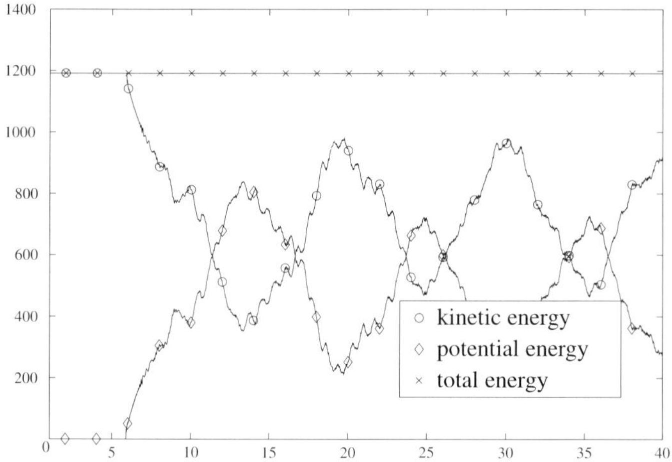

Fig. 6. Energy versus time ($\Delta t = 0.01$).

5 Conclusions

The main new contribution of the present work lies in the design of the algorithmic contact forces within the framework of the NTS contact element. In particular, the newly-proposed parametrization of the NTS contact constraints in terms of appropriate invariants along with the use of the notion of a discrete gradient are the main features which facilitate the design of an energy-momentum scheme. In the developments presented herein we have tacitly assumed that active nodes of contact have been properly detected.

In recent years the mortar method has emerged as an alternative to the NTS approach for the spatial resolution of the contact problem, see, for example, Puso and Laursen [15] and Hüeber and Wohlmuth [10]. Our present approach can as well be

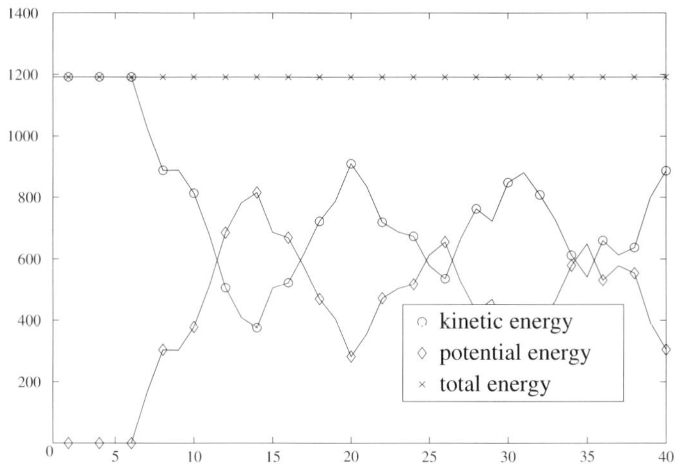

Fig. 7. Energy versus time ($\Delta t = 1.0$).

applied to the mortar description of contact. This will be the subject of a follow-up paper.

References

1. S.S. Antman. *Nonlinear Problems of Elasticity*, 2nd edition. Springer-Verlag, 2005.
2. F. Armero and E. Petöcz. Formulation and analysis of conserving algorithms for frictionless dynamic contact/impact problems. *Comput. Methods Appl. Mech. Engrg.*, 158:269–300, 1998.
3. P. Betsch. The discrete null space method for the energy consistent integration of constrained mechanical systems. Part I: Holonomic constraints. *Comput. Methods Appl. Mech. Engrg.*, 194(50–52):5159–5190, 2005.
4. P. Betsch and P. Steinmann. Conservation properties of a time FE method. Part II: Time-stepping schemes for nonlinear elastodynamics. *Int. J. Numer. Methods Eng.*, 50:1931–1955, 2001.
5. P. Betsch and P. Steinmann. Conservation properties of a time FE method. Part III: Mechanical systems with holonomic constraints. *Int. J. Numer. Methods Eng.*, 53:2271–2304, 2002.
6. C.W. Gear, G.K. Gupta, and B.J. Leimkuhler. Automatic integration of the Euler-Lagrange equations with constraints. *J. Comp. Appl. Math.*, 12:77–90, 1985.
7. O. Gonzalez. Time integration and discrete Hamiltonian systems. *J. Nonlinear Sci.*, 6:449–467, 1996.
8. O. Gonzalez. Mechanical systems subject to holonomic constraints: Differential-algebraic formulations and conservative integration. *Physica D*, 132:165–174, 1999.

9. O. Gonzalez. Exact energy and momentum conserving algorithms for general models in nonlinear elasticity. *Comput. Methods Appl. Mech. Engrg.*, 190:1763–1783, 2000.
10. S. Hüeber and B.I. Wohlmuth. Mortar methods for contact problems. In P. Wriggers and U. Nackenhorst (Eds), *Analysis and Simulation of Contact Problems*, Lecture Notes in Applied and Computational Mechanics, Vol. 27 Springer-Verlag, 2006, pp. 39–47.
11. T.A. Laursen. *Computational Contact and Impact Mechanics*. Springer-Verlag, 2002.
12. T.A. Laursen and V. Chawla. Design of energy conserving algorithms for frictionless dynamic contact problems. *Int. J. Numer. Methods Eng.*, 40:863–886, 1997.
13. T.A. Laursen and G.R. Love. Improved implicit integrators for transient impact problems – Geometric admissibility withing the conserving framework. *Int. J. Numer. Methods Eng.*, 53:245–274, 2002.
14. J.E. Marsden and T.S. Ratiu. *Introduction to Mechanics and Symmetry*, 2nd edition. Springer-Verlag, 1999.
15. M.A. Puso and T.A. Laursen. A mortar segment-to-segment contact method for large deformation solid mechanics. *Comput. Methods Appl. Mech. Engrg.*, 193:601–629, 2004.
16. J.C. Simo. Numerical analysis and simulation of plasticity. In P.G. Ciarlet and J.L. Lions (Eds), *Handbook of Numerical Analysis*, Vol. VI. Elsevier Science, 1998, pp. 183–499.
17. J.C. Simo and O Gonzalez. Recent results on the numerical integration of infinite-dimensional Hamiltonian systems. In T.J.R. Hughes, E. Onate, and O.C. Zienkiewicz (Eds), *Recent Developments in Finite Element Analysis*. CIMNE, Barcelona, 1994, pp. 255–271.
18. J.C. Simo and N. Tarnow. The discrete energy-momentum method. Conserving algorithms for nonlinear elastodynamics. *Z. angew. Math. Phys. (ZAMP)*, 43:757–792, 1992.
19. C. Truesdell and W. Noll. *The Non-Linear Field Theories of Mechanics*, 3rd edition. Springer-Verlag, 2004.
20. P. Wriggers. *Computational Contact Mechanics*. John Wiley & Sons, 2002.
21. P. Wriggers, T. Vu Van, and E. Stein. Finite element formulation of large deformation impact-contact problems with friction. *Computers & Structures*, 36:2049–2071, 1993.

Symmetrization of Various Friction Models Based on an Augmented Lagrangian Approach

Alexander Konyukhov and Karl Schweizerhof

Institut für Mechanik, Universität Karlsruhe, Englerstraße 2, D-76131 Karlsruhe, Germany
E-mail: {konyukhov, schweizerhof}@ifm.uni-karlsruhe.de

Abstract. The standard implementation of the classical Coulomb friction model together with the Newton iterative method for the finite element method leads to non-symmetric tangent matrices for sliding zones of contact surfaces. This fact is known in literature as consequence of the non-associativity of the friction law. Considering anisotropic models for friction, especially including coupling of adhesion and friction, leads to additional non-symmetries due to anisotropy. Since, non-symmetry of matrices is a non-desirable feature of most engineering problems, various proposals for symmetrization are known in computational mechanics. A further suggestion is made in this contribution. The covariant approach for both isotropic and anisotropic frictional contact problems leads to a very simple structure of the tangent matrices. This allows to obtain very robust tangent matrices within the symmetrized Augmented Lagrangian method. In the current contribution, the nested Uzawa algorithm is applied for symmetrization within the Augmented Lagrangian approach for an anisotropic friction model including adhesion and friction. The numerical examples show the good convergence behavior for various problems such as small and large sliding problems.

Key words: Anisotropic friction and adhesion, covariant description, Augmented Lagrangian method, symmetrization.

1 Introduction

The penalty method for frictional contact problems [1,2] currently is among the most popular schemes in finite element packages, leads to the satisfaction of the contact constraints, such as non-penetration and sticking conditions, only within a certain tolerance. This tolerance is defined by the penalty parameters for both normal and tangential direction. As is known, the classical method of Lagrangian multipliers leads to an exact satisfaction of contact constraints, however, one should take care of the number of multipliers due to the often overstiff behavior of contact interfaces, see e.g. in [3]. Additional degrees of freedoms for the contact tractions are often

Peter Wriggers and Udo Nackenhorst (eds.), IUTAM Symposium on Computational Contact Mechanics, 97–111.

mentioned among the disadvantages of this method. Recently, various combinations of the Mortar method have been developed in e.g. [4–6] allowing to overcome overstiff behavior especially in the case where contact tractions are computed point-wise at integration points. In addition, good results for the patch-test have been shown also in [3] exploiting the, so-called, segment-to-segment approach, coinciding with the Mortar method with penalty descriptions of the contact traction. So-called dual Lagrange multipliers have been developed for non-frictional contact in Wohlmuth [7] allowing to condense degrees of freedom for contact traction. Using this approach, the frictional constraints should be carefully treated as a sequence of the Tresca type friction model, see [8]. This approach is similar to the Augmented Lagrangian approach allowing to satisfy the contact constraints for non-penetration and sticking within the nested algorithm. The method is described theoretically in Bertsekas [9] and Fortin and Glowinski [10]. Pietrzak and Curnier [11] developed the Augmented Lagrangian approach for frictional contact including the corresponding saddle point functional. Laursen and Simo [12] proposed a symmetrization procedure based on the nested Uzawa algorithm. This approach is based on symmetrization of the corresponding tangent matrices for the solution of the equilibrium equations, while the contact constraints have been still iteratively satisfied within the external loop with a specified tolerance.

In the current contribution, we will describe the symmetric Augmented Lagrangian method for the coupled anisotropic contact model including anisotropy for friction and adhesion as developed in Konyukhov and Schweizerhof [13, 14]. The geometrical structure of the corresponding tangent matrices allows us to construct very simple symmetric tangent matrices for the anisotropic case. The isotropic case can then be defined as a reduction of the anisotropic case. Good convergence rates are illustrated in the numerical examples.

2 Covariant Description of the Coupled Anisotropic Friction Model

We shortly present here the main details of the coupled interface contact model including anisotropy for adhesion and friction and refer to [13, 14] for further details and corresponding derivations.

At the beginning, a local surface coordinate system is introduced as

$$\mathbf{r}_s(\xi^1, \xi^2, \xi^3) = \boldsymbol{\rho}(\xi^1, \xi^2) + \mathbf{n}\xi^3, \tag{1}$$

where ξ^1, ξ^2 are two convective coordinates and responsible for the tangential contact interaction. The third coordinate ξ^3 is the value of the penetration and is used to define the properties of the normal interaction

$$\xi^3 = (\mathbf{r}_s - \boldsymbol{\rho}) \cdot \mathbf{n}. \tag{2}$$

The vector of contact tractions **R** is defined as a covariant vector and, therefore, is expressed via the contravariant basis vectors $\boldsymbol{\rho}^i$ and **n** in the coordinate system (1) as sum of the tangential and normal components

$$\mathbf{R} = \mathbf{T} + \mathbf{N} = T_i \boldsymbol{\rho}^i + N\mathbf{n}. \tag{3}$$

For the generalization into anisotropy for adhesion, the tangential traction vector $T_i \boldsymbol{\rho}^i$ and the normal traction vector $N\mathbf{n}$ are assumed to be decoupled. Therefore, the generalized constitutive equations for tangential tractions are taken in rate form as

$$\frac{d\mathbf{T}}{dt} = \mathbf{B}(\mathbf{v}_s - \mathbf{v}), \tag{4}$$

where **B** is the anisotropic adhesion tensor. The covariant differentiation operations are involved throughout whenever the rate form is given.

The constitutive equation for normal traction is given in closed form and can be viewed as a simple penalty regularization procedure for this normal traction N

$$N = \epsilon_N \xi^3, \tag{5}$$

where ϵ_N is a parameter of normal compliance, or a penalty parameter.

The anisotropy for friction is chosen as a model of Coulomb type involving proportionality of the frictional force to the normal traction N. The corresponding yield function is written as

$$\Phi = \sqrt{f^{ij} T_i T_j} - |N| = \sqrt{\mathbf{T} \cdot \mathbf{F}\mathbf{T}} - |N|, \tag{6}$$

where $\mathbf{F} = f^{ij} \boldsymbol{\rho}_i \boldsymbol{\rho}_j$ is the anisotropic friction tensor.

2.1 Incremental Formulation of the Coupled Anisotropic Model

Though, initially the model is formulated in the continuous rate form, it has been transformed into incremental form for the final computational model via the application of the backward Euler scheme.

(i) The full displacement vector $\Delta\boldsymbol{\xi} = \boldsymbol{\xi}^{(n+1)} - \boldsymbol{\xi}^{(n)}$ is decomposed additively into an elastic increment $\Delta\boldsymbol{\xi}^{el}$ and into a sliding increment $\Delta\boldsymbol{\xi}^{sl}$:

$$\Delta\boldsymbol{\xi} = \Delta\boldsymbol{\xi}^{el} + \Delta\boldsymbol{\xi}^{sl}, \tag{7}$$

where both vectors are defined in the surface metrics, namely,

$$\Delta\boldsymbol{\xi} := \Delta\xi^i \boldsymbol{\rho}_i = (\xi^i_{(n+1)} - \xi^i_{(n)})\boldsymbol{\rho}_i. \tag{8}$$

(ii) The trial elastic, or adhesion force $\mathbf{T}^{\text{tr}}_{(n+1)}$ is computed via the incremental evolution equations:

$$\mathbf{T}^{\text{tr}}_{(n+1)} - \mathbf{T}_{(n)} = \mathbf{B}^{(n+1)}(\boldsymbol{\xi}^{el}_{(n+1)} - \boldsymbol{\xi}^{el}_{(n)}). \tag{9}$$

(iii) The final result if the tangential traction $\mathbf{T}$ is the elastic one (belongs to the adhesion region), or the plastic one if provided by the yield condition of the Coulomb type in each load step becomes:

$$\begin{aligned}\Phi^{\text{tr}} &:= \sqrt{\mathbf{T}^{\text{tr}}_{(n+1)} \cdot \mathbf{F}^{(n+1)}\mathbf{T}^{\text{tr}}_{(n+1)}} - |N_{(n+1)}| \\ &= \sqrt{f^{ij}T^{\text{tr}}_{i\,(n+1)}T^{\text{tr}}_{j\,(n+1)}} - |N_{(n+1)}|,\end{aligned} \tag{10}$$

- If $\Phi^{\text{tr}} < 0$ then the trial force is a real sticking force $\mathbf{T} = \mathbf{T}^{\text{tr}}$.
- If $\Phi^{\text{tr}} \geq 0$ then the sliding force must be obtained via the maximum of the energy dissipation function given in the incremental form.

(iv) All contact parameters such as sliding traction and sliding distance should be derived via the principle of the maximum dissipation

$$D^{(n+1)} := \Delta\boldsymbol{\xi}^{sl} \cdot \mathbf{T}^{sl}_{(n+1)} = \Delta\xi^{i}_{sl}T^{sl}_{i\,(n+1)}, \quad D^{(n+1)} \longrightarrow \max. \tag{11}$$

Using the necessary optimization conditions for the functional $D^{(n+1)}$ together with the Kuhn–Tucker conditions from (iii) the closed form for the sliding force $\mathbf{T}^{sl}$ is obtained as:

$$\mathbf{T}^{sl} = -\frac{\mathbf{BFT}^{\text{tr}}}{\sqrt{\mathbf{BFT}^{\text{tr}} \cdot \mathbf{FBFT}^{\text{tr}}}}|N|. \tag{12}$$

3 Linearization Process and Structure of Matrices

Since the frictional problem is nonlinear, an iterative solution based on a Newton scheme should be applied. The important part of the implementation is then a consistent tangent matrix which differs for sticking and sliding according to the return-mapping scheme. We present here the results with particular focus on the structure of the matrices especially their symmetry. According to numerical experiences reported in [15, 16]. we will also exclude the curvature parts of the matrices.

3.1 Linearization of the Normal Part δW_c^N

We denote $D(f)$ as a linearization operator acting on a functional f in the covariant form. Thus, linearization of the virtual work of the contact normal traction N is given as:

$$D(\delta W_c^N) = D(\int_s N\delta\xi^3 ds) = -\underbrace{\int_s \epsilon_N(\delta\mathbf{r}_s - \delta\boldsymbol{\rho})\cdot(\mathbf{n}\otimes\mathbf{n})(\mathbf{v}_s - \mathbf{v})ds}_{\text{leads to } \mathbf{K}_N^m} \tag{13}$$

$$-\underbrace{\int_s \epsilon_N \xi^3 \left(\delta\boldsymbol{\rho}_{,j}\cdot a^{ij}(\mathbf{n}\otimes\boldsymbol{\rho}_i)(\mathbf{v}_s - \mathbf{v}) + (\delta\mathbf{r}_s - \delta\boldsymbol{\rho})\cdot a^{ij}(\boldsymbol{\rho}_j\otimes\mathbf{n})\mathbf{v}_{,i}\right)ds,}_{\text{leads to } \mathbf{K}_N^r} \tag{13a}$$

where a^{ij} are contravariant components of the metric tensor for the master surface and $\boldsymbol{\rho}_j = \partial\boldsymbol{\rho}/\partial\xi^j$, $j = 1, 2$ are covariant basis (tangent) vectors. Here, the first term (resp. the second term) after approximation of the geometry leads to the main part of the contact matrix $\mathbf{K}_N^m$ (resp. the rotational part of the contact matrix $\mathbf{K}_N^r$).

All parts can be algorithmically implemented for any order of approximation. In order to do this only the operator $\mathbf{A}$ for the approximation of the surface geometry has to be introduced. The derivatives with respect to convective coordinates $\mathbf{A}_\xi$ are also necessary. Thus, the approximation can be written as

$$\mathbf{r}_s - \boldsymbol{\rho} = \mathbf{A}\{\mathbf{x}\}, \quad \boldsymbol{\rho}_\xi = \mathbf{A}_\xi\{\mathbf{x}\}, \tag{14}$$

where $\mathbf{x}$ is a nodal vector for the standard FE implementation, or a control points (knots) vector for a CAD approximation. This leads e.g. to the following structure of the contact matrix $\mathbf{K}_N^m$

$$\mathbf{K}_N^m = -\int_s \epsilon_N \mathbf{A}^T\cdot(\mathbf{n}\otimes\mathbf{n})\mathbf{A}ds, \tag{15}$$

where the integral is computed via the set of Gauss points defined on the slave segment and penetrating into the master surface (so-called penalty based Mortar method, see Fischer and Wriggers [6]).

3.2 Linearization of the Tangential Part δW_c^T: Sticking Case

The sticking case is understood as a case where the tangential traction remains in the elastic region and, therefore, is computed via the evolution equations (ii). Linearization leads to

$$D(\delta W_c^T) = D(\int_s T_i^{\text{tr}}\delta\xi^i ds) = \tag{16}$$

$$= \underbrace{\int_s (\delta\mathbf{r}_s - \delta\boldsymbol{\rho})\cdot\mathbf{B}(\mathbf{v}_s - \mathbf{v})ds -}_{\text{leads to } \mathbf{K}_{T,st}^m}$$

$$\underbrace{-\int_s T_i\left((\delta\mathbf{r}_s - \delta\boldsymbol{\rho})\cdot a^{il}a^{jk}\,\boldsymbol{\rho}_k\otimes\boldsymbol{\rho}_l\,\mathbf{v}_j + \delta\boldsymbol{\rho}_{,j}\cdot a^{ik}a^{jl}\,\boldsymbol{\rho}_k\otimes\boldsymbol{\rho}_l\,(\mathbf{v}_s - \mathbf{v})\right)ds,}_{\text{leads to } \mathbf{K}_{T,st}^r}$$

3.3 Linearization of the Tangential Part δW_c^T: Sliding Case

The sliding case is understood as a case where the yield condition (iii) is fulfilled and, therefore, the tangential traction is computed as a sliding traction given in Equation (12). The corresponding linearization leads to

$$D(\delta W_c^T) = D(\int_s T_i^{sl} \delta\xi^i ds) = \tag{17}$$

$$= \underbrace{\int_s \left((\delta \mathbf{r}_s - \delta \boldsymbol{\rho}) \cdot \frac{\epsilon_N \mathbf{BFT}^{\text{tr}} \otimes \mathbf{n}}{\sqrt{\mathbf{BFT}^{\text{tr}} \cdot \mathbf{FBFT}^{\text{tr}}}} (\mathbf{v}_s - \mathbf{v}) \right) ds}_{\text{leads to } \mathbf{K}_{T,sl}^{m,1}}$$

$$- \underbrace{\int_s \left((\delta \mathbf{r}_s - \delta \boldsymbol{\rho}) \cdot \frac{|N|\, \mathbf{BFB}}{\sqrt{\mathbf{BFT}^{\text{tr}} \cdot \mathbf{FBFT}^{\text{tr}}}} (\mathbf{v}_s - \mathbf{v}) \right) ds}_{\text{leads to } \mathbf{K}_{T,sl}^{m,2}}$$

$$+ \underbrace{\int_s \left((\delta \mathbf{r}_s - \delta \boldsymbol{\rho}) \cdot \frac{|N|\, \mathbf{BFT}^{\text{tr}} \otimes (\mathbf{BFB})^T \mathbf{FBFT}^{\text{tr}}}{\sqrt{(\mathbf{BFT}^{\text{tr}} \cdot \mathbf{FBFT}^{\text{tr}})^3}} (\mathbf{v}_s - \mathbf{v}) \right) ds}_{\text{leads to } \mathbf{K}_{T,sl}^{m,3}}$$

$$- \underbrace{\int_s T_i^{sl} \left[(\delta \mathbf{r}_s - \delta \boldsymbol{\rho}) \cdot a^{il} a^{jk}\, \boldsymbol{\rho}_k \otimes \boldsymbol{\rho}_l\, \mathbf{v}_j + \delta \boldsymbol{\rho}_{,j} \cdot a^{ik} a^{jl}\, \boldsymbol{\rho}_k \otimes \boldsymbol{\rho}_l\, (\mathbf{v}_s - \mathbf{v}) \right] ds.}_{\text{leads to } \mathbf{K}_{T,sl}^{r}}$$

Now we can summarize the results concerning the symmetry of the necessary tangent matrices. As expected, all parts concerning non-frictional and sticking contact, namely $\mathbf{K}_N^m$, $\mathbf{K}_N^r$, $\mathbf{K}_{T,st}^m$ and $\mathbf{K}_{T,st}^r$ are symmetric. Attention should be paid to the sliding tangent matrix, because it contains both symmetric and non-symmetric parts:

$\mathbf{K}_{T,sl}^{m,1}$ is the first main part due to the coupling of the normal and the sliding tractions. It appears due to linearization of the normal traction N and it is non-symmetric.

$\mathbf{K}_{T,sl}^{m,2}$ is the second main part. It appears due to linearization of the normal trial tangential traction $\mathbf{T}^{\text{tr}}$ and preserves symmetry.

$\mathbf{K}_{T,sl}^{m,3}$ is the third main part and non-symmetric. It appears due to linearization of the complex term $1/\sqrt{\mathbf{BFT}^{\text{tr}} \cdot \mathbf{FBFT}^{\text{tr}}}$ reflecting the coupling of anisotropy for adhesion and friction.

Finally, $\mathbf{K}_{T,sl}^{r}$ is the rotational, symmetric part reflecting rotation of the master segment, for more details see [17].

4 Augmented Lagrangian Method and Symmetric Uzawa Algorithm

The main advantage of the Augmented Lagrangian in comparison to the penalty method is the possibility to select a value of the penalty parameter leading to well conditioned tangent matrices together with the enforcement of the constraint conditions (non-penetration and sticking) within a specified tolerance. The method is constructed as a nested algorithm known as Uzawa algorithm and possesses linear convergence. The algorithm using the full consistent tangent matrix is known as exact Uzawa algorithm, while the inexact algorithm is exploiting a somehow simplified matrix, see results on convergence in Stadler [18]. Laursen and Simo [12] developed a symmetrized Uzawa algorithm, where symmetric matrices for the sliding case have been obtained under the assumption that the normal traction remains constant for the solution of the equilibrium equations within an internal loop. This preserves the quadratic rate of convergence for the internal loop. The correct value is enforced then within the external loop where a linear rate of convergence is preserved.

4.1 Limitations of the Augmented Lagrangian Approach for the Coupled Anisotropic Model

The complexity for a generalization of the discussed algorithm into anisotropic friction especially including anisotropic adhesion is that the sliding tangent matrix is fully non-symmetric, moreover, the assumption of isotropy together with the constant normal traction would lead only to the rotational part of the tangent matrix. Since, this part has no influence for small displacement problems as well as in the case when the master segment has no rotation, we obtain only zero matrices. This makes it impossible to obtain any solution. Fortunately, the covariant approach allows to estimate the influence of the matrices part by part. Thus, we can construct a symmetric algorithm and analyze it numerically.

Another problem for the coupled anisotropy is that the sticking case is defined when the slave contact point lays inside the elliptic adhesion domain, see Figure 1. However, as is known from numerical results the ratio $a/b = \mu_1\varepsilon_2/\mu_2\varepsilon_1$ does not influence the convergence result in the penalty based approach and can lead to correct kinematics even for large sliding problems. An enforcement to put a slave contact point inside the adhesion ellipse via the Augmented Lagrangian approach necessarily leads to dis-convergence in cases as $a/b \to 0$, or $a/b \to \infty$. In computation due to the linear convergence of the Augmented Lagrangian method the global number of iteration is proportional to the ratio a/b for $a > b$, e.g. a simply computable case with $a/b = 10$ would lead to a 10-times increase of the global number of iterations for the Augmented Lagrangian method. Thus, one should judge the coupled anisotropic model as an interface model for tangential traction, rather than a penalty based approach.

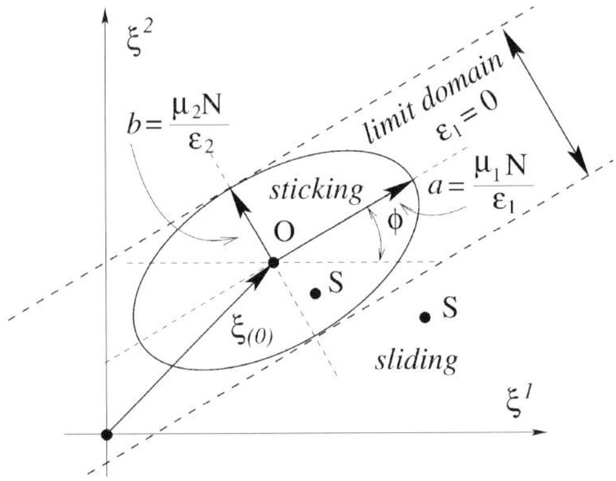

Fig. 1. Allowable elastic region (adhesion domain).

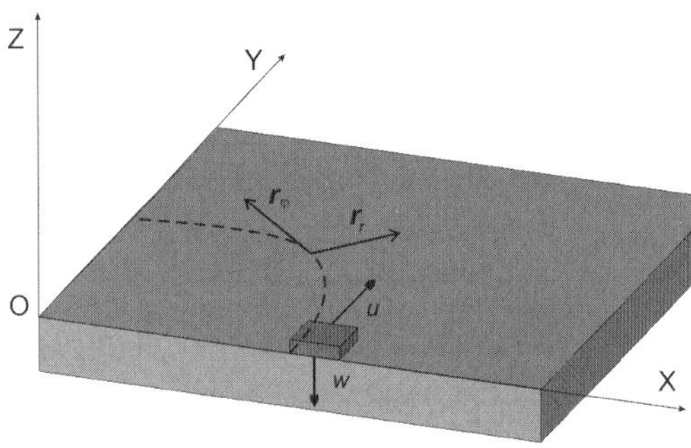

Fig. 2. Geometry and loading for the case of polar orthotropy.

Summarizing the discussion, we can define *computable cases* for the Augmented Lagrangian method in application to the coupled anisotropic model for:

1. *Small sliding problems with low anisotropy.* Small sliding can be numerically defined as a case where it is important to compute the distribution of the sticking-sliding zone. The low anisotropy is defined then as a case with $a/b \approx 1$.
   In this case, Lagrangian multipliers for both normal N, and tangential traction $\mathbf{T}$ are augmented. The convergence for tangential displacements is checked by the proximity to the initial contact point:

$$\sqrt{a_{ij}(\xi^i_{(I)} - \xi^i_{(K-1)})(\xi^j_{(I)} - \xi^j_{(K-1)})} \le \epsilon_T, \tag{18}$$

   where ϵ_T is a specified tolerance for tangential displacements. The convergence for normal displacements is checked as in the standard isotropic case by the normal penetration

$$|\xi^3| \leq \epsilon_N, \tag{19}$$

where ϵ_N is a specified tolerance for the normal penetration.

2. *Large sliding problem with arbitrary anisotropy.* Large sliding can be numerically defined as a case where sliding is reached within a single load step. In this case, the global kinematical behavior of a contacting body is of interest rather than the distribution of the sticking-sliding zone.
   In this case, the Lagrangian multiplier only for the normal traction N needs to be updated because the tolerance of the sticking condition will normally not influence the tolerance of the computed trajectory. Thus, convergence is checked only for the normal displacement.

The symmetrized algorithm in accordance with the inexact Uzawa approach which appears to be numerically effective for the current anisotropic problem is constructed as follows. The internal loop, No. 3 in Tables 1 and 2, serves for the solution of the equilibrium equations. The normal traction $N^{K,L,I}$ is computed via the augmented scheme and updated inside loop 2. The trial tangential traction $\mathbf{T}^{K,L,I}$ for *small sliding problems* is computed via the augmented scheme. The update according to the scheme $\Delta\lambda_T^L = \Delta\lambda_T^{L-1} + \mathbf{B}(\boldsymbol{\xi}_{(I)} - \boldsymbol{\xi}_{(K-1)})$ inside loop 2 allows to enforce sticking conditions similar to the normal penetration. The modification for *large sliding problems* is as follows: the sticking condition is satisfied in the penalty form, then the trial tangential traction $\mathbf{T}^{K,L,I}$ is computed as

Table 1. Update scheme for normal traction N.

1. Loop over applied in load incremental steps: $K = K + 1$
   initialization of Lagrange multiplier $\lambda_N^0 = 0$
   - 2. Loop over augmented multipliers: $L = L + 1$
     - 3. Iterative solution of global equilibrium equations: with normal force
       $$N^{K,L,I} = \lambda_N^{L-1} + \varepsilon_N \xi_I^3$$
       penetration ξ_I^3 is computed in each iteration I.

     Update multipliers
     $$\lambda_N^L = N^{K,L,I}$$
     Convergence is checked by the non-penetration condition:
     $$|\xi_I^3| \leq \epsilon_N, \; for \; N^{K,L,I} < 0,$$
     where ϵ_N - specified tolerance for normal displacements

Table 2. Update scheme for trial tangential traction $\mathbf{T}$.

1. Loop over load applied in incremental steps: $K = K + 1$
   initial condition for tangential traction $\mathbf{T}^{0,0,0} = 0$
   initialization of Lagrange multiplier $\Delta\lambda_T^0 = 0$

   2. Loop over augmented multipliers: $L = L + 1$

      3. Iterative solution of
         global equilibrium equations:
         with tangential traction

         $$\mathbf{T}^{K,L,I} = \mathbf{T}^{K-1,L,I} + \Delta\lambda_T^{L-1} + \mathbf{B}(\xi_{(I)} - \xi_{(K-1)})$$

         $\xi_{(I)}$ is a projection point in each iteration I,
         $\xi_{(K-1)}$ is a projection point in the previous load step.

   Update multipliers

   $$\Delta\lambda_T^L = \Delta\lambda_T^{L-1} + \mathbf{B}(\xi_{(I)} - \xi_{(K-1)})$$

   Convergence is checked by proximity
   to the initial sticking point:

   $$\sqrt{a_{ij}(\xi^i_{(I)} - \xi^i_{(K-1)})(\xi^j_{(I)} - \xi^j_{(K-1)})} \leq \epsilon_T,$$

   where ϵ_T - specified tolerance for tangential displacements

$\mathbf{T}^{K,L,I} = \mathbf{T}^{K-1,L,I} + \mathbf{B}(\xi_{(I)} - \xi_{(K-1)})$. The Lagrange multiplier for the tangential traction vector $\Delta\lambda_T^{L-1}$ is not introduced and, therefore, an update loop 2 does not exist.

An important modification for symmetrization should be done for the return-mapping scheme, see Table 3, where the sliding force instead of Equation (12) is computed via the augmented multiplier for the normal traction λ_N^{L-1} as

$$\mathbf{T}^{sl} = -\frac{\mathbf{BFT}^{\text{tr}}}{\sqrt{\mathbf{BFT}^{\text{tr}} \cdot \mathbf{FBFT}^{\text{tr}}}} |\lambda_N^{L-1}|. \tag{20}$$

The yield function (6) is respectively modified as

$$\Phi_\lambda = \sqrt{\mathbf{T}_{\text{tr}}^{K,L,I} \cdot \mathbf{FT}_{\text{tr}}^{K,L,I}} - |\lambda_N^{L-1}|. \tag{21}$$

This leads to a constant sliding force for the internal loop and, therefore, the first main part of the sliding tangent matrix is zero $\mathbf{K}_{T,sl}^{m,1} = 0$. Then, the full tangent matrix for the sliding case becomes:

$$\mathbf{K}_{\mathrm{sl}}^{\mathrm{full}} = \mathbf{K}_N^m + \mathbf{K}_N^r + \mathbf{K}_{T,\mathrm{sl}}^{m,2} + \mathbf{K}_{T,\mathrm{sl}}^{m,3} + \mathbf{K}_{T,\mathrm{sl}}^r. \tag{22}$$

The part $\mathbf{K}_{T,\mathrm{sl}}^{m,3}$ is still non-symmetric due to anisotropy. The matrix is fully symmetric only for isotropic friction. As it was found in numerical computations we can exclude this part with only a small loss of efficiency. This finally leads to the following matrix in the sliding case according to the inexact Uzawa algorithm:

$$\mathbf{K}_{\mathrm{sl}}^{\mathrm{full}} = \mathbf{K}_N^m + \mathbf{K}_N^r + \mathbf{K}_{T,\mathrm{sl}}^{m,2} + \mathbf{K}_{T,\mathrm{sl}}^r. \tag{23}$$

The tangent matrix for sticking remains symmetric:

$$\mathbf{K}_{\mathrm{st}}^{\mathrm{full}} = \mathbf{K}_N^m + \mathbf{K}_N^r + \mathbf{K}_{T,\mathrm{st}}^m + \mathbf{K}_{T,\mathrm{st}}^r. \tag{24}$$

Table 3. Return-mapping scheme for the symmetric Augmented Lagrangian method.

1. Loop over load applied in incremental steps: $K = K + 1$
   initial condition for tangential traction $\mathbf{T}^{0,0,0} = 0$
   initialization of Lagrange multiplier $\Delta\lambda_T^0 = 0$
   initialization of Lagrange multiplier $\Delta\lambda_N^0 = 0$

   2. Loop over load augmented multipliers: $L = L + 1$

      3. Iterative solution of
         global equilibrium equations:
         a) compute trial tangential traction $\mathbf{T}_{\mathrm{tr}}^{K,L,I}$ (Table 2)

         b) compute trial yield function Φ_λ

         $$\Phi_\lambda = \sqrt{\mathbf{T}_{\mathrm{tr}}^{K,L,I} \cdot \mathbf{F}\mathbf{T}_{\mathrm{tr}}^{K,L,I}} - |\lambda_N^{L-1}|.$$

         c) return-mapping: real tangential traction $\mathbf{T}^{K,L,I}$

         $$\mathbf{T}^{K,L,I} = \begin{cases} \mathbf{T}_{\mathrm{tr}}^{K,L,I} & if\ \Phi_\lambda < 0 \\ \mathbf{T}^{sl} = -\dfrac{\hat{\mathbf{T}}}{\sqrt{\hat{\mathbf{T}} \cdot \mathbf{F}\hat{\mathbf{T}}}} |\lambda_N^{L-1}| & if\ \Phi_\lambda \geq 0 \end{cases} \tag{25}$$

      where $\hat{\mathbf{T}} = \mathbf{B}\mathbf{F}\mathbf{T}_{\mathrm{u}}^{K,L,I}$

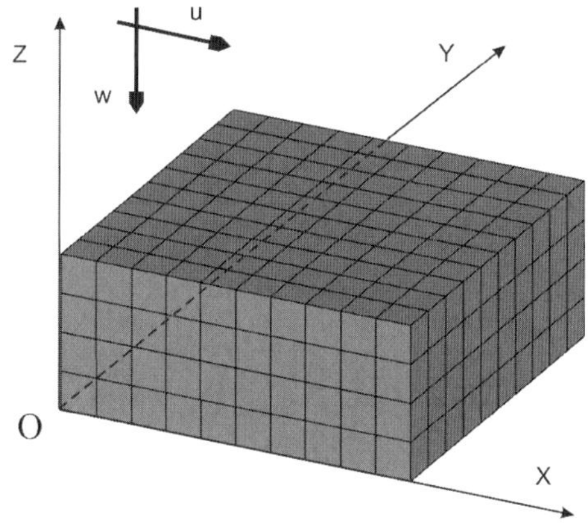

Fig. 3. Geometry and loading of the rectangular in plane block.

5 Numerical Examples

Two cases with different anisotropy have been selected to illustrate the convergence for the proposed approach: constant orthotropy and polar orthotropy on the plane.

5.1 Small Sliding Problem: Constant Orthotropy

We present here an example, which has been analyzed for the penalty based approach in [14]. The rectangular block, see Figure 3, is considered on an orthotropic plane. The dimensions of the block are $10 \times 10 \times 4$ with linear elastic properties: Young's modulus $E = 2.10 \cdot 10^4$ and Poisson ratio $\nu = 0.3$, assumed within a consistent dimension system.

The case of constant orthotropy is defined by the spectral representation of both the adhesion tensor $\mathbf{B}$ and the friction tensor $\mathbf{F}$ as follows:

$$\mathbf{B} = [b^i_j] = -\begin{bmatrix} \varepsilon_1 \cos^2\alpha + \varepsilon_2 \sin^2\alpha & (\varepsilon_1 - \varepsilon_2)\sin\alpha\cos\alpha \\ (\varepsilon_1 - \varepsilon_2)\sin\alpha\cos\alpha & \varepsilon_1 \sin^2\alpha + \varepsilon_2 \cos^2\alpha \end{bmatrix}. \tag{26}$$

$$\mathbf{F} = [f^i_k] = \begin{bmatrix} \dfrac{1}{\mu_1^2}\cos^2\beta + \dfrac{1}{\mu_2^2}\sin^2\beta & \left(\dfrac{1}{\mu_1^2} - \dfrac{1}{\mu_2^2}\right)\sin\beta\cos\beta \\ \left(\dfrac{1}{\mu_1^2} - \dfrac{1}{\mu_2^2}\right)\sin\beta\cos\beta & \dfrac{1}{\mu_1^2}\sin^2\beta + \dfrac{1}{\mu_2^2}\cos^2\beta \end{bmatrix}. \tag{27}$$

The following parameters are taken for the computations: normal penalty parameter: $\varepsilon_N = 2.1 \cdot 10^4$; adhesion parameters: $\varepsilon_1 = 3.0 \cdot 10^4$, $\varepsilon_1 = 2.0 \cdot 10^4$; friction coefficients: $\mu_1 = 3.0$, $\mu_2 = 2.0$; tolerance for penetration: $\epsilon_N = 1.0 \cdot 10^{-5}$;

Table 4. Constant orthotropy on a plane. Convergence results for the symmetric Augmented Lagrangian method.

Aug. No.	Number of eq. iter.	Tolerance
1	7	$1.8742 \cdot 10^{-3}$
2	14	$6.0419 \cdot 10^{-4}$
3	12	$1.7536 \cdot 10^{-4}$
4	12	$4.0497 \cdot 10^{-5}$
5	12	$3.5972 \cdot 10^{-6}$
	$\sum 56$	

tolerance for tangential displacement: $\epsilon_T = 1.0 \cdot 10^{-5}$; orthotropic angles: $\alpha = 45^o, \beta = 45^o$. This leads to a, so-called, geometrical isotropic case because of the ratio $\varepsilon_1\mu_2/\varepsilon_2\mu_1 = 1.0$.

The block (see Figure 3) is located on the XOY plane and loaded by vertically prescribed displacements on the upper surface $w = 1.0 \cdot 10^{-2}$. The penalty based approach gives convergence in 4 iterations, but both normal and tangential displacements inside the sticking region do not satisfy the prescribed tolerance. The results for the Augmented scheme are presented in Table 4. The method shows linear convergence as can be seen from the *Tolerance* column.

5.2 Large Sliding Problem: Polar Orthotropy

For the case with large sliding, we consider the block on a plane with polar orthotropy see [14]. An elastic block with dimensions $1 \times 1 \times 0.25$ and mesh $4 \times 4 \times 1$ is positioned on a rigid block, see Figure 2. Linear elastic material is assumed within a consistent dimension system: Young's modulus $E = 2.10 \cdot 10^4$; Poisson ratio $\nu = 0.3$. The loading is applied sequentially by prescribing displacements on the upper surface in $(1+n)$ steps: (1) vertical loading with $w = 1.0 \cdot 10^{-2}$, (2) n steps with horizontal displacement increments $\Delta u = 1.0 \cdot 10^{-2}$ along the X axis. The frictional tensor is isotropic with $\mu_1 = \mu_2 = 0.2$, but the adhesion tensor has the following structure in the Cartesian coordinate system:

$$\mathbf{B} = -\frac{1}{x^2+y^2}\begin{bmatrix} \varepsilon_r x^2 + \varepsilon_\varphi y^2 & (\varepsilon_r - \varepsilon_\varphi)xy \\ (\varepsilon_r - \varepsilon_\varphi)xy & \varepsilon_r y^2 + \varepsilon_\varphi x^2 \end{bmatrix}. \tag{28}$$

Parameters for the adhesion tensor are chosen as $\varepsilon_r = 1000$, $\varepsilon_\varphi = 0.0$, in order to obtain the circular trajectory of the sliding block. Now only the normal tractions N are updated and a tolerance for penetration $\epsilon_N = 1.0 \cdot 10^{-5}$ is chosen. The gap for the penalty based approach is computed as $\xi^3 = 7.26 \cdot 10^{-3}$ if the normal penalty $\varepsilon_N = 2.1 \cdot 10^4$ is chosen. The total number of iterations (4556) compared with the penalty approach (1834) is influenced only by the normal penetration. The final coordinates of the nodal point $X = 0.292$, $Y = 4.996$ are compared with the results focused with the penalty approach $X = 0.306$, $Y = 4.994$ in order to show that the main kinematical effect of the anisotropic surfaces is preserved for the Augmented Lagrangian method. A comparison of both trajectories leads to a slight difference comparable to the value of the initial penetration and the initial sticking displacements.

6 Conclusions

In this contribution a symmetrization of the stiffness matrix has been developed within the Augmented Lagrangian method for anisotropic contact surfaces including both anisotropy for adhesion and anisotropy for friction domains. It is shown that in general a fully coupled model necessarily leads to a fully non-symmetric matrix in the case of sliding, but the covariant approach allows to estimate the structure of the tangent matrix part by part and, therefore, allows to construct a symmetric matrix used in accordance with the inexact Uzawa algorithm. However, as is shown, the Augmented Lagrangian method cannot be directly applied to arbitrary anisotropic surfaces due to convergence problems. Thus, to create a robust algorithm, contact problems have to be subdivided into *small sliding problems* with low anisotropy where the distribution of the sticking-sliding zone is of interest, and into *large sliding problems* where the trajectory of the sliding body is of interest.

For *the small sliding problems*, both the normal, and the tangential contact tractions are augmented within the nested Uzawa algorithm. This makes it possible to enforce both normal and tangential sticking displacements to satisfy a prescribed tolerance. For *the large sliding problems* only the normal traction is augmented leading to the enforcement of only the normal displacements to satisfy a prescribe tolerance. In both cases the return-mapping scheme is exploited to obtain the real sliding tractions. Numerical examples including constant as well as nonlinear orthotropy, e.g. a polar orthotropy showed the effectiveness of the proposed approach.

References

1. Wriggers P (2002) *Computational Contact Mechanics*. John Wiley & Sons.
2. Laursen TA (2002) *Computational Contact and Impact Mechanics. Fundamentals of Modeling Interfacial Phenomena in Nonlinear Finite Element Analysis*. Springer, New York.
3. Harnau M, Konyukhov A, Schweizerhof K (2005) Algorithmic aspects in large deformation contact analysis using 'Solid-Shell' elements. *Comput Struct*, 83:1804–1823.
4. Puso MA, Laursen TA (2004) A mortar segment-to-segment contact method for large deformation solid mechanics. *Comp Meth Appl Mech Engrg*, 193:601–629.
5. Puso MA, Laursen TA (2004) A mortar segment-to-segment frictional contact method for large deformations. *Comp Meth Appl Mech Engrg*, 193:4891–4913.

6. Fischer KA, Wriggers P (2005) Frictionless 2D Contact formulations for finite deformations based on the mortar method. *Comput Mech*, 36:226-244.
7. Wohlmuth B (2000) A mortar finite element method using dual spaces for the Lagrange multiplier. *SIAM J Numer Anal*, 38:989–1012.
8. Hüeber S, Stadler G, Wohlmuth B (2006) A primal-dual active set algorithm for three-dimensional contact problems with Coulomb friction. Preprint, Inst Angewandte Anal. und Num. Simulation, Stuttgart.
9. Bertsekas, D.P. (1982) *Constrained Optimization and Lagrange Multiplier Methods*. Academic Press, New York.
10. Fortin M., Glowinski, R. (1983) *Augmented Lagrangian Methods: Applications to the Numerical Solutions of Boundary-Value Problems*. North-Holland, Amsterdam.
11. Pietrzak G, Curnier A (1999) Large deformation frictional contact mechanics: Continuum formulation and augmented Lagrangian treatment. *Comp Meth Appl Mech Engrg*, 177:351–381.
12. Laursen TA, Simo JC (1993) A continuum-based finite element formulation for the implicit solution of multibody, large deformation-frictional contact problems. *Int J Numer Methods Engrg*, 36:3451–3485.
13. Konyukhov A, Schweizerhof K (2006) Covariant description of contact interfaces considering anisotropy for adhesion and friction: Part 1. Formulation and analysis of the computational model. *Comp Meth Appl Mech Engrg*, 196:103–117.
14. Konyukhov A, Schweizerhof K (2006) Covariant description of contact interfaces considering anisotropy for adhesion and friction. Part 2. Linearization, finite element implementation and numerical analysis of the model. *Comp Meth Appl Mech Engrg*, 196:289–303.
15. Konyukhov A, Schweizerhof K (2004) Contact formulation via a velocity description allowing efficiency improvements in frictionless contact analysis. *Comput Mech*, 33:165–173.
16. Konyukhov A, Schweizerhof K (2005) Covariant description for frictional contact problems. *Comput Mech*, 35:190–213.
17. Konyukhov A, Schweizerhof K (2006) A special focus on 2D formulations for contact problems using a covariant description. *Int J Numer Methods Engrg*, 66:1432–1465.
18. Stadler G (2004) Infinite-dimensional semi-smooth Newton and Augmented Lagrangian methods for friction and contact problems in elasticity. PhD Thesis, University of Graz, Graz.

Adaptive BEM for Contact Problems with Friction

Alexey Chernov[1] and Ernst P. Stephan[2]

[1] *Seminar for Applied Mathematics, ETH-Zentrum, CH-8092 Zürich, Switzerland*
E-mail: achernov@math.ethz.ch
[2] *Institut für Angewandte Mathematik, Leibniz Universität Hannover, D-30167 Hannover, Germany*
E-mail: stephan@ifam.uni-hannover.de

Abstract. We describe two methods for the numerical solution of two-body frictional contact problems based on pure boundary integral formulations, and corresponding automatic mesh refinement procedures. For the h-version of the penalty BEM we use local reliable and efficient a posteriori error estimators of the residual type [1]. For the hp-version of the mortar BEM we use heuristic local hp a posteriori residual error indicators and a three-step refinement procedure [1].

Key words: A posteriori error estimate, adaptive mesh refinement, contact, Tresca friction, Steklov–Poincaré operator.

1 Introduction

Frictional contact problems are very important for a wide range of problems in solid mechanics and engineering. Usually the zone of contact is relatively small and the surface contact forces are very high, which cause large errors for the numerical simulations if uniform meshes are used. This problem can be solved by appropriate mesh refinement near the contact interface. Often the contact zone is not known in advance and only a guess for a zone of possible contact is available. Therefore, automatic adaptive refinement procedures are very important, since they allow to find and eliminate by mesh refinement current local sources of the error. The local error contributions of each mesh element are computed, based only on the known data, i.e. the Galerkin solution, the boundary conditions and the volume forces.

We use the boundary element method (BEM) for modelling the linear elastic response of the material. Here, no degrees of freedom in the interior of the physical domain are needed. Using properties of the boundary integral operators, a pure boundary integral formulation is obtained for frictional contact problems. Based on [1]

Peter Wriggers and Udo Nackenhorst (eds.), IUTAM Symposium on Computational Contact Mechanics, 113–122.

we use penalty and mortar methods for the numerical solutions of one- and two-body contact problems and present error indicators suitable for adaptive refinement procedures. Furthermore, we describe refinement algorithms and give illustrative numerical examples.

2 Contact Problems with Tresca Friction

We consider two non-overlapping polygonal domains Ω^i, $i = 1, 2$ with Lipschitz boundaries Γ^i. Each Γ^i consists of three mutually disjoint measurable parts Γ^i_D, Γ^i_N and Γ^i_C, where Dirichlet, Neumann and frictional contact conditions are prescribed. In order to specify the contact conditions, we assume that there exists a bijective mapping $b : \Gamma^1_C \to \Gamma^2_C$ which identifies the "opposite" points of the both potential contact areas and along which the distance between the bodies is measured [4]. Now, we can identify $\Gamma_C := \Gamma^1_C$ and Γ^2_C in terms of b. Let $g(\boldsymbol{x}), \boldsymbol{x} \in \Gamma_C$ be the distance between the bodies. We define a jump of $h \in C(\Gamma^1_C \cup \Gamma^2_C)$ across Γ_C pointwise as $[h](\boldsymbol{x}) := h(\boldsymbol{x}) - h(b(\boldsymbol{x})), \boldsymbol{x} \in \Gamma_C$, where the normal or the tangential displacement ($u_n := \boldsymbol{u} \cdot \boldsymbol{n}$ or $u_t := \boldsymbol{u} \cdot \boldsymbol{t}$) is at h. Equilibrium across Γ_C implies that the sum of tractions on the sides Γ^1_C and Γ^2_C vanishes, therefore the scalar normal and tangential boundary stresses $\sigma_n := \boldsymbol{n}^1 \cdot \sigma(\boldsymbol{u}^1) \cdot \boldsymbol{n}^1$ and $\sigma_t := \boldsymbol{t}^1 \cdot \sigma(\boldsymbol{u}^1) \cdot \boldsymbol{t}^1$ are a.e. well defined. Here $\boldsymbol{n}^1$ and $\boldsymbol{t}^1$ are normal and tangential vectors on Γ^1_C. The stress tensor σ is determined by $\boldsymbol{u}$ with Hooke's law and under small strain assumptions there holds with $\lambda > 0$, $\mu > 0$

$$\sigma(\boldsymbol{u}) = \lambda \operatorname{tr}\varepsilon(\boldsymbol{u}) + 2\mu\varepsilon(\boldsymbol{u}), \qquad \varepsilon(\boldsymbol{u}) = (\nabla \boldsymbol{u} + \nabla \boldsymbol{u}^T)/2.$$

The classical formulation of the problem is given by

$$\begin{aligned}
\operatorname{div}\sigma(\boldsymbol{u}) &= 0 && \text{in } \Omega := \Omega^1 \cup \Omega^2,\\
\boldsymbol{u} &= 0 && \text{on } \Gamma_D := \Gamma^1_D \cup \Gamma^1_D,\\
\sigma(\boldsymbol{u}) \cdot \boldsymbol{n} &= \hat{\boldsymbol{t}} && \text{on } \Gamma_N := \Gamma^1_N \cup \Gamma^2_N,\\
&\left.\begin{aligned}\sigma_n \le 0, \quad [u_n] \le g, \quad \sigma_n([u_n] - g) = 0,\\ |\sigma_t| \le \mathcal{F}, \quad \sigma_t[u_t] + \mathcal{F}|[u_t]| = 0,\end{aligned}\right\} && \text{on } \Gamma_C.
\end{aligned} \tag{1}$$

The so-called given friction function $\mathcal{F} \ge 0$ defines pointwise the sticking threshold of the bodies, i.e. as it can be seen from (1), if the absolute value of the tangential stress does not reach the given friction $|\sigma_t| < \mathcal{F}$ then $[u_t] = 0$, and $[u_t] \neq 0$ is only possible if $|\sigma_t| = \mathcal{F}$. In the more general case of Coulomb's friction law $\mathcal{F}$ is proportional to the normal stress $\mathcal{F} := \mu_f \sigma_n$, where μ_f is the friction coefficient. In order to derive the weak formulation for (1) we assume that $\mathcal{F} \in L_2(\Gamma_C)$, $\hat{\boldsymbol{t}} \in [H^{-1/2}(\Gamma_N)]^2$. For the definition of the Sobolev spaces $H^r(\Gamma)$, $\tilde{H}^r(\Gamma)$ with $r \in \mathbb{R}$ and their norms, see [8]. We shall use corresponding bold face symbols for spaces with vector-valued elements, e.g. $\boldsymbol{H}^1(\Omega) := [H^1(\Omega)]^2$. We define

$$H_D^1(\Omega) := \{u \in H^1(\Omega^1 \cup \Omega^2), u = 0 \text{ on } \Gamma_D\}.$$

The variational formulation for (1) is obtained in a standard way by using the Gauss theorem in Ω_1 and Ω_2. We obtain

$$\int_\Omega \sigma(u) : \varepsilon(v)\, dx = \int_{\Gamma_N \cup \Gamma_C} (\sigma(u) \cdot n) \cdot v\, ds, \quad \forall v \in H_D^1(\Omega). \tag{2}$$

Applying the contact conditions with Tresca friction leads to the uniquely solvable [6] variational inequality in $\Omega^1 \cup \Omega^2$:

Find $u \in \mathcal{K}_\Omega := \{u \in H_D^1(\Omega), [u_n] \le g \text{ on } \Gamma_C\}$:

$$\int_\Omega \sigma(u) : \varepsilon(v-u)\, dx + \int_{\Gamma_C} \mathcal{F}(|[v_t]| - |[u_t]|)\, ds \ge \int_{\Gamma_N} \hat{t} \cdot (v-u)\, ds, \quad \forall v \in \mathcal{K}_\Omega. \tag{3}$$

Next, we introduce the Dirichlet-to-Neumann map on $\Gamma := \Gamma^1 \cup \Gamma^2$

$$S : H^{1/2}(\Gamma) \to H^{-1/2}(\Gamma)$$
$$u|_\Gamma \mapsto \sigma(u)|_\Gamma \cdot n.$$

Inserting $S(u|_\Gamma) = \sigma(u)|_\Gamma \cdot n$ in (2) yields

$$\int_\Omega \sigma(u) : \varepsilon(v)\, dx = \int_\Gamma S(u|_\Gamma) \cdot v\, ds, \quad \forall v \in H_D^1(\Omega). \tag{4}$$

It is well-known [9] that the Steklov–Poincaré operator S can be computed via

$$S := W + (K' + 1/2)V^{-1}(K + 1/2)$$

with the help of the boundary integral operators

$$V\phi(x) = \int_\Gamma G(x, y)\phi(y)\, ds_y, \qquad K\phi(x) = \int_\Gamma \mathcal{T}_{n_y} G(x, y)^T \phi(y)\, ds_y,$$

$$K'\phi(x) = \int_\Gamma \mathcal{T}_{n_x} G(x, y)\phi(y)\, ds_y, \quad W\phi(x) = -\mathcal{T}_{n_x} \int_\Gamma \mathcal{T}_{n_y} G(x, y)^T \phi(y)\, ds_y,$$

where $G(x, y)$ is the fundamental solution of the Lamé operator and $\mathcal{T}_{n_z} v = (\sigma(v)|_\Gamma \cdot n_z)$ is the conormal derivative of $v(z)$ with respect to $z = x, y$ [11].

Due to relation (4) we can perform the boundary reduction of the domain formulation (3) and give a pure boundary integral formulation of the initial problem (1). Since $\tilde{H}^{1/2}(\Gamma \setminus \Gamma_D) = \{u|_\Gamma : u \in H_D^1(\Omega)\}$ we obtain the following constraint boundary variational inequality:

Find $u \in \mathcal{K} := \{u \in \tilde{H}^{1/2}(\Gamma \setminus \Gamma_D), [u_n] \le g \text{ on } \Gamma_C)\}$:

$$\int_\Gamma (Su) \cdot (v-u)\, ds + \int_{\Gamma_C} \mathcal{F}(|[v_t]| - |[u_t]|)\, ds \ge \int_{\Gamma_N} \hat{t} \cdot (v-u)\, ds, \quad \forall v \in \mathcal{K}, \tag{5}$$

which is uniquely solvable [1]. Note that the bilinear form of the above formulation contains the non-local operator S and the Gâteaux non-differentiable friction functional.

3 Penalty Method

Applying penalty methods to (3) is a standard approach for FE-simulations of contact problems, see e.g. [12]. We approximate the constraint boundary variational inequality (5) by the penalty boundary variational formulation:

Find $\boldsymbol{u}^\varepsilon \in \tilde{\boldsymbol{H}}^{1/2}(\Gamma \setminus \Gamma_D)$:

$$\int_\Gamma (S\boldsymbol{u}^\varepsilon) \cdot \boldsymbol{v}\, ds - \int_{\Gamma_C} (p_n^\varepsilon(u_n^\varepsilon)[v_n] + p_t^\varepsilon(u_n^\varepsilon)[v_t])\, ds = \int_{\Gamma_N} \hat{\boldsymbol{t}} \cdot \boldsymbol{v}\, ds, \qquad \forall \boldsymbol{v} \in \tilde{\boldsymbol{H}}^{1/2}(\Gamma \setminus \Gamma_D), \tag{6}$$

with the pointwise definition

$$p_n^\varepsilon(u_n^\varepsilon) := -\frac{([u_n^\varepsilon] - g)^+}{\varepsilon_n}, \qquad p_t^\varepsilon(u_t^\varepsilon) := -\frac{\mathcal{F}[u_t^\varepsilon]^*}{\varepsilon_t}, \tag{7}$$

where $\varepsilon_n, \varepsilon_t > 0$ are penalty parameters and $a^* := \operatorname{sign}(a)\min(\varepsilon_t, |a|)$, $a^+ := \max(a, 0)$ for some $a \in \mathbb{R}$. Note that (6) is formulated on the unconstrained space $\tilde{\boldsymbol{H}}^{1/2}(\Gamma \setminus \Gamma_D)$.

In the frictionless case ($\mathcal{F} \equiv 0$) of the one-body problem ($\boldsymbol{u}^2 \equiv 0$) for the hp-version of BEM we show in [2] that the error between the solution $\boldsymbol{u}$ of the original formulation (5) and the corresponding Galerkin solution $\boldsymbol{U}^\varepsilon$ of (6) satisfies the a priori estimate

$$\|\boldsymbol{u} - \boldsymbol{U}^\varepsilon\|_{\boldsymbol{H}^{1/2}(\Gamma)} \le C\left(\frac{h}{p}\|\boldsymbol{u}\|_{\boldsymbol{H}^{3/2}(\Gamma)} + \left(\varepsilon_n + \frac{h}{p}\right)\|S\boldsymbol{u}\|_{\boldsymbol{H}^{1/2}(\Gamma)}\right)$$

with $C > 0$ independent of h, p and $\boldsymbol{u}$ and $\varepsilon_n \sim (h/p)^{1-\nu}$ for arbitrary fixed $\nu \in (0; 1)$.

Let $\boldsymbol{\psi}^\varepsilon = S\boldsymbol{u}^\varepsilon$ be the traction, $\boldsymbol{\Psi}^\varepsilon$ its Galerkin approximation and P_n^ε, P_t^ε be the Galerkin approximations of p_n^ε and p_t^ε respectively. We define the norm

$$\begin{aligned} |||\boldsymbol{u}^\varepsilon - \boldsymbol{U}^\varepsilon||| := \Big(&||\boldsymbol{u}^\varepsilon - \boldsymbol{U}^\varepsilon||^2_{\boldsymbol{H}^{1/2}(\Gamma)} + ||\boldsymbol{\psi}^\varepsilon - \boldsymbol{\Psi}^\varepsilon||^2_{\boldsymbol{H}^{-1/2}(\Gamma)} \\ &+ ||{\varepsilon_n}^{1/2}(p_n^\varepsilon - P_n^\varepsilon)||^2_{L_2(\Gamma_C)} + ||{\varepsilon_t}^{1/2}\mathcal{F}^{-1/2}(p_t^\varepsilon - P_t^\varepsilon)||^2_{L_2(\Gamma_C)}\Big)^{1/2}. \end{aligned} \tag{8}$$

The following residual a posteriori error estimate is proven in [1] for the h-version of BEM on quasiuniform meshes $\mathcal{T}_h$ if $\varepsilon_n \ge \tilde{C}h$, $\varepsilon_t \ge \tilde{C}\mathcal{F}h$ for some constant $\tilde{C} > 0$

$$c\sum_{I \in \mathcal{T}_h} \eta_h^2(I) \le |||\boldsymbol{u}^\varepsilon - \boldsymbol{U}^\varepsilon|||^2 \le C\sum_{I \in \mathcal{T}_h} \eta_h^2(I)$$

with the local error indicators

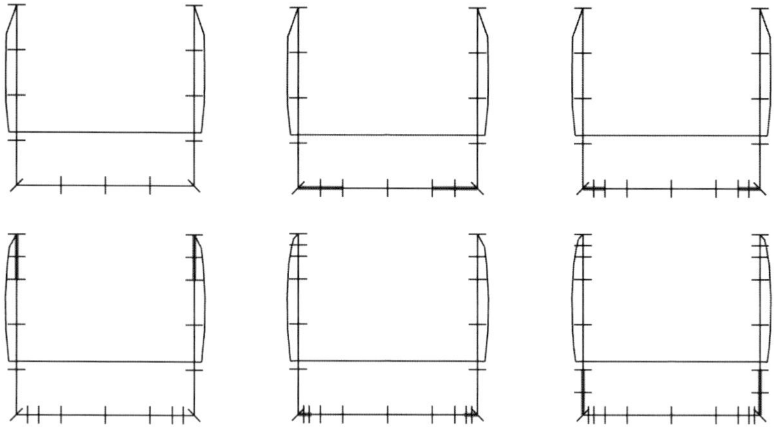

Fig. 1. Sequence of the adaptively generated meshes and deformed geometries (value of the displacement is multiplied by 10^4).

$$\eta_h^2(I) := h_I \|\hat{\boldsymbol{t}} - \hat{S}\boldsymbol{U}^\varepsilon\|^2_{L_2(I\cap\Gamma_N)} + h_I \|P_n^\varepsilon \boldsymbol{n} + P_t^\varepsilon \boldsymbol{t} - \hat{S}\boldsymbol{U}^\varepsilon\|^2_{L_2(I\cap\Gamma_C)} \\ + h_I \left\| \frac{\partial}{\partial s}(V\boldsymbol{\Psi}^\varepsilon - (K + 1/2)\boldsymbol{U}^\varepsilon) \right\|^2_{L_2(I)}. \qquad (9)$$

Here $\hat{S}$ is the discretization of S (see e.g. [2]), $I \in \mathcal{T}_h$ is the mesh element and c, C are positive constants independent of the mesh size h.

We give a numerical example of a frictional contact problem between the two-dimensional elastic body, occupying $\Omega := [-1, 1]^2$, and a rigid horizontal obstacle $\gamma := [-1, 1] \times \{-1 + d\}$, which is pushed upwards with $d = 0.6 \cdot 10^{-4}$. The body is fixed along the upper horizontal boundary, i.e. $\boldsymbol{u}^\varepsilon = 0$ on $\Gamma_D := [-1, 1] \times \{1\}$, while the remaining part of the boundary might be in contact with the obstacle γ, i.e. $\Gamma_C := \partial\Omega \setminus \Gamma_D$ and $\Gamma_N := \emptyset$. We choose Young modulus $E = 266926.0$, Poisson ratio $\nu = 0.29$ and the Tresca friction $\mathcal{F} = 0.1$.

The uniform initial mesh of 16 elements is used. Continuous piecewise linear approximation is used for the displacement $\boldsymbol{u}^\varepsilon$ and discontinuous piecewise constant approximation is used for the traction $\boldsymbol{\psi}^\varepsilon$. We use the local error indicator (9) in the following mesh refinement strategy. We compute the maximum value $\eta_h^{\max}$ of the local error indicators $\eta_h(I)$ over all mesh elements $I \in \mathcal{T}_h$, and compare it with $\eta_h(I)$. Then the element I is halved if its error indicator $\eta_h(I)$ is larger than 90% of $\eta_h^{\max}$, otherwise it remains without changes.

We show the resulting sequence of meshes and the corresponding deformed configurations in Figure 1. The deformed configuration is plotted for the displacement, multiplied with 10^4, to make the deformation visible. The labelling in the mesh is used for the elements being refined within the current refinement step. Note that the

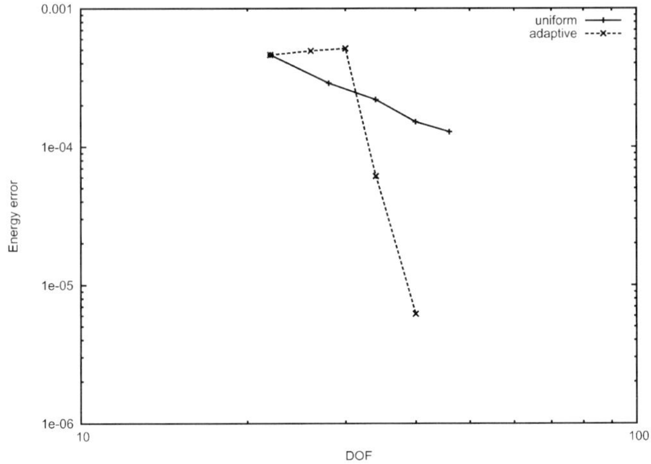

Fig. 2. $\mathrm{err}(\boldsymbol{U}^{\varepsilon}) := \big| \,|||\boldsymbol{U}^{\varepsilon}||| - |||\boldsymbol{U}^{\varepsilon}_{190}|||\, \big|$ for uniform and adaptive mesh refinement.

elements having symmetric positions are refined simultaneously (in the same refinement step), which is caused by the symmetry of the problem geometry. Most of the refinement happens in the *current* zone of contact and near the corners of Ω.

We compare the decay of the error in the norm $|||\cdot|||$ for uniform and adaptive refinements. As a reference norm, we take $|||\boldsymbol{U}^{\varepsilon}_{190}|||$, where $\boldsymbol{U}^{\varepsilon}_{190}$ is the solution, obtained on the uniform mesh with 190 degrees of freedom. The plots of $\mathrm{err}(\boldsymbol{U}^{\varepsilon}) := \big| \,|||\boldsymbol{U}^{\varepsilon}||| - |||\boldsymbol{U}^{\varepsilon}_{190}|||\, \big|$ are presented in Figure 2. The adaptive refinement procedure provides significantly better results than the uniform refinement with the same number of the degrees of freedom.

4 Mortar Method and *hp*-Adaptive Error Control

An alternative technique for approximately solving the variational inequality (5) is the mortar method [3]. In contrast to the penalty method, no intermediate formulation is needed and the discrete version of the variational inequality (5) is solved. In order to obtain it, we decompose each $\Gamma^i, i = 1,2$ into a finite family of the straight line segments $\mathcal{T}_h^i$ such that each segment belongs to Γ_D^i, or Γ_N^i, or Γ_C^i. Based on $\mathcal{T}_h^i, i = 1,2$, we define the space of globally continuous piecewise polynomial functions $\mathcal{V}_{hp}^i$ (discretization of the displacement) and the space of globally discontinuous piecewise polynomial functions $\mathcal{W}_{hp}^i$ (discretization of the traction). Let $\mathcal{P}_{p_I}(I)$ be the space of polynomials on I of degree at most p_I. Then we demand $\boldsymbol{U}|_I \in \mathcal{P}_{p_I}(I)$ and $\boldsymbol{\Psi}|_I \in \mathcal{P}_{p_I-1}(I)$ for arbitrary $\boldsymbol{U} \in \mathcal{V}_{hp}^i$ and $\boldsymbol{\Psi} \in \mathcal{W}_{hp}^i$. Note that the mesh nodes and the polynomial degrees do not match in general across

Γ_C, which is strongly desirable in many applications [12]. On the other hand, the non-matching property makes discretization of the convex set of admissible solutions $\mathcal{K}$ in (5) more complex, since the condition $u_n^1 - u_n^2 \equiv [u_n] \leq g$ is difficult to realize in each $\boldsymbol{x} \in \Gamma_C$. We impose this condition in a weak sense. In order to define discrete contact conditions we introduce auxiliary spaces of normal traces on Γ_C^i by $\mathcal{N}_{hp}^i := \{W = \boldsymbol{U} \cdot \boldsymbol{n}^i|_{\Gamma_C} : \boldsymbol{U} \in \boldsymbol{\mathcal{V}}_{hp}^i\}$, and the mortar space $\mathcal{M}_{hp}^1 := \{\Psi \in \mathcal{N}_{hp}^1 : \Psi \in \mathcal{P}_{p_I-1}(I), \text{ if } I \cap \partial\Gamma_C \neq \emptyset\}$, without loss of generality associated with $\mathcal{T}_h^1$. We define the hp-mortar projection operator (e.g. [3, 10]) $\pi_{hp}^1 : H^{1/2}(\Gamma_C) \to \mathcal{N}_{hp}^1$ by

$$\begin{aligned} \pi_{hp}^1 \varphi &= \varphi && \text{in } \partial\Gamma_C, \\ \int_{\Gamma_C} (\varphi - \pi_{hp}^1 \varphi)\Psi^1 \, ds &= 0 && \forall \Psi^1 \in \mathcal{M}_{hp}^1, \end{aligned} \tag{10}$$

Let G_{hp}^i be the set of Gauss–Lobatto nodes associated with the elements of $\mathcal{T}_h^i$. Let $\boldsymbol{\mathcal{V}}_{hp} := \boldsymbol{\mathcal{V}}_{hp}^1 \times \boldsymbol{\mathcal{V}}_{hp}^2$. Now we give a discrete counterpart of (5):

Find $\boldsymbol{U} \in \mathcal{K}_{hp} := \left\{\boldsymbol{U} \in \boldsymbol{\mathcal{V}}_{hp} : \pi_{hp}^1[U_n](\boldsymbol{x}) \leq g(\boldsymbol{x}) \;\; \forall \boldsymbol{x} \in G_{hp}^1 \cap \Gamma_C\right\}$:

$$\begin{aligned} \int_\Gamma (\hat{S}\boldsymbol{U}) \cdot (\boldsymbol{\Phi} - \boldsymbol{U}) \, ds + \int_{\Gamma_C} \mathcal{F}(|\pi_{hp}^1[\Phi_t]| - |\pi_{hp}^1[U_t]|) \, ds \\ \geq \int_{\Gamma_N} \hat{\boldsymbol{t}} \cdot (\boldsymbol{\Phi} - \boldsymbol{U}) \, ds, \quad \forall \boldsymbol{\Phi} \in \mathcal{K}_{hp}, \end{aligned} \tag{11}$$

Note that in general $\mathcal{K}_{hp} \not\subset \mathcal{K}$. In [3] we proved that on quasiuniform meshes

$$\|\boldsymbol{u} - \boldsymbol{U}\|_{\boldsymbol{H}^{1/2}(\Gamma)} \leq C \left(\frac{h}{p}\right)^{1/4} \|\boldsymbol{u}\|_{\boldsymbol{H}^{3/2}(\Gamma)}$$

under additional assumptions on the discretization parameters. In [1] two different solution algorithms for the mortar method are given: the Dirichlet-to-Neumann iterations, based on the variational inequality (5), and the Uzawa algorithm, based on a boundary saddle point formulation equivalent to (5), see [1, 5], whose discretization (12) is given below. Furthermore, a heuristic local a posteriori error indicator is motivated for the saddle point formulation. In order to formulate (12) we define the following spaces of continuous piecewise polynomial functions

$$\begin{aligned} Y_{hp}^1 &:= \left\{P \in C(\Gamma_C) : \forall I \in \mathcal{T}_{hp}^1 \cap \Gamma_C, P \in \mathcal{P}_{p_I}(I)\right\}, \\ M_{n,hp} &:= \left\{P_n \in Y_{hp}^1 : P_n(x) \geq 0, \, \forall x \in G_{hp}^1 \cap \Gamma_C\right\}, \\ M_{t,hp} &:= \left\{P_t \in Y_{hp}^1 : |P_t(x)| \leq \mathcal{F}(x), \, \forall x \in G_{hp}^1 \cap \Gamma_C\right\}. \end{aligned}$$

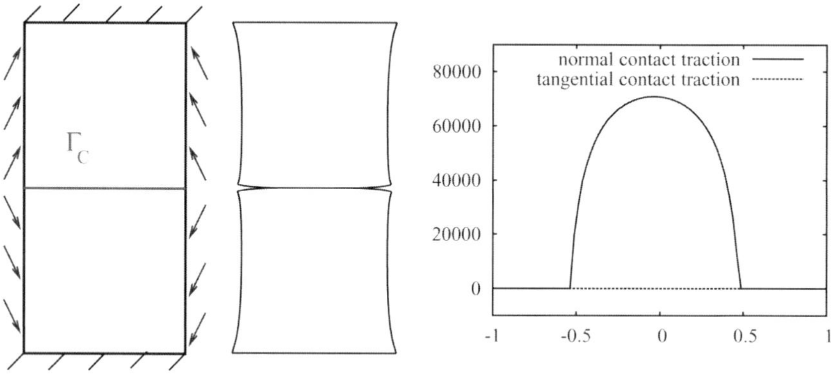

Fig. 3. Model problem, deformed configuration and contact traction.

Note that the inequality constraints are enforced only on the discrete set of the Gauss–Lobatto points G^1_{hp}. Then for $g = 0$ the discretized version of the saddle point formulation reads:

Find $\boldsymbol{U} \in \boldsymbol{\mathcal{V}}_{hp}$, $\boldsymbol{P} \in \boldsymbol{\mathcal{M}}_{hp} := M_{n,hp} \times M_{t,hp}$ such that

$$\begin{aligned} \int_\Gamma (\hat{S}\boldsymbol{U}) \cdot \boldsymbol{\Phi}\, ds + \int_{\Gamma_C} \boldsymbol{P} \cdot \boldsymbol{\Phi}\, ds &= \int_{\Gamma_N} \hat{\boldsymbol{t}} \cdot \boldsymbol{\Phi}\, ds, \qquad \forall \boldsymbol{\Phi} \in \boldsymbol{\mathcal{V}}_{hp}, \\ \int_{\Gamma_C} (\boldsymbol{Q} - \boldsymbol{P}) \cdot \boldsymbol{U}\, ds &\le 0, \qquad \forall \boldsymbol{Q} \in \boldsymbol{\mathcal{M}}_{hp}. \end{aligned} \tag{12}$$

We use the following heuristic local a posteriori error indicator $\eta_{hp}(I) = (\eta^2_C(I) + \eta^2_N(I) + \eta^2_{\hat{S}}(I))^{1/2}$, motivated in [1], where

$$\begin{aligned} \eta^2_C(I) &:= \frac{h_I}{p_I} \|[\boldsymbol{P}]\|^2_{L_2(I\cap\Gamma_C)} + \left(\frac{h_I}{p_I}\right)^{-1} \|[U_n]^+\|^2_{L_2(I\cap\Gamma_C)} \\ &\qquad + \frac{h_I}{p_I} \|(-\boldsymbol{P}) - \hat{S}\boldsymbol{U}\|^2_{L_2(I\cap\Gamma_C)}, \end{aligned}$$

$$\eta^2_N(I) := \frac{h_I}{p_I} \|\hat{\boldsymbol{t}} - \hat{S}\boldsymbol{U}\|^2_{L_2(I\cap\Gamma_N)}, \qquad \eta^2_{\hat{S}}(I) := \frac{h_I}{p_I} \|V\boldsymbol{\Psi} - (K + 1/2)\boldsymbol{U}\|^2_{L_2(I)}$$

and $\boldsymbol{U} \in \boldsymbol{\mathcal{V}}_{hp}$, $\boldsymbol{P} \in \boldsymbol{\mathcal{M}}_{hp}$ solve (12). The error indicator $\eta_{hp}(I)$ is similar to the heuristic error indicator applied by Krause and Wohlmuth [7] with the h-version of mortar FEM to contact problems.

In our numerical example we consider two elastic bodies $\Omega^1 := [-1; 1] \times [0; 2]$ and $\Omega^2 := [-1; 1] \times [-2; 0]$. The bodies are fixed along $\Gamma^1_D := [-1; 1] \times \{2\}$ and $\Gamma^2_D := [-1; 1] \times \{-2\}$ respectively. Furthermore, $\Gamma^1_C = \Gamma^2_C = [-1; 1] \times \{0\}$ and $\Gamma^i_N := \Gamma^i \setminus (\Gamma^i_D \cup \Gamma^i_C)$, $i = 1, 2$. The uniform load is distributed along Γ^i_N as shown in Figure 3, such that non-zero contact tractions appear on Γ_C. The tangential traction vanishes since the configuration is symmetric.

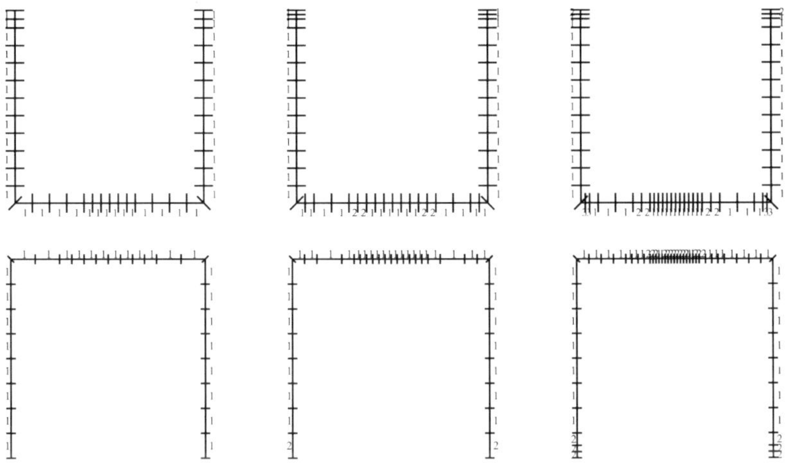

Fig. 4. Adaptively generated meshes and polynomial degrees after 3, 6 and 9 refinement steps.

For the automatic mesh refinement procedure we use the three-step hp-adaptive algorithm, used e.g. by Maischak and Stephan [9]. First, we compute the maximal error indicator value $\eta_{hp}^{\max}$ of the local error indicators $\eta_{hp}(I)$ over all mesh elements I, and compare it with $\eta_{hp}(I)$. Then the element I is halved (h-refinement step) if its indicator value $\eta_{hp}(I)$ is larger than 90% of $\eta_{hp}^{\max}$, the polynomial degree p_I is increased by 1 (p-refinement step) if $\eta_{hp}(I)$ is between 85% and 90% of $\eta_{hp}^{\max}$, and nothing is done otherwise. The sequence of automatically generated meshes and polynomial degrees obtained with our approach is shown in Figure 4.

Acknowledgement

The work of the second author has been supported by DFG Grant No. STE 573/7-1.

References

1. Chernov A (2006) Nonconforming boundary elements and finite elements for interface and contact problems with friction – hp-version for mortar, penalty and Nitsche's methods. PhD Thesis, Universität Hannover.
2. Chernov A, Maischak M, Stephan E P (2007) A priori error estimates for hp penalty BEM for contact problems in elasticity. Submitted to *Comput Methods Appl Mech Engrg*.
3. Chernov A, Maischak M, Stephan E P (2007) hp-mortar boundary element method for two-body contact problems with friction. Submitted to *Math Methods Appl Sci*.
4. Haslinger J, Hlaváček I (1980) Contact between elastic bodies. I. Continuous problems. *Appl Mat*, 25:324–347.

5. Haslinger J, Hlaváček I, Nečas J (1996) Numerical methods for unilateral problems in solid mechanics. In *Handbook of Numerical Analysis*, Vol. IV. North-Holland, Amsterdam, pp 313–485.
6. Jarušek J (1983) Contact problems with bounded friction coercive case. *Czechoslovak Math J*, 33(108):237–261.
7. Krause R, Wohlmuth B I (2002) A Dirichlet-Neumann type algorithm for contact problems with friction. *Comput Visual Sci*, 5:139–148.
8. Lions J-L, Magenes E (1972) *Non-Homogeneous Boundary Value Problems and Applications*, Vol. I. Springer-Verlag, New York.
9. Maischak M, Stephan E P (2007) Adaptive hp-versions of boundary element methods for elastic contact problems. *Comp Mech*, 39:597–607.
10. Seshaiyer P, Suri M (2000) Uniform *hp* convergence results for the mortar finite element method. *Math Comp*, 69:521–546.
11. Stephan E P (2004) Coupling of boundary element methods and finite element methods. *Encyclopedia of Comp Mech*, 1:375-412.
12. Wriggers P (2002) *Computational Contact Mechanics*. John Wiley & Sons.

Recent Extensions of Mortar-Based Contact Formulations: Lubrication Modeling and Parallel Implementations

Tod A. Laursen, Eui Joong Kim and Bin Yang

Computational Mechanics Laboratory, Department of Civil and Environmental Engineering, Duke University, Durham, NC 27708, U.S.A.
E-mail: laursen@duke.edu

Abstract. Mortar element methods have recently been successfully applied to large deformation dry-frictional contact problems. In this contribution, we summarize two recent extensions of the mortar based contact approach: solution of lubricated contact problems between deformable solid bodies, and implementation of mortar methods in parallel equation solving frameworks.

Lubricated contact problems, in which thin fluid films exist between solid contact surfaces, are widely found in engineering and sciences. Based on Reynolds equations, different numerical methods have been developed for solving lubricated contact problems between rigid or linearly elastic solid bodies (see, for example, Christopherson [2] and Hamrock et al. [6]). In our approach, the fluid film thickness is computed from a least squares projection based on dual basis functions [12]. The free boundary problem for the fluid phase is regularized with a penalty method. The solid and the fluid phase unknowns are solved in a fully coupled system of equations, based on the linearization of the weak forms of the governing equations for the solid and the fluid phases. The performance and potential of the proposed method is demonstrated here in a two dimensional example of elastohydrodynamic lubrication.

Another important issue for mortar based contact formulations lies in determining their effectiveness within parallel computing architectures. In general, contact problems result in potentially ill-conditioned and highly unsymmetric stiffness matrices, so that direct methods have been preferred (particularly in highly nonlinear large deformation applications). However, direct methods require memory and computation time up to $\mathcal{O}(n^{1.5})$ and $\mathcal{O}(n^2)$, respectively, so that their applicability to very large scale contact simulations appears to be limited. Iterative methods require memory and computation time of $\mathcal{O}(n)$ and $\mathcal{O}(n^{1.5})$, making their use in large-scale numerical simulations tempting. We consider here a conceptually simple implementation which uses iterative methods to solve the linear systems generated by consistent linearization of the mortar contact formulation, on distributed memory machines, and examine performance in a representative numerical example.

Key words: Lubrication, friction, contact, mortar formulation, parallel computation.

Peter Wriggers and Udo Nackenhorst (eds.), IUTAM Symposium on Computational Contact Mechanics, 123–146.

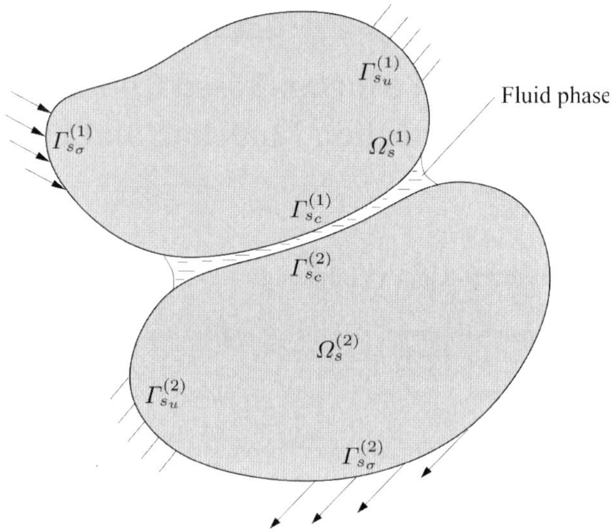

Fig. 1. Notation for two body lubricated contact problems.

1 Mortar Formulation of Lubricated Contact

We consider in this first section the lubricated contact between two elastic bodies, and for simplicity, consider the case where all deformations are small, so that linearized kinematics apply. The problem at hand is depicted in Figure 1; two solid bodies occupy the regions denoted by $\Omega_s^{(1)}$ and $\Omega_s^{(2)}$. The surfaces of these bodies $\partial\Omega_s^{(i)}$ are divided into $\Gamma_{s_u}^{(i)}$, where displacements are prescribed; $\Gamma_{s_\sigma}^{(i)}$, where tractions are prescribed; and $\Gamma_{s_c}^{(i)}$, where the lubricated contact will take place. We (arbitrarily) denote $\Gamma_{s_c}^{(1)}$ and $\Gamma_{s_c}^{(2)}$ as slave and master surfaces, respectively, although these terminologies will not retain their same meanings as in traditional strategies. As usual, we assume that the Dirichlet, Neumann, and contact boundaries of each body are nonintersecting, and that their union comprises the entirety of $\partial\Omega_s^{(i)}$ for each body i.

The solid bodies will be assumed to be governed by the usual equations of linear elasticity; these equations (as well as their finite element treatment) are standard and will not be treated in any depth here. In lubricated contact problems, the thicknesses of any lubricant films that developed will be assumed to be much smaller than other characteristic dimensions of the problem, so that the fluid response can be assumed to be governed by the Reynolds equation. This representation is derived from the Navier–Stokes and continuity equations (see, e.g., [5]), and is defined over a surface domain giving the topology of the lubricated region, which here will be taken as the slave side of a slave-master contact patch. As shown schematically in Figure 2, the lubrication domain Ω_f, exclusive of the dry contact region Ω_d,

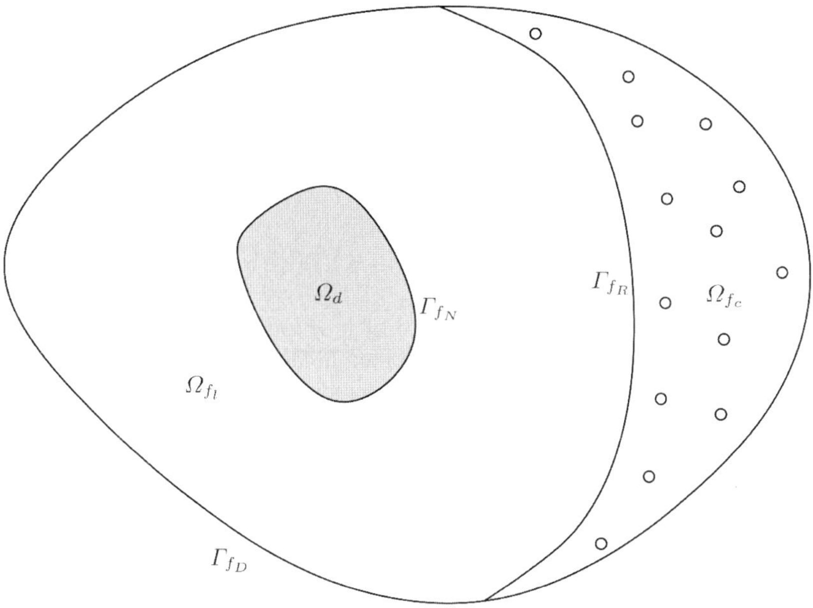

Fig. 2. Notation for the lubrication domain.

may be divided into two different subdomains, Ω_{f_l} and Ω_{f_c}. Ω_{f_l} is the subdomain where the lubricant is continuous and is in compression, while Ω_{f_c} is the subdomain where the lubricant has cavitations and is ruptured (we assume that the fluid film can only support a negligible amount of tension before rupture). Γ_{f_D} is the surface where Dirichlet boundary conditions are prescribed for the Reynolds equation, while Γ_{f_N} is the boundary between the dry contact domain and the lubrication domain and where Neumann boundary conditions are prescribed. Finally, Γ_{f_R}, which is called the Reynolds boundary, is the boundary between Ω_{f_l} and Ω_{f_c}; solution for this boundary is a key desired attribute of the algorithm to be described.

Based on these considerations, we have in general that

$$\Omega_f = \Gamma_{s_c}^{(1)} \setminus \Omega_d, \tag{1}$$

and

$$\Omega_f = \Omega_{f_l} \cup \Omega_{f_c}, \tag{2}$$

although in this work dry contact is not considered for simplification of the formulation. As a result, we have

$$\Omega_f = \Gamma_{s_c}^{(1)}. \tag{3}$$

In this formulation of small deformation lubricated problems, the primary unknowns will be the displacement fields $\boldsymbol{u}^{(i)}$ $(i = 1, 2)$ in the two solid bodies, and the pressure field p in the fluid film. With an eye toward finite element treatments of the problem, we define solution and weighting spaces $\mathcal{C}^{(i)}$ and $\mathcal{V}^{(i)}$, consisting of solutions $\boldsymbol{u}^{(i)}$ and their variations $\overset{*}{\boldsymbol{u}}{}^{(i)}$ according to

$$\mathcal{C}^{(i)} = \left\{ \boldsymbol{u}^{(i)} : \bar{\Omega}_s^{(i)} \rightarrow \mathbb{R}^{n_{dm}} \middle| \boldsymbol{u}^{(i)} \in H^1(\Omega_s^{(i)}), \boldsymbol{u}^{(i)} = \bar{\boldsymbol{u}}^{(i)} \text{ in } \Gamma_{s_u}^{(i)} \right\} \tag{4}$$

and

$$\mathcal{V}^{(i)} = \left\{ \overset{*}{\boldsymbol{u}}{}^{(i)} : \bar{\Omega}_s^{(i)} \rightarrow \mathbb{R}^{n_{dm}} \middle| \overset{*}{\boldsymbol{u}}{}^{(i)} \in H^1(\Omega_s^{(i)}), \overset{*}{\boldsymbol{u}}{}^{(i)} = \boldsymbol{0} \text{ in } \Gamma_{s_u}^{(i)} \right\}, \tag{5}$$

where $H^1(\cdot)$ represents the Sobolev space of functions with square integrable derivatives. The fluid phase response is described in terms of solution and weighting spaces $\mathcal{P}$ and $\mathcal{Q}$, consisting of solutions p and their variations $\overset{*}{p}$ according to

$$\mathcal{P} = \left\{ p : \bar{\Omega}_f \rightarrow \mathbb{R}^{n_{dm}-1} \middle| p \in H^1(\Omega_f), p = 0 \text{ in } \Gamma_{f_D} \right\} \tag{6}$$

and

$$\mathcal{Q} = \left\{ \overset{*}{p} : \bar{\Omega}_f \rightarrow \mathbb{R}^{n_{dm}-1} \middle| \overset{*}{p} \in H^1(\Omega_f), \overset{*}{p} = 0 \text{ in } \Gamma_{f_D} \right\}. \tag{7}$$

1.1 Fluid Phase Equations

As stated before, the pressure distribution in the regions of thin film lubrication will be described in terms of the Reynolds equation (see [5]) and is written as

$$\frac{\partial(\rho h)}{\partial t} + \tilde{\nabla} \cdot \left(-\frac{\rho h^3}{12\mu} \tilde{\nabla} p + \frac{\rho(\tilde{\boldsymbol{V}}^{(1)} + \tilde{\boldsymbol{V}}^{(2)})}{2} h \right) = 0, \text{ in } \Omega_{f_l}, \tag{8}$$

where ρ is the current mass density of the lubricant, h is the fluid film thickness (coupling strongly to the deformation of the solid phase), μ is the viscosity of the lubricant, $\tilde{\boldsymbol{V}}^{(1)}$ and $\tilde{\boldsymbol{V}}^{(2)}$ are the shear velocity fields at either side of the fluid layer, and p is the fluid pressure. Equation (8) is typically taken as an equation governing the spatial distribution of the fluid pressure p; we adopt that approach here as well. In general, the fluid viscosity μ is frequently taken to have a direct dependence on fluid pressure. Here, we consider a so-called Barus form of the viscosity (see, e.g., [5]), such that

$$\mu = \mu_0 e^{\alpha p}, \tag{9}$$

where μ_0 is the viscosity when $p = 0$ and α is a property of the lubricating fluid.

As mentioned previously, we make the choice here to use the slave surface $\Gamma_{s_c}^{(1)}$ as the topology over which the fluid equations will be described. Effectively, the Reynolds equation given by (8) prescribes the distribution of pressures over this surface, which will provide the normal component of the "contact" tractions acting on the solids on either side of the interface. In the first term of Equation (8), $\partial/\partial t$ is the spatial time derivative, while $\tilde{\nabla}$ is the surface gradient operator on the slave surface, defined as

$$\tilde{\nabla} = \frac{\partial}{\partial \xi^\alpha} \boldsymbol{\tau}^\alpha, \tag{10}$$

where $\boldsymbol{\tau}^\alpha$ denotes the contravariant basis vectors on the slave surface. $\tilde{\boldsymbol{V}}^{(1)}$ and $\tilde{\boldsymbol{V}}^{(2)}$ are conveniently described as:

$$\tilde{\boldsymbol{V}}^{(i)} = \boldsymbol{V}^{(i)} - \left(\boldsymbol{V}^{(i)} \cdot \boldsymbol{n}\right) \boldsymbol{n} = \tilde{V}_\alpha^{(i)} \boldsymbol{\tau}^\alpha, \ (i{=}1{,}2), \tag{11}$$

where $\boldsymbol{n}$ is the surface normal associated with $\Gamma_{s_c}^{(1)}$.

The volume flow rate is defined as:

$$\tilde{\boldsymbol{q}} = -\frac{h^3}{12\mu} \tilde{\nabla} p + \frac{\tilde{\boldsymbol{V}}^{(1)} + \tilde{\boldsymbol{V}}^{(2)}}{2} h; \tag{12}$$

examination of (12) shows that $\tilde{\boldsymbol{q}}$ may also be conveniently described as a vector field on the slave surface. On the Neumann boundary Γ_{f_N}, the film thickness h goes to zero, leading to

$$\tilde{\boldsymbol{q}} = \boldsymbol{0} \text{ on } \Gamma_{f_N}. \tag{13}$$

The Dirichlet boundary condition may be prescribed on its boundary as

$$p = 0 \text{ on } \Gamma_{f_D}. \tag{14}$$

Finally, in the cavitation domain Ω_{f_c}, the fluid pressure is close to the atmosphere pressure, generally taken to be negligible compared to the fluid pressure in the lubrication domain Ω_{f_l}. Consequently, we have

$$p = 0 \text{ in } \Omega_{f_c}. \tag{15}$$

To satisfy the continuity of mass flow across the boundary Γ_{f_R} (also known as free boundary), the Reynolds boundary condition is defined as

$$p = 0 \text{ and } \frac{\partial p}{\partial \boldsymbol{n}} = \tilde{\nabla} p \cdot \tilde{\boldsymbol{n}} = 0, \text{ on } \Gamma_{f_R}, \tag{16}$$

where $\tilde{\boldsymbol{n}}$ is the outward normal vector of Γ_{f_R}. Based on (16), (15) and (12), one can easily prove that the mass flow continuity is satisfied across the Reynolds boundary. However, the position of the Reynolds boundary is in general an unknown to be solved as part of the problem, and a major obstacle in formulating lubrication problems.

In this work, we shall assume steady state flow and incompressibility (constant density) for the fluid phase, so that one obtains the simplified Reynolds equation

$$\tilde{\nabla}\cdot\left(-\frac{h^3}{12\mu}\tilde{\nabla}p+\frac{\tilde{\boldsymbol{V}}^{(1)}+\tilde{\boldsymbol{V}}^{(2)}}{2}h\right)=0 \text{ in } \Omega_{f_l}. \tag{17}$$

In formulating the lubrication equations to enable determination of the Reynolds boundary as part of the solution, we turn now to an approach discussed in earlier literature by Rohde and McAllister [10] and Cryer [3]. As per these authors, the lubrication problem summarized in (17), (14), (13) and (16) can be represented in terms of an equivalent complementary problem. We note that in the lubricated region we have

$$p\geq 0 \text{ and } \tilde{\nabla}\cdot\left(-\frac{h^3}{12\mu}\tilde{\nabla}p+\frac{\tilde{\boldsymbol{V}}^{(1)}+\tilde{\boldsymbol{V}}^{(2)}}{2}h\right)=0 \text{ in } \Omega_{f_l}, \tag{18}$$

while in the cavitated region we may write

$$p=0 \text{ and } \tilde{\nabla}\cdot\left(-\frac{h^3}{12\mu}\tilde{\nabla}p+\frac{\tilde{\boldsymbol{V}}^{(1)}+\tilde{\boldsymbol{V}}^{(2)}}{2}h\right)\geq 0 \text{ in } \Omega_{f_c}. \tag{19}$$

Therefore, in the whole fluid phase domain Ω_f (encompassing both lubrication and cavitation regions), we may write

$$p\left[\tilde{\nabla}\cdot\left(-\frac{h^3}{12\mu}\tilde{\nabla}p+\frac{\tilde{\boldsymbol{V}}^{(1)}+\tilde{\boldsymbol{V}}^{(2)}}{2}h\right)\right]=0, \text{ in } \Omega_f, \tag{20}$$

subject to the inequality constraints

$$p\geq 0 \text{ in } \Omega_f, \tag{21}$$

and

$$\tilde{\nabla}\cdot\left(-\frac{h^3}{12\mu}\tilde{\nabla}p+\frac{\tilde{\boldsymbol{V}}^{(1)}+\tilde{\boldsymbol{V}}^{(2)}}{2}h\right)\geq 0 \text{ in } \Omega_f. \tag{22}$$

with p and $\tilde{\boldsymbol{q}}$ satisfing boundary conditions (14) and (13).

Thus, the solution to the original lubrication problem summarized by (17), (15), (14), (13) and (16) can be obtained by solving the equivalent problem (20)–(22). Different approaches have been proposed to solve this problem. In this work, the complementary problem is penalty regularized, and then a weak form is set up to solve the problem which exploits the mortar formalism to discretize the fluid domain equations.

1.2 Weak Form of the Lubrication Problem and Mortar Discretization

The weak form of the penalty regularized Reynolds equations may be written as

$$G^f(\boldsymbol{u}, p, \overset{*}{p}) := \int_{\Omega_f} \overset{*}{p} \left[\tilde{\nabla} \cdot \left(-\frac{h^3}{12\mu} \tilde{\nabla} p + \frac{\tilde{\boldsymbol{V}}^{(1)} + \tilde{\boldsymbol{V}}^{(2)}}{2} h \right) - \varepsilon_p \langle -p \rangle \right] d\Omega = 0, \tag{23}$$

where ε_p is a regularization parameter treating the complementarity conditions associated with resolution of the Reynolds boundary. Note that G^f is also a function of $\boldsymbol{u}$ because the film thickness h is dependent on the displacement of the solid phase. Using integration by parts, one obtains

$$\begin{aligned} G^f(\boldsymbol{u}, p, \overset{*}{p}) := &- \int_{\Omega_f} \left(-\frac{h^3}{12\mu} \tilde{\nabla} p + \frac{\tilde{\boldsymbol{V}}^{(1)} + \tilde{\boldsymbol{V}}^{(2)}}{2} h \right) \cdot \tilde{\nabla} \overset{*}{p} d\Omega \\ &- \int_{\Omega_f} \varepsilon_p \langle -p \rangle \overset{*}{p} d\Omega + \int_{\Gamma_{f_N}} \overset{*}{p} \tilde{\boldsymbol{n}} \cdot \tilde{\boldsymbol{q}} d\Gamma = 0, \end{aligned} \tag{24}$$

where

$$\tilde{\nabla} p = \frac{\partial p}{\partial \xi^\alpha} \boldsymbol{\tau}^\alpha = p_{,\alpha} \boldsymbol{\tau}^\alpha. \tag{25}$$

Since the flux $\tilde{\boldsymbol{q}}$ satisfies the Neumann boundary condition (13), one can rearrange terms in (24) and obtain

$$\begin{aligned} G^f(\boldsymbol{u}, p, \overset{*}{p}) := &\int_{\Omega_f} \frac{h^3}{12\mu} \tilde{\nabla} p \cdot \tilde{\nabla} \overset{*}{p} d\Omega - \int_{\Omega_f} \varepsilon_p \langle -p \rangle \overset{*}{p} d\Omega \\ &- \int_{\Omega_f} \left(\frac{\tilde{\boldsymbol{V}}^{(1)} + \tilde{\boldsymbol{V}}^{(2)}}{2} h \right) \cdot \tilde{\nabla} \overset{*}{p} d\Omega = 0. \end{aligned} \tag{26}$$

This is a highly nonlinear equation because of the dependence of h on the solid phase deformations and because of the nonsmoothness caused by the Macauley bracket. In general, we use consistent linearization within a global Newton–Raphson framework to solve these equations monolithically with the solid phase equations.

With an eye to such a monolithic formulation, the lubricated contact problem can be expressed via:

Find $\boldsymbol{u} \in \mathcal{C}$ and $p \in \mathcal{P}$, such that

$$G^{\text{int,ext}}(\boldsymbol{u}, \overset{*}{\boldsymbol{u}}) + G^c(\boldsymbol{u}, p, \overset{*}{\boldsymbol{u}}) = 0 \; \forall \overset{*}{\boldsymbol{u}} \in \mathcal{V} \tag{27}$$

and

$$G^f(\boldsymbol{u}, p, \overset{*}{p}) - 0 \; \forall \overset{*}{p} \in \mathcal{Q}, \tag{28}$$

where $G^{\text{int,ext}}(\boldsymbol{u}, \overset{*}{\boldsymbol{u}})$ refers to the usual virtual work associated with the contacting solid phases (both internal and external, excluding the virtual work G^c arising from tractions generated by the lubrication conditions). Equations (27) and (28) are coupled by the traction fields applied on the solid phase by the lubricant (included in G^c) and the variation of the film shape with the deformations of the solid phase.

The crux of the approach proposed here is that we consider the use of mortar approximations as a natural way to describe the lubricated contact interaction over areas of contact. In general, the spatial discretization of the bodies $\Omega_s^{(i)}$ is achieved by using the finite set of elements $\mathcal{E}^{(i)^h}$:

$$\Omega_s^{(i)^h} = \bigcup_{\forall e \in \mathcal{E}^{(i)^h}} \Omega_{s_e}^{(i)}. \tag{29}$$

where $\Omega_{s_e}^{(i)}$ denotes the subdomain occupied by element e. The discretization of contact surfaces $\Gamma_{s_c}^{(i)^h}$ occurs over subsets of $\partial\Omega_s^{(i)^h}$. Note that the slave surface $\Gamma_{s_c}^{(1)^h}$ coincides with the lubrication domain Ω_f^h if dry contact is not considered.

Finite dimensional subspaces of solution and weighting spaces in (4) and (5) are denoted $\mathcal{C}^{(i)^h}$ and $\mathcal{V}^{(i)^h}$, and may be defined via

$$\begin{aligned}\mathcal{C}^{(i)^h} = \Big\{ & \boldsymbol{u}^{(i)^h} : \bar{\Omega}_s^{(i)^h} \to \mathbb{R}^{n_{dm}} \,|\, \boldsymbol{u}^{(i)^h} \in C^0(\Omega^{(i)^h}), \\ & \forall e \in \mathcal{E}^{(i)^h}, \boldsymbol{u}^{(i)^h}(\Omega_e^{(i)^h}) \in \mathbb{P}_N(\Omega_e^{(i)^h}), \boldsymbol{u}^{(i)^h} = \bar{\boldsymbol{u}}^{(i)} \text{ in } \Gamma_{s_u}^{(i)^h} \Big\}\end{aligned} \tag{30}$$

and

$$\begin{aligned}\mathcal{V}^{(i)^h} = \Big\{ & \overset{*}{\boldsymbol{u}}{}^{(i)^h} : \bar{\Omega}^{(i)^h} \to \mathbb{R}^{n_{dm}} \,|\, \overset{*}{\boldsymbol{u}}{}^{(i)^h} \in C^0(\Omega^{(i)^h}), \\ & \forall e \in \mathcal{E}^{(i)^h}, \overset{*}{\boldsymbol{u}}{}^{(i)^h}(\Omega_e^{(i)^h}) \in \mathbb{P}_N(\Omega_e^{(i)^h}), \overset{*}{\boldsymbol{u}}{}^{(i)^h} = \boldsymbol{0} \text{ in } \Gamma_{s_u}^{(i)^h} \Big\},\end{aligned} \tag{31}$$

where $\mathbb{P}_N(\Omega_e^{(i)^h})$ is the set of all polynomials on $\Omega_e^{(i)^h}$ of order $\leq N$. The mortar and nonmortar fields, $\boldsymbol{u}^{(i)^h}(\Gamma_{s_c}^{(i)^h}) \subset \mathcal{X}^{(i)^h}$, and the variations of these fields, $\overset{*}{\boldsymbol{u}}{}^{(i)^h}(\Gamma_{s_c}^{(i)^h}) \in \mathcal{W}^{(i)^h}$, are subsets of $\mathcal{C}^{(i)^h}$ and $\mathcal{V}^{(i)^h}$ and are obtained by restriction of (30) and (31) to contact surfaces $\Gamma_{s_c}^{(i)^h}$. The discretized mortar multiplier space, which is physically interpreted as the space containing contact tractions, is defined on the nonmortar side (i.e., the slave surface) as

$$\mathcal{M}^h = \Big\{ \boldsymbol{\lambda}^h | \boldsymbol{\lambda}^h \in C^0(\Gamma_{s_c}^{(1)^h}); \forall e \in \mathcal{G}^h, \boldsymbol{\lambda}^h(\Gamma_{s_{ce}}^{(1)^h}) \in \mathbb{P}_N(\Gamma_{s_{ce}}^{(1)^h}), \Big\}, \tag{32}$$

where $\mathcal{G}^h$ is the set of nonmortar element edges making up the slave surface.

Similarly, finite dimensional subspaces of solution and weighting spaces in (6) and (7) are denoted $\mathcal{P}^h$ and $\mathcal{Q}^h$, and may be defined via

$$\mathcal{P}^h = \left\{ p^h : \bar{\Omega}_f^h \rightarrow \mathbb{R}^{n_{dm-1}} \middle| p^h \in C^0(\Omega_f^h), \right.$$
$$\left. \forall e \in \mathcal{G}^h,\, p^h(\Omega_{f_e}^h) \in \mathbb{P}_N(\Omega_{f_e}^h),\, p^h = 0 \text{ in } \Gamma_{f_D}^h \right\} \tag{33}$$

and

$$\mathcal{Q}^h = \left\{ \overset{*}{p}^h : \bar{\Omega}_f^h \rightarrow \mathbb{R}^{n_{dm-1}} \middle| \overset{*}{p}^h \in C^0(\Omega_f^h), \right.$$
$$\left. \forall e \in \mathcal{G}^h,\, \overset{*}{p}^h(\Omega_{f_e}^h) \in \mathbb{P}_N(\Omega_{f_e}^h),\, \overset{*}{p}^h = 0 \text{ in } \Gamma_{f_D}^h \right\}. \tag{34}$$

Notably, $\mathcal{Q}^h$, $\mathcal{P}^h$ and $\mathcal{M}^h$ are defined on the same set of elements.

With these definitions, we may now define the discretized version of contact virtual work by developing shape function expansions for the above contact surface fields and substituting into (51). The discretized contact virtual work can be expressed as

$$G^c(\boldsymbol{u}^h, p^h, \overset{*}{\boldsymbol{u}}^h) = -\sum_A \sum_B \sum_C \boldsymbol{\lambda}_A \cdot \left[n_{AB}^{(1)} \overset{*}{\boldsymbol{u}}_B^{(1)} - n_{AC}^{(2)} \overset{*}{\boldsymbol{u}}_C^{(2)} \right], \tag{35}$$

where $\boldsymbol{\lambda}_A$ is the contact traction, and where the mortar integrals $n_{AB}^{(1)}$ and $n_{AC}^{(2)}$ are expressed as

$$n_{AB}^{(1)} = \int_{\Gamma_{s_c}^{(1)h}} N_A^{(1)}\left(\boldsymbol{\xi}^{(1)}(\boldsymbol{X})\right) N_B^{(1)}\left(\boldsymbol{\xi}^{(1)}(\boldsymbol{X})\right) d\Gamma, \tag{36}$$

and

$$n_{AC}^{(2)} = \int_{\Gamma_{s_c}^{(1)h}} N_A^{(1)}\left(\boldsymbol{\xi}^{(1)}(\boldsymbol{X})\right) N_C^{(2)}\left(\boldsymbol{\xi}^{(2)}\left(\bar{\boldsymbol{Y}}(\boldsymbol{X})\right)\right) d\Gamma. \tag{37}$$

The normal and tangential portions of the contact traction, generated by the fluid but acting on the solid phases, are now exposed by splitting $\boldsymbol{\lambda}_A$ into normal and frictional parts

$$\boldsymbol{\lambda}_A = \boldsymbol{\lambda}_{A_N} + \boldsymbol{\lambda}_{A_T}. \tag{38}$$

The normal and tangential (i.e. shear stress applied by the fluid film) tractions in (38) are defined as

$$\boldsymbol{\lambda}_{A_N} = -p_A \boldsymbol{n}_A \tag{39}$$

and

$$\boldsymbol{\lambda}_{A_T} = -\frac{\mu_A}{h_A}\left(\tilde{\boldsymbol{V}}_A^{(1)} - \tilde{\boldsymbol{V}}_A^{(2)}\right) - \frac{h_A}{2}\tilde{\nabla} p_A, \tag{40}$$

where

$$\mu_A = \mu_0 e^{\alpha p_A}. \tag{41}$$

$\tilde{\nabla} p_A$ is the surface gradient of pressure at the slave node A, which introduces another difficulty for lubricated contact problems since the pressure gradient is not continuous at each node in general. Accordingly, a smoothing procedure has to be adopted to get $\tilde{\nabla} p_A$. Although this is a technical detail which will not be treated in depth here, we use a least squares smoothing procedure in conjunction with the dual mortar basis (see [11]) to efficiently compute the pressure gradient at each slave node.

To obtain the numerical formulation of the fluid equations, we first note that the fluid pressure field and its variation may be interpolated as

$$\begin{aligned} p^h(\boldsymbol{X}) &= \sum_{J\in\eta} N_J^{(1)}\left(\boldsymbol{\xi}^{(1)}(\boldsymbol{X})\right) p_J \\ &= \sum_{J\in\eta-\eta_g} N_J^{(1)}\left(\boldsymbol{\xi}^{(1)}(\boldsymbol{X})\right) p_J + \sum_{J\in\eta_g} N_J^{(1)}\left(\boldsymbol{\xi}^{(1)}(\boldsymbol{X})\right) g_J \end{aligned} \tag{42}$$

$$\overset{*}{p}(\boldsymbol{X}) = \sum_{I\in\eta-\eta_g} N_I^{(1)}\left(\boldsymbol{\xi}^{(1)}(\boldsymbol{X})\right) \overset{*}{p}_I, \tag{43}$$

where η is the set of all nodes on the discretized slave surface, η_g is the set of nodes where Dirichlet boundary conditions are prescribed, and g_J is the boundary value of p at node J. Since $p = 0$ at the Dirichlet boundary, we have

$$p^h(\boldsymbol{X}) = \sum_{J\in\eta-\eta_g} N_J^{(1)}\left(\boldsymbol{\xi}^{(1)}(\boldsymbol{X})\right) p_J. \tag{44}$$

Substitution of (44) and (43) into the discretized form of Equation (26) gives

$$\begin{aligned} &G^f(\boldsymbol{u}^h, p^h, \overset{*}{p}{}^h) \\ &= \sum_{I\in\eta-\eta_g} \overset{*}{p}_I \Bigg\{ \sum_{J\in\eta-\eta_g} \left(\int_{\Omega_f^h} \frac{(h^h(\boldsymbol{X}))^3}{12\mu} N_{I,\alpha}^{(1)}\left(\boldsymbol{\xi}^{(1)}(\boldsymbol{X})\right) N_{J,\beta}^{(1)}\left(\boldsymbol{\xi}^{(1)}(\boldsymbol{X})\right) m^{\alpha\beta} d\Omega \right) p_J \\ &- \int_{\Omega_f^h} N_I^{(1)}\left(\boldsymbol{\xi}^{(1)}(\boldsymbol{X})\right) \varepsilon_p \langle - \sum_{J\in\eta-\eta_g} N_J^{(1)}\left(\boldsymbol{\xi}^{(1)}(\boldsymbol{X})\right) p_J \rangle d\Omega \\ &- \int_{\Omega_f^h} N_{I,\alpha}^{(1)}\left(\boldsymbol{\xi}^{(1)}(\boldsymbol{X})\right) \boldsymbol{\tau}^\alpha(\boldsymbol{X}) \cdot \frac{\tilde{\boldsymbol{V}}^{(1)^h}(\boldsymbol{X}) + \tilde{\boldsymbol{V}}^{(2)^h}(\boldsymbol{X})}{2} h^h(\boldsymbol{X}) d\Omega \Bigg\} = 0, \end{aligned} \tag{45}$$

where $h^h(\boldsymbol{X})$, $\tilde{\boldsymbol{V}}^{(1)^h}(\boldsymbol{X})$ and $\tilde{\boldsymbol{V}}^{(2)^h}(\boldsymbol{X})$ are discretized forms of film thickness and surface velocities. They are defined as:

$$h^h(\boldsymbol{X}) = \sum_{A=1}^{n_s} N_A^{(1)}\left(\boldsymbol{\xi}^{(1)}(\boldsymbol{X})\right) h_A, \tag{46}$$

$$\tilde{\boldsymbol{V}}^{(i)^h}(\boldsymbol{X}) = \sum_{A=1}^{n_s} N_A^{(1)}\left(\boldsymbol{\xi}^{(1)}(\boldsymbol{X})\right)\tilde{\boldsymbol{V}}_A^{(i)}, \quad i = 1, 2, \tag{47}$$

where $\tilde{\boldsymbol{V}}_A^{(1)}$ is defined as:

$$\tilde{\boldsymbol{V}}_A^{(1)} = \boldsymbol{V}_A^{(1)} \cdot \boldsymbol{\tau}_{A_\alpha} \otimes \boldsymbol{\tau}_A^\alpha. \tag{48}$$

In (48), $\boldsymbol{V}_A^{(1)}$ is the spacial velocity of the slave node A. $\boldsymbol{\tau}_{A_\alpha}$ and $\boldsymbol{\tau}_A^\alpha$ are covariant and contravariant basis vectors, and $\tilde{\boldsymbol{V}}_A^{(2)}$ is obtained through mortar projection onto the slave surface, using the same dual basis mentioned above for the pressure gradients. $m^{\alpha\beta}$ in (45) are the so called contravariant components of the metric tensor which are defined as

$$m^{\alpha\beta} = \boldsymbol{\tau}^\alpha \cdot \boldsymbol{\tau}^\beta, \tag{49}$$

where $\boldsymbol{\tau}^\alpha$ and $\boldsymbol{\tau}^\beta$ are surface tangential contravariant basis vectors.

This brief presentation omits the details of how these approximation strategies lead to a numerical representation of the boundary value problem, but these details are standard and consistent with mortar approximation strategies found in other, slightly more conventional settings (see, e.g., [8]). We also employ consistent linearization of the resulting simultaneous equations for the nodal displacements and fluid pressures, enabling the monolithic solution approach we have mentioned previously.

1.3 Numerical Example

An elastohydrodynamically lubricated line contact problem is now explored to highlight the performance of the proposed approach. As illustrated in Figure 3, we consider a thin layer of fluid film between an elastic roller (cylinder) and a rigid flat surface. The rigid surface is moving with a constant speed v, which generates pressure inside the fluid film. The pressure in the fluid film will deform the elastic cylinder, and the deformation causes the change in the fluid film profile.

This problem has been tested by many researchers with different numerical methods (see, for instance, the works by Hamrock and coworkers [6, 7]). However, in most of these research works, the elastic deformation of the cylinder was evaluated by a semi-analytical integral equation, and the fluid phase was solved by either finite difference or finite element methods. Here, as mentioned earlier, we consider a monolithic numerical approach to the problem, which if effective will make the method much more broadly applicable to lubrication problems in solid mechanics.

In this example, both a constant viscosity and Barus viscous law are considered. As mentioned in [7], the elastic deformations of the cylinder are much larger than the fluid film thickness. Any inaccuracy in the film thickness will cause large pressure oscillations, especially for Barus viscous materials under very high loadings. The material for the cylinder taken to be isotropic and linearly elastic,

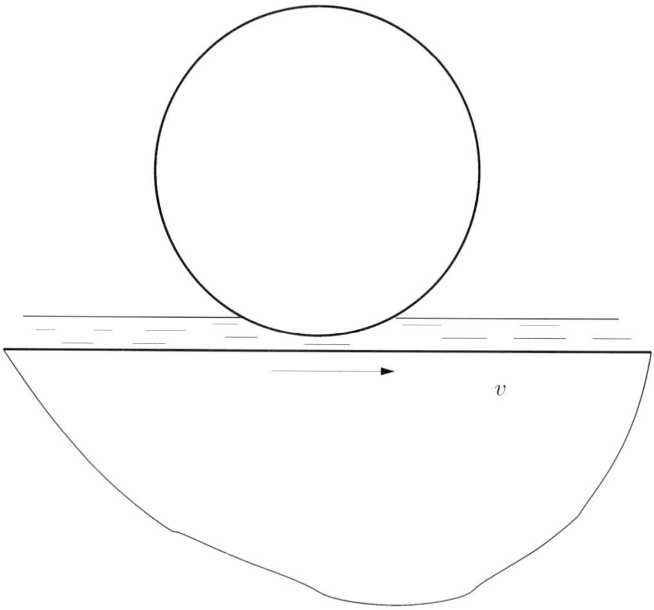

Fig. 3. An elastohydrodynamically lubricated line contact problem.

with $E = 2.1 \times 10^5$ MPa and $\nu = 0.3$. The initial viscosity for the fluid is $\mu_0 = 4.0 \times 10^{-8}$ MPa·s and the piezoviscous constant α in (9) is chosen as 0.005 MPa^{-1}. The speed v of the ground is 1157.5 mm/s. This is a two dimensional problem since side leakage is ignored. The load is applied as prescribed displacement at the top of the half cylinder until the total reaction force reaches the desired value. The final loads (reaction forces in the vertical direction) in this example are all about 374 N. Only half of the cylinder is discretized and the finite element mesh is shown in Figure 4. As seen in Figure 4(b), the mesh at the contact region is significantly refined. The rigid flat surface is defined as the slave surface and the cylinder surface is defined as the master surface. The (one dimensional) mesh for the fluid phase is inherited from the rigid flat surface. The radius of the cylinder is 10 mm and the width of the contact region on the flat block surface is 0.698 mm.

Figures 5 and 6 plot the pressure profiles for lubricated contact problems with and without shear stress, respectively. The pressure profile in Figure 5 is similar to the results obtained by the semi-analytical approach adopted by other researchers (because shear stresses were ignored in these research works when calculating deformations of the solid phases). However, Figure 6 shows the obvious influence of the shear stress on the pressure profile, indicating the importance of these terms in the predictions obtained. Another effect of the shear stress is that it results in more

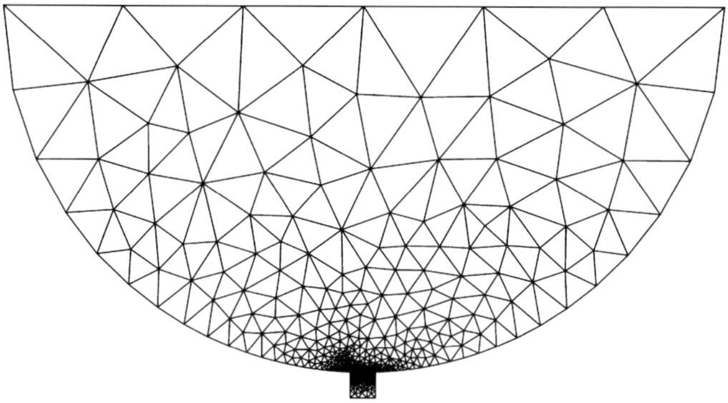

(a)

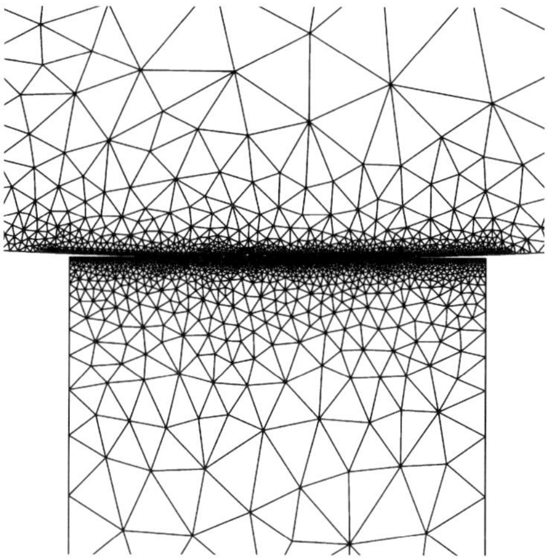

(b)

Fig. 4. Finite element mesh for the elastohydrodynamically lubricated contact problem: (a) entire mesh and (b) enlarged view of the contact region.

oscillations in the pressure profile as can be seen in Figure 6. These oscillations can be reduced if a more refined mesh is used, as plotted in Figure 7.

2 A Simple Strategy for Parallel Treatment of Mortar Contact

For the purposes of discussing parallel implementation of mortar-based interface constraints, we switch our attention now to the two body dry contact problem, de-

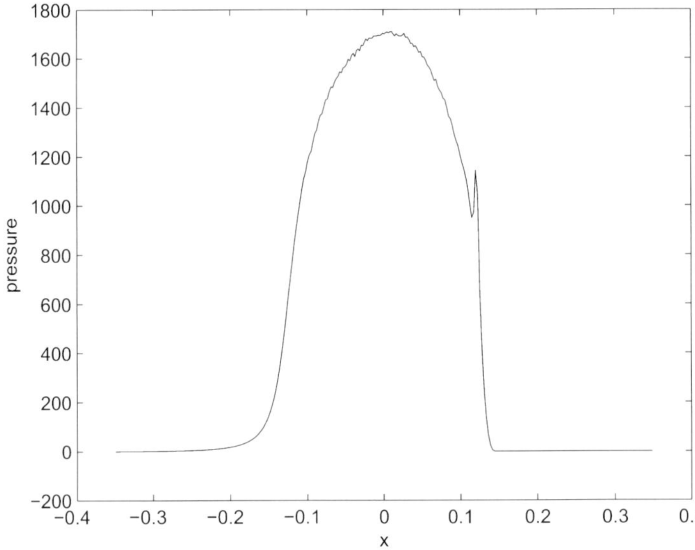

Fig. 5. Computed pressure distribution for the lubricated contact problem, obtained without consideration of shear stress.

picted schematically in Figure 8. We remove the restriction of small deformations in this section, and consider the case where two deformable bodies $\Omega^{(1)}$ and $\Omega^{(2)}$ are expected to come into contact. We select the surfaces $\Gamma_c^{(i)}$, $i = 1, 2$ to include all potential points of contact. At any time t, the virtual work for the system can be expressed in terms of the deformation mapping $\boldsymbol{\varphi}$ and the admissible variations $\overset{*}{\boldsymbol{\varphi}}$ as

$$\begin{aligned} G(\boldsymbol{\varphi}, \overset{*}{\boldsymbol{\varphi}}) :&= \sum_{i=1}^{2} G^{(i)}(\boldsymbol{\varphi}^{(i)}, \overset{*}{\boldsymbol{\varphi}}{}^{(i)}) \\ &= \sum_{i=1}^{2} \Bigg\{ \int_{\Omega^{(i)}} \left[\rho_0^{(i)} \overset{*}{\boldsymbol{\varphi}}{}^{(i)} \cdot \boldsymbol{A}^{(i)} + Grad \overset{*}{\boldsymbol{\varphi}} : \boldsymbol{P}^{(i)} \right] d\Omega \\ &\qquad - \int_{\Omega^{(i)}} \overset{*}{\boldsymbol{\varphi}}{}^{(i)} \cdot \boldsymbol{F}^{(i)} d\Omega - \int_{\Gamma_\sigma^{(i)}} \overset{*}{\boldsymbol{\varphi}}{}^{(i)} \cdot \bar{\boldsymbol{T}}^{(i)} d\Gamma \Bigg\} \\ &\qquad - \sum_{i=1}^{2} \int_{\Gamma_c^{(i)}} \overset{*}{\boldsymbol{\varphi}}{}^{(i)} \cdot \boldsymbol{t}^{(i)} d\Gamma = 0 \\ &= G^{\text{int,ext}}(\boldsymbol{\varphi}, \overset{*}{\boldsymbol{\varphi}}) + G^c(\boldsymbol{\varphi}, \overset{*}{\boldsymbol{\varphi}}) = 0, \end{aligned} \tag{50}$$

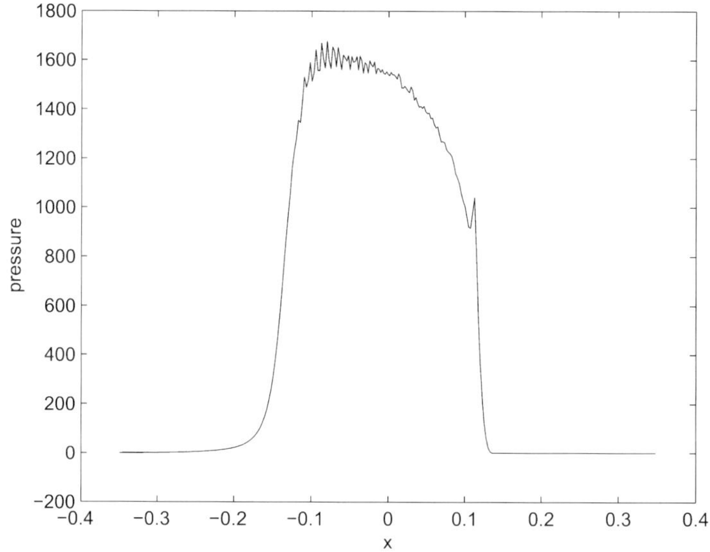

Fig. 6. Computed pressure distribution for the lubricated contact problem, with shear stress included.

where $G^{\text{int,ext}}(\boldsymbol{\varphi}, \overset{*}{\boldsymbol{\varphi}})$ refers to the virtual work arising from the internal and external forces in the two bodies, and $G^c(\boldsymbol{\varphi}, \overset{*}{\boldsymbol{\varphi}})$ denotes the virtual work associated with the contact tractions. The notation $\boldsymbol{A}^{(i)}$ has been employed to denote the material acceleration field in body (i) (in the event that inertial effects are present), $\rho_0^{(i)}$ denotes the reference density, $\bar{\boldsymbol{T}}^{(i)}$ are the prescribed tractions, $\boldsymbol{P}^{(i)}$ is the first Piola–Kirchhoff stress tensor, and $\boldsymbol{F}^{(i)}$ denotes the body force. By transforming to the spatial configuration, the contact virtual work can be expressed as:

$$G^c(\boldsymbol{\varphi}, \overset{*}{\boldsymbol{\varphi}}) := -\int_{\gamma_c^{(1)}} \boldsymbol{\lambda}^{(1)}(\boldsymbol{X}, t) \cdot \left(\overset{*}{\boldsymbol{\varphi}}{}^{(1)}(\boldsymbol{X}, t) - \overset{*}{\boldsymbol{\varphi}}{}^{(2)}(\bar{\boldsymbol{Y}}, t) \right) d\gamma, \tag{51}$$

where $\boldsymbol{\varphi}^{(2)}(\bar{\boldsymbol{Y}})$ is the current position of the contact point for $\boldsymbol{X}$.

By convention, the contact traction $\boldsymbol{\lambda}$ is defined on the nonmortar surface, notationally identified here as $\gamma_c^{(1)}$, according to

$$\boldsymbol{\lambda}^h(\boldsymbol{X}) = \sum_{A=1}^{n_n} N_A^1\left(\boldsymbol{\xi}^1(\boldsymbol{X})\right)\boldsymbol{\lambda}_A. \tag{52}$$

Incorporation of (52) and the shape function expansions for the variations into (51) gives

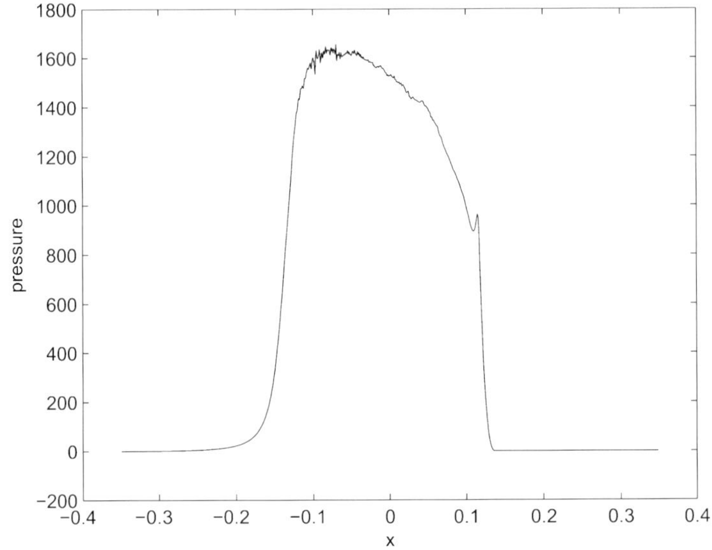

Fig. 7. Pressure distribution for the lubricated contact problem with shear stress included, using a finer mesh than depicted in Figure 6.

$$G^c(\boldsymbol{\varphi}^h, \overset{*}{\boldsymbol{\varphi}}{}^h) = -\sum_A \sum_{B,C} \boldsymbol{\lambda}_A \cdot \left(n^1_{AB} \overset{*}{\boldsymbol{\varphi}}{}^1_B - n^2_{AC} \overset{*}{\boldsymbol{\varphi}}{}^2_C \right) \tag{53}$$

where n^1_{AB} and n^2_{AC} are called mortar integrals. Efficient computation of these is the crux of an effective mortar-based contact algorithm, particularly in large deformation contact analysis. These integrals are defined as follows:

$$\begin{aligned} n^1_{AB} &= \int_{\gamma_c^{(1)}} N^1_A(\boldsymbol{\xi}^1(\boldsymbol{X})) N^1_B(\boldsymbol{\xi}^1(\boldsymbol{X})) d\gamma \\ n^2_{AC} &= \int_{\gamma_c^{(1)}} N^1_A(\boldsymbol{\xi}^1(\boldsymbol{X})) N^2_C(\boldsymbol{\xi}^2(\overline{\boldsymbol{Y}})) d\gamma, \end{aligned} \tag{54}$$

and computation of them involves evaluation of shape function inner products computed over the nonmortar side, $\gamma_c^{(1)}$. We note their similarity to the mortar integrals summarized in (36) and (37) for the lubricated contact problem, but with one important difference: since our focus here is large deformation contact analysis, these integrals must be computed in the spatial configuration and are therefore dependent upon the solution. Implementation of this idea requires efficient searching algorithms that find intersections between opposing element surfaces (these intersections are line segments in two dimensions, and polygonal surfaces in three dimensions); details are omitted here but the interested reader should refer to [13, 14] for more details.

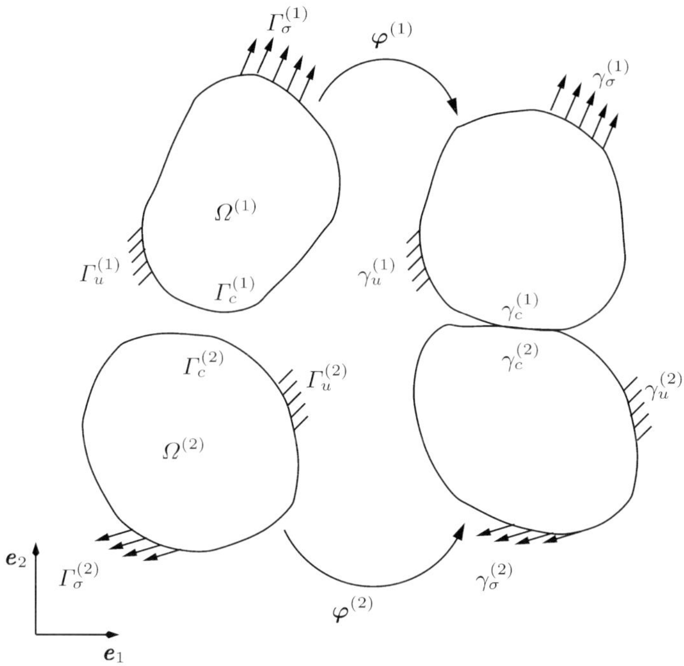

Fig. 8. Notation for the two body large deformation contact problem.

Another important feature of a mortar contact implementation is the decomposition of the surface tractions $\boldsymbol{\lambda}_A$ into normal and tangential components via

$$\boldsymbol{\lambda}_A = \lambda_{A_N} \boldsymbol{n}_A + \boldsymbol{\lambda}_{A_T} \tag{55}$$

where $\boldsymbol{n}_A$ is a locally averaged surface normal vector based at node A, and where the tangential traction $\boldsymbol{\lambda}_{A_T}$ is expressed in a tangent plane orthogonal to this normal vector. The contact pressure λ_{A_N}, is subject to the Kuhn–Tucker conditions

$$\begin{aligned} \lambda_{A_N} &\geq 0 \\ g_{A_N} &\leq 0 \\ \lambda_{A_N} g_{A_N} &= 0 \end{aligned} \tag{56}$$

where g_{A_N} is the scalar gap function at node A, defined via

$$\begin{aligned} g_{A_N} &= \boldsymbol{g}_A \cdot \boldsymbol{n}_A, \\ \boldsymbol{g}_A &:= \sum_B n^1_{AB} \boldsymbol{\varphi}^1_B - \sum_C n^2_{AC} \boldsymbol{\varphi}^2_C. \end{aligned} \tag{57}$$

As is usually the case in contact analysis, these constraints can be enforced using Lagrange multiplier, penalty and augmented Lagrangian approaches; examples of this can be seen in [8, 9, 14].

The frictional traction is most easily computed using a penalty regularization procedure, which enables its determination using a trial-state, return map algorithm. This procedure can be summarized (for example) in two dimensions as

$$\boldsymbol{\lambda}_{A_{T_{n+1}}} = \begin{cases} \boldsymbol{\lambda}^{\text{trial}}_{A_{T_{n+1}}} = \lambda^{\text{trial}}_{A_{T_{n+1}}} \boldsymbol{\tau}_A, & \text{if } \Phi^{\text{trial}}_{n+1} \leq 0, \text{ stick,} \\ \mu \|\boldsymbol{\lambda}_{A_N}\| \boldsymbol{\tau}_A, & \text{otherwise, slip,} \end{cases} \tag{58}$$

where

$$\lambda^{\text{trial}}_{T_{A_{n+1}}} = \lambda_{T_{A_n}} - \epsilon_T \boldsymbol{\tau}_A \cdot \left[\sum_C (n^2_{AC_{n+1}} - n^2_{AC_n}) \boldsymbol{\varphi}^2_C - \sum_B (n^1_{AB_{n+1}} - n^1_{AB_n}) \boldsymbol{\varphi}^1_B \right], \tag{59}$$

and

$$\Phi^{\text{trial}}_{n+1} = \|\boldsymbol{\lambda}^{\text{trial}}_T\| - \mu \|\boldsymbol{\lambda}_N\| \tag{60}$$

where ϵ_T is the tangential penalty parameter and the subscript $n+1$ means a state associated with the current iteration of the $n+1$st load (or time) step, and n is to be associated with the nth converged load step. It is important to note in (59) that the increments of motion in the tangent plane can be expressed in terms of increments in the mortar integrals themselves, giving us the important property of frame indifference and also facilitating a convenient data structure for contact histories. More details on this point can be found in [14].

2.1 Parallel Implementation of Implicit Mortar Formulations

We consider here an implicit equation solving framework, such that full linearization is performed in a consistent way to obtain the contact stiffness in a Newton–Raphson framework. Full details of this linearization can be found, for example, in [9].

In this study, we introduce a very simple approach to parallel implementation, consisting of a standard partitioning of the finite element discretization (FE), plus the assignment of an entire contact element (CE), containing all information pertaining to contact interaction, to each processor. Accordingly, each processor stores the data for its partition of the FE mesh, as well as the entire CE, as shown in Figure 9. The CE (indicated by a thick solid outline on the boundary of bodies Ω_1 and Ω_2 in Figure 9(a), for a schematic two dimensional case) is typically a one dimensional manifold in a two dimensional problem and a two dimensional manifold in a three dimensional problem. This simple hybrid partitioning strategy is motivated by the desire to avoid load imbalance, and the observation that for the problems we have studied, the computation time associated with contact searching and stiffness assembly usually amount to less than 5% of total computation time, with very

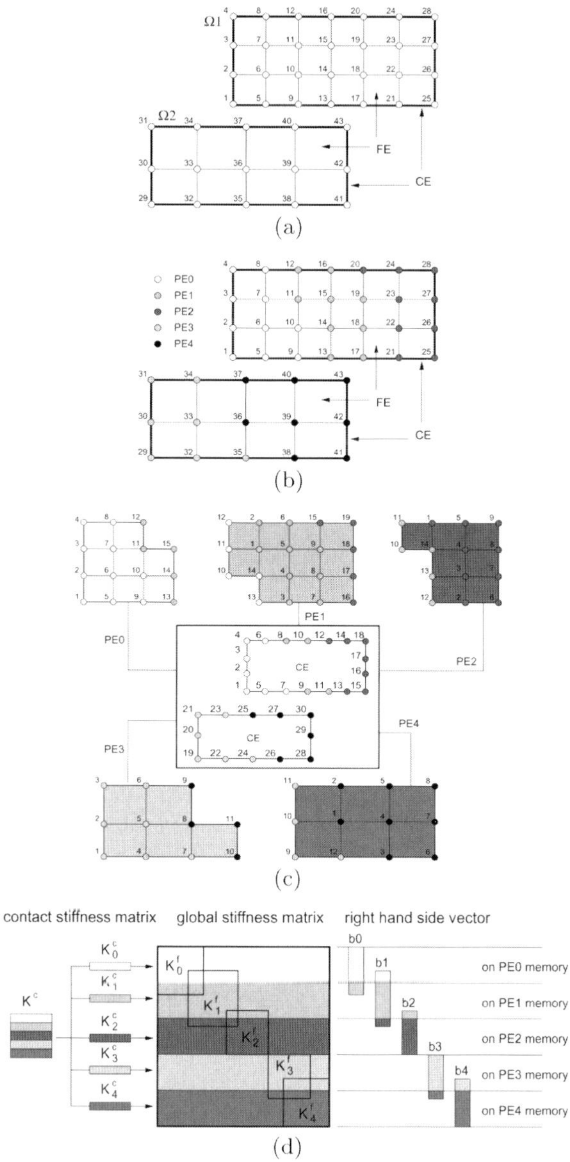

Fig. 9. Simple example for hybrid parallel mortar contact algorithm: (a) entire finite element domain + entire contact element; (b) node-wise domain partitioning with nonoverlap; (c) partitioned FE based on node partition + entire CE for each processor; (d) assembling global stiffness matrix from contact stiffness matrix ($\boldsymbol{K}^C$) and local FE stiffness matrix ($\boldsymbol{K}_i^f$, $i :=$ subdomain ID) and right hand side vector ($\boldsymbol{b}_i$).

modest memory complexity associated with the CE (particularly when large n are considered).

Continuing to refer to Figure 9(a), the conceptual steps in developing the parallel mortar contact algorithm may now be described. During problem initialization, the following steps are to be performed:

1. Partition nodes on a nonoverlapping basis, without regard to the contact interaction, so that partitioned nodes are stored on each corresponding processor and labeled accordingly (Figure 9(b)). Each processor stores only its portion of the global solution vector $\boldsymbol{d}$ after parallel solution of each Newton–Raphson linearization of the global equilibrium equations.
2. Partition finite elements, based on the nodal partition. For each processor, find all finite elements that have a node on that processor and store those elements. Nodes not belonging to that processor in these FEs are stored as *shared*. Once the finite element partitioning is done, nodal IDs are renumbered from global to local as shown in Figure 9(c).
3. Store contact information on each processor.

Then, during problem execution, the following steps take place during each global equilibrium iteration:

1. Construct stiffness matrices and right hand side (RHS) vector. On each processor, we construct the local FE stiffness matrix and RHS vector, as well as the contact stiffness matrix for the entire CE, as shown in Figure 9(d).
2. Assemble stiffness matrices and RHS vector. Based on mapping between local and global IDs, the local stiffness matrix and local RHS vector are assembled into the global system. If the equation number of an element in the local FE stiffness matrix does not fall into that processor, then those data are sent to the neighbor processor containing that equation. Similarly, each processor assembles contact stiffness and right hand sides to the global stiffness matrix for those components whose equation number falls within its range; otherwise, these are discarded.
3. After solution of the linearized global system using a parallel solver (e.g. PETSc), perform data update through interprocessor communication, as shown in Figure 10. For example, processor PE0 sends data for node 6, 7, 8 and 9 to PE1 and receives data for nodes 11, 12, 13, 14 and 15 from PE1. At the the same time, PE1 sends and receives these data to and from PE0. For update of the contact data, each processor updates its part of the CE and shares the CE data to facilitate the next contact calculation. In our implementation, the master processor (e.g. PE0) gathers the data for the CE from the other processors and broadcasts the data to the other processors.

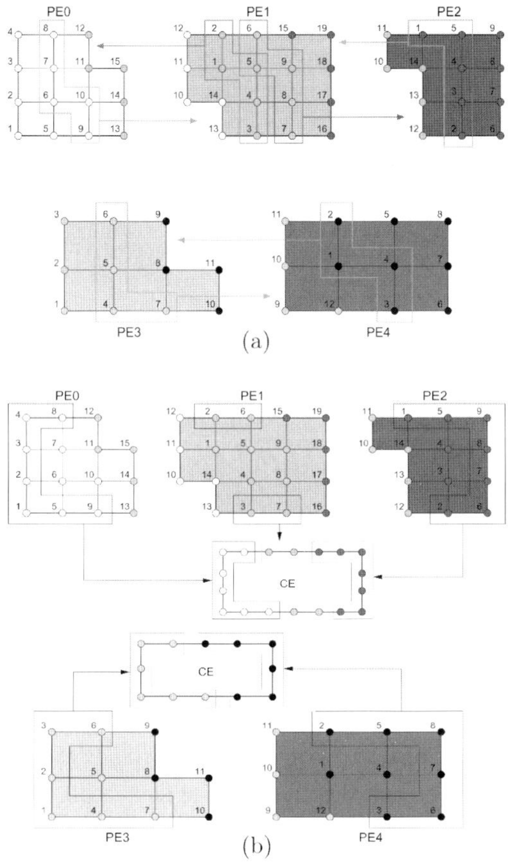

Fig. 10. Data communication in hybrid parallel mortar contact algorithm. (a) data communication for FE; (b) contact data (nodal data on CE) update from each processor.

2.2 Numerical Example

To demonstrate typical performance, we now briefly summarize a numerical example. Here, we use PETSc [1] for iterative solution of successive linearized systems. Among the Krylov iterative solvers available in PETSc, GMRES(30) was used with block Jacobi and ILU(p) as a preconditioner. The Compressed Sparse Row (CSR) format was used for storage of the sparse matrix, enabling memory savings and reduction of computation time. The results presented include simulations performed on two machines: an Intel dual Xeon based Linux cluster (CSEM, 2.8GHz-Processor, 2GByte-Memory, 512 KB-L2 cache) at Duke university, and an

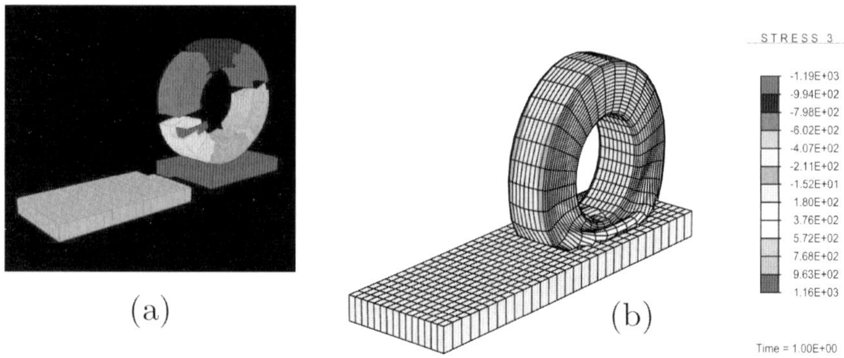

Fig. 11. Tire crushing simulation: (a) 8-subpartitioned domain; and (b) stress (σ_{zz}) distribution at $t = 1.0$ sec.

HP-Compaq AlphaServer (AKA, Lemieux with 1 GHz-Processor, 4GByte-Memory) at the Pittsburgh Supercomputing Center.

In our implementations of mortar contact using Newton–Raphson iteration and direct solvers (such as, for example, [4]), tolerances in the relative residual norm of around 10^{-15} are routinely achieved. We thus use this same tolerance for the iterative solvers, and check the fixed-size scalability of our approach over two problem scales, considering a tire crush simulation as an example problem (see Figure 11). In an initial, small to medium sized mesh, we have 95,223 nodes and 63,200 hexahedral elements. The model is composed of three different parts: (1) the rim, (2) the tire, composed of a neo-Hookean material, and (3) an idealized ground. The coefficient of friction for contacting surfaces was taken as $\mu = 0.5$. The bottom surface of the ground elements is fixed, while the displacement of the rim is prescribed downward in a proportional manner. An example of the finite element partition with 8 subdomains is shown in Figure 11(a).

Fifty time steps were used to model the deformation of the tire. Within the Newton–Raphson iteration scheme, a convergence criterion in terms of the energy norm was set as 1.0×10^{-18}; the number of unknowns in this initial mesh was 248,946. The contour of stress (σ_{zz}) at time 1.0 sec is shown in Figure 11(b), while Table 1 shows the performance evaluation in this first model, with parallel efficiencies noted in parentheses; results are given in the table on both parallel computers tested.

To examine performance in larger problems, we generated a larger version of the model with 302,400 nodes and 431,305 hexahedral elements; the number of unknowns in this case is 1,136,292. Due to some difficulties with file I/O on the CSEM cluster, this calculation was performed on Lemieux only. In this case, the number of time steps was defined as 20 (less than half of the value in the medium-scale case) for efficient use of limited computational resources on Lemieux. Table 2 shows the per-

Table 1. Fixed-size scalability for medium-scale tire crush simulation (time in hours).

PEs	CSEM		LEMIEUX	
	solver	total	solver	total
4	35.74 (1.00)	37.36 (1.00)	50.56 (1.00)	52.70 (1.00)
8	19.34 (0.92)	20.64 (0.91)	26.56 (0.95)	28.33 (0.93)
16	12.87 (0.69)	14.08 (0.66)	14.86 (0.85)	16.15 (0.82)
32	7.01 (0.64)	7.73 (0.60)	8.54 (0.74)	9.30 (0.71)
64	5.40 (0.41)	6.10 (0.38)	5.31 (0.60)	6.10 (0.54)

Table 2. Fixed-size scalability for large-scale tire crush simulation in Lemieux (time in hours).

PEs	solver (S)		rest (R)		**total (S+R)**	
	time	efficiency	time	efficiency	time	efficiency
64	8.45	1.00	2.69	1.00	11.14	1.00
128	5.87	0.72	2.04	0.66	7.91	0.70
256	3.90	0.54	1.42	0.47	5.32	0.52
512	1.98	0.53	0.82	0.41	2.80	0.50

formance results for this large-scale model. Since the granularity of the large-scale model is about 3.5 times smaller than the medium-scale model, the scalability in the large-scale case indicates worse performance than in medium-scale case. However, our parallel implementation as applied to this larger scale case shows high parallel efficiency up to 512 processors.

3 Summary and Conclusions

In this paper, we have discussed two recent extensions of the mortar formulations as applied to contact and interface mechanics: lubricated contact formulations on one hand, and parallel implementations of solid-to-solid, large deformation contact on the other. Numerical examples have been presented for each application area. Future work will continue to explore the enhanced range of interface mechanics applications that may be opened up through consideration of this numerical framework.

Acknowledgments

We would like to thank CSEM at Duke university for allowing us to use the CSEM cluster. Supercomputing resources on HP-Compaq AlphaServer at Pittsburgh Supercomputing Center wer e granted under award MSS060001P. Additionally, this work was supported in part by Sandia National Laboratories under Contract 21757. This support is gratefully acknowledged.

References

1. S. Balay, K. Buschelman, V. Eijkhout, W.D. Gropp, D. Kaushik, M.G. Knepley, L. C. McInnes, B. F. Smith, and H. Zhang. PETSc users manual. Technical Report ANL-95/11 – Revision 2.1.5, Argonne National Laboratory, 2004.
2. D.G. Christopherson. An analysis of metal forming processes using large deformation elastic-plastic formulations. *Proceeding of the Institution of Mechanical Engineers*, 146:126–135, 1941.
3. C.W. Cryer. The method of christopherson for solving free boundary problems for ininite journal bearings by means of finite differences. *Mathematics of Computation*, 25:435–443, 1971.
4. A. Gupta. *WSMP: Watson Sparse Matrix Package, Part-I,II (User's Guide)*. IBM T.J. Watson Research Center, 2000.
5. B. J. Hamrock. *Fundamentals of Fluid Film Lubrication*. NASA, Washington, D.C., 1991.
6. B.J. Hamrock, P. Pan, and R.T. Lee. Pressure spikes in elastohydrodynamically lubricated conjunctions. *ASME Journal of Tribology*, 110:279–284, 1988.
7. L.G. Houpert and B.J. Hamrock. Fast approach for calculating film thicknesses and pressures in elastohydrodynamically lubricated contacts at high loads. *ASME Journal of Tribology*, 108:411–420, 1986.
8. S. Hueber and B.I. Wohlmuth. A primal-dual active set strategy for non-linear multibody contact problems. *Computer Methods in Applied Mechanics and Engineering*, 194:3147–3166, 2005.
9. M.A. Puso and T.A. Laursen. A mortar segment-to-segment contact method for large deformation solid mechanics. *Computer Methods in Applied Mechanics and Engineering*, 193:601–629, 2004.
10. S.M. Rohde and G.T. McAllister. A variational formulation for a class of free boundary problems arising in hydrodynamic lubrication. *International Journal of Engineering Science*, 13:841–850, 1975.
11. B.I. Wohlmuth. A mortar finite elment method using dual spaces for the lagrange multiplier. *SIAM Journal on Numericl Analysis*, 38:989–1012, 2000.
12. B.I. Wohlmuth. *Discretization Methods and Iterative Solvers Based on Domain Decomposition*. Springer-Verlag, Heidelberg, 2001.
13. B. Yang and T.A. Laursen. A contact searching algorithm for large deformation mortar formulation. *Computational Mechanics*, 2007, to appear.
14. B. Yang, T.A. Laursen, and X.N. Meng. Two dimensional mortar contact methods for large deformation frictional sliding. *International Journal for Numerical Methods in Engineering*, 62:1183–1225, 2005.

A Coupled Contact and Decohesion Analysis of Laminated Beams

Alberto Carpinteri, Marco Paggi and Giorgio Zavarise

Department of Structural and Geotechnical Engineering, Politecnico di Torino, Corso Duca degli Abruzzi 24, 10129 Torino, Italy
E-mail: marco.paggi@polito.it

Abstract. The problem of interface decohesion in laminated beams is addressed with reference to the double cantilever beam (DCB) geometry. The paper deals with the analysis of the influence of non-uniform bonding properties or interfacial defects on the crack propagation process and its stability. In spite of the relative simplicity of the considered case study, which is under pure Mode I deformation, a contact problem takes place due to the decohesion. The occurrence of contact makes the problem highly non-linear, since the extension of the contact area is *a priori* unknown. The finite element method with a generalized interface constitutive law is used to enforce the contact constraint along the interface and to model the progress of the delamination process.

Key words: Contact, decohesion, cohesive models, laminated beams, finite elements.

1 Introduction

Nowadays the problem of debonding occurring at the interface between different materials is the subject of extensive researches. Most of the contributions in this field have focused on the mathematical and numerical modeling of the non-linear mechanical response of the interface [1–3]. In this framework, cohesive models are commonly adopted to set up a local interface constitutive law. The origin of such laws can be traced back to the work of Dugdale [4] and Barenblatt [5], where tractions on zones near the crack tip were used to model plastic yielding in ductile materials. Extensions to the analysis of the non-linear fracture process zones in quasi-brittle materials were also proposed by Hillerborg [6], whereas Carpinteri firstly applied a cohesive formulation to the study of ductile-brittle transition and snap-back instability in concrete [7–11]. More recently, we have proposed an interpretation of size-scale effects in metal matrix composites and their connection with superplasticity by modeling the debonding between fibers and matrix with a cohesive formulation [12].

Peter Wriggers and Udo Nackenhorst (eds.), IUTAM Symposium on Computational Contact Mechanics, 147–156.

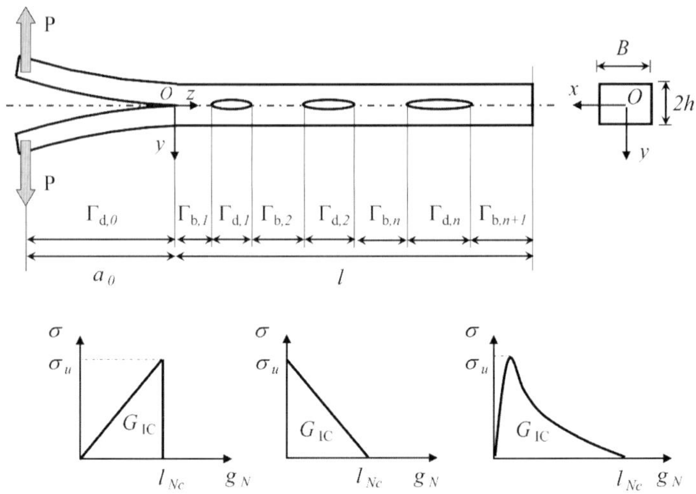

Fig. 1. Scheme of a DCB with multiple microcracks ahead of the macrocrack tip and shape of the cohesive laws compared in this study.

According to this approach, the localized damaged zone can be modeled as a pair of surfaces with no volume between them. The local stress state can be replaced by an equivalent traction acting on the surfaces, which non-linearly depends on the amount of displacement discontinuity. The input parameters of the cohesive models, i.e. the peak stress, σ_u, and the interface fracture energy, G_{IC}, are usually estimated according to conventional mechanical tests designed to evaluate the properties of the interface [13–16]. Among them, the debonding double cantilever beam geometry (DCB) is widely adopted for the evaluation of the Mode I fracture energy (see Figure 1 for a scheme of the tested geometry and refer to [17] for more details on the testing procedure).

Most of the analytical models available in the literature for the analysis of these problems are based on the theory of a beam on an elastic foundation [18]. Uniform bonding properties are usually assumed in [19–22]. According to this approach, the unilateral contact condition along the interface is not fulfilled, thus admitting a penetration between the two arms of the cantilever beam. This simplification is usually considered acceptable in case of uniform bonding, since these penetrations are thought to be recovered by the finite thickness of the adhesive. In addition, if the influence of shear deformation is taken into account, the effect of the closure of the gap is expected to be less important [23]. On the other hand, for other problems involving crack interaction, like the one addressed in this paper, the imposition of the contact inequalities is crucial (see e.g. [24] for the problem of vibration of laminated beams with multiple cracks).

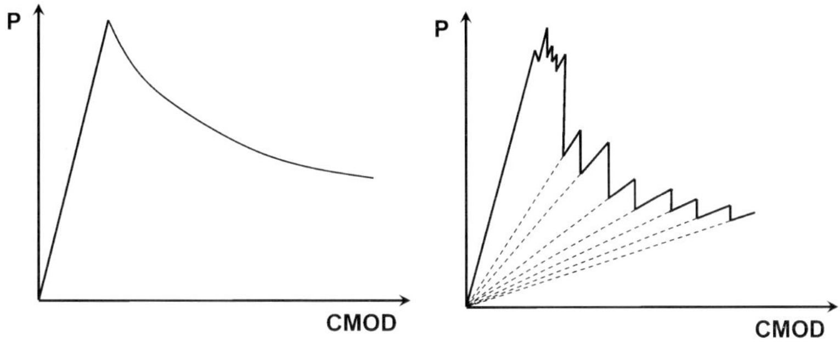

(a) Uniform bonding. (b) Nonuniform bonding.

Fig. 2. Comparison between typical mechanical responses of a DCB test, depending on the uniformity of the bonding properties along the interface (experimental data adapted from [29]). The parameter CMOD denotes the crack mouth opening displacement.

Concerning the numerical models, the finite element method is usually employed with a proper discretization of the candidate contact zones with interface elements. In general, much attention is given to the cohesive law, whereas a minor attention is provided to the contact problem, which is not explicitly addressed [25–27]. In fact, in most of the commercial FE programs, a penalty-like relation is directly added to the cohesive formulation to deal with the problem of crack closure, and only in a few cases the contact unilateral constraint is properly modeled [28].

As far as the problem of decohesion in laminated beams with interfacial defects is concerned, this study is motivated by the fact that adherend surface preparation plays a critical role in developing bonded joints. Inadequate surface roughening, environmental effects, peel ply chemical contamination, and other factors (both mechanical and chemical) can prevent adhesives from bonding properly to composites, resulting in interfacial failures and localized regions of extremely weak bonding [29]. As a consequence, the mechanical response of the DCB test expressed in terms of pulling force, P, vs. crack mouth opening displacement, CMOD, results in an extremely jagged curve instead of a smooth one (see Figure 2).

In this context, aiming at modeling such unstable mechanical responses, a finite element model is proposed with the aid of a generalized interface constitutive law. This approach permits to enforce the unilateral contact condition along the interface, as well as to model the progress of the delamination process.

2 Finite Element Model

To deal with the above-described problem, interfaces are modeled according to a contact formulation which can handle cohesive forces. Contact constraints can be enforced using a modified penalty method. Depending on the contact status, an automatic switching procedure is adopted in order to choose between cohesive and contact models. According to the node-to-segment contact strategy [30], it is possible to compute for each node along the master contact surface the normal gap that is a measure of the relative displacement of the two arms of the DCB. The generalized interface constitutive law, proposed in [1, 31], is particularly suitable for the analysis of coupled contact and decohesion problems occurring in interface mechanical problems.

When opening takes place, interface tractions are computed according to a cohesive formulation, which is a function, respectively, of the gap measured in the normal direction, g_N, of a critical value of the opening displacement, l_{Nc}, and of the material strength, σ_u (see Figure 1):

$$\sigma_N = \sigma_N(g_N, l_{Nc}, \sigma_u). \tag{1}$$

A linear softening cohesive law is usually adopted when the problem of cohesive crack propagation takes place inside a structural component without the presence of preexisting material discontinuities (see e.g. [8–11]). In such a case, the cohesive tractions transmitted along the crack faces are always decreasing functions of the crack opening displacements, starting from σ_u at $g_N = 0$, down to zero when the opening is equal to the critical value, i.e. $g_N = l_{Nc}$. On the other hand, when an interface already exists, this line is usually the weakest link and the crack path is *a priori* known. In these situations the cohesive models proposed in the literature consider cohesive tractions increasing from zero up to a maximum, σ_u, in correspondence of a given interface opening, and then decreasing down to zero when $g_N = l_{Nc}$ (see e.g. the polynomial model proposed in [32]). In some cases, a simplified tension cut-off relationship is used to make the problem analytically treatable.

In the numerical scheme, when the inequality $g_N < l_{Nc}$ holds, the cohesive tractions computed as a function of the interface displacements are multiplied by the corresponding area of influence belonging to each Gauss point along the interface element. To do it, an enhanced formulation of the node-to-segment contact FE formulation has to be used [33].

During closure, i.e. for $g_N \leq 0$, contact forces deriving from the interaction of the interfaces can be expressed with a penalty-like model [34]:

$$F_N = C g_N^m, \tag{2}$$

where the parameters C and m depend on the deformation assumptions of the profiles in contact. Instead of making an attempt to list the variety of approaches in the literature, we refer the reader to the survey available in [31, 35, 36]. Here we simply adopt a penalty formulation which is recovered by setting the penalty parameter m in Equation (2) equal to the unity.

Both in case of decohesion and contact, contributions of cohesive and contact forces are added to the global virtual work equation:

$$\delta W = \mathbb{A}(F_N \delta g_N), \tag{3}$$

where the symbol $\mathbb{A}$ denotes an assembly operator for all the interface nodes either inside the process zone or in contact.

Within the analysis, a main difficulty due to the contact constraints is related to the fact that the debonded regions are *a priori* unknown, and the corresponding boundary value problem has to be solved with an iterative method. The Newton–Raphson solution procedures, commonly used for solving nonlinear problems, require the determination of the tangent stiffness matrix. Consistent linearization of the equation set (3) leads to:

$$\Delta\delta W = \frac{\partial F_N}{\partial g_N} \Delta g_N \delta g_N + F_N \Delta\delta g_N, \tag{4}$$

where the symbols δ and Δ denote, respectively, variations and linearizations. In Equation (4) the normal forces are given by either the cohesive formulation or the contact model. Linearizations and variations of the normal gap can be found in [37] for the contact model and in [1] for the cohesive law, as well as the discretized version of these expressions for a direct implementation in the finite element formulation.

3 Numerical Examples

As a case study, let us consider a DCB test with the following geometrical and mechanical parameters (see Figure 1): $a_0 = 30$ mm, $l = 70$ mm, $h = 1.5$ mm, $B = 1$ mm, $E = 69$ GPa, $\sigma_u = 57$ MPa, $G_{\mathrm{IC}} = 10$ kN/m. We consider the existence of internal debonded regions, each one having a length equal to that of the bonded portions, and periodically distributed along the interface. Under these conditions, the length of the regions $\Gamma_{d,i}$ and $\Gamma_{b,i}$ is equal to $l_d = l/(2n+1)$.

The application of our generalized interface constitutive law implemented in the finite element code FEAP (courtesy of Professor Taylor) permits to analyze the decohesion problem taking into account the whole non-linear interface behavior. This fact is very well evidenced in Figure 3a, where the deformed meshes for the DCB test with and without taking into consideration the contact constraints are compared in the case of $n = 0$ for a small zone behind the macrocrack tip. Clearly, when contact is disregarded, interpenetration of the two arms of the beam takes place, which is a physically unacceptable situation.

Figure 3b shows a comparison between the finite element results and the analytical prediction determined according to the formulation in [38] for a totally bonded interface, i.e. for $n = 0$. In the analytical formulation a simplified tension cut-off cohesive model is used to obtain a closed-form solution. The finite element results take into account the effect of contact along the interface and consider three different cohesive laws: the tension cut-off relationship between cohesive tractions and opening displacements, a linear softening law [8–11], and a polynomial relationship [32].

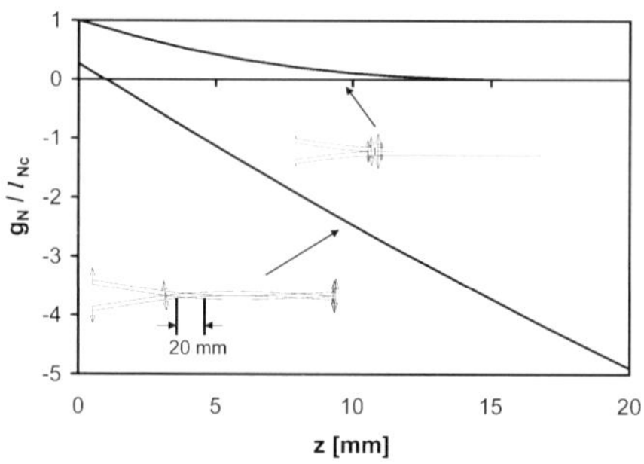

(a) Vertical displacements with or without modeling the contact closure constraints. Reaction forces are marked with arrows.

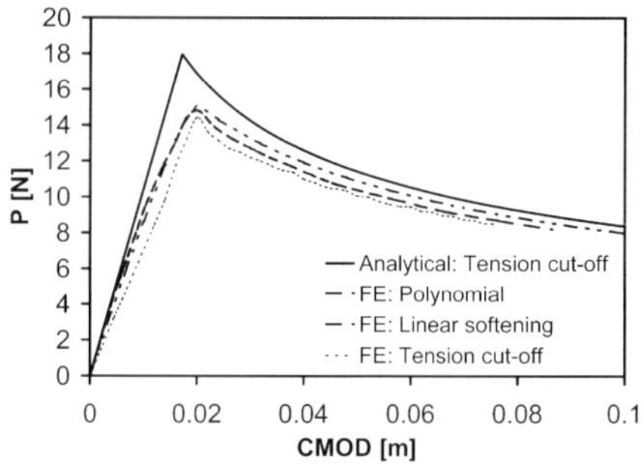

(b) Comparison between analytical and finite element solutions for a DCB test with totally bonded interface.

Fig. 3. Results comparison.

The cohesive parameters σ_u and G_{IC} are the same for all the models. The good agreement between the computed P vs. CMOD curves obtained using the three cohesive models demonstrates that the shape of the cohesive law does not play a major role.

On the other hand, the comparison between the analytical and the finite element solution in case of a totally bonded interface shows that the effect of contact results into a reduction of the critical force required for the onset of decohesion. This effect can be explained by noting that the fulfillment of the unilateral contact condition along the interface results into a higher value of the opening displacement at the real macrocrack tip, $z = 0$, as compared to the case when contact is disregarded (see Figure 3a). Compressive stresses in the region close to the macrocrack tip are also shown in the contour plot of the vertical stresses in Figure 4.

The finite element results for a DCB test with 5 and 10 defects are shown in Figure 5a. The peak load corresponding to the onset of delamination and the initial stiffness of the composite beam are both decreasing functions of the number of defects. The initial stiffness of the composite is measured as the slope of the load vs. CMOD curve before reaching the peak load.

Finally, it has to be pointed out that the finite element algorithm permits to follow the whole delamination process. Under displacement control, when the first bonded portion of the interface fails, a sudden reduction in the reaction force, P, takes place in correspondence of the microcrack, when a jump in the macrocrack length occurs from $a_0 + l_d$ to $a_0 + 2l_d$ (see points A and B in Figure 5b). After that, another bonded portion of the beam is involved and the reaction force progressively increases (see points B and C in Figure 5b). The same curve BC could be followed in case of a DCB with initial crack length equal to $a_0 + 2l_d$ (see the dashed line connecting the origin of the plane with the point B in Figure 5b). Exactly the same behavior is repeated when the other defects are encountered.

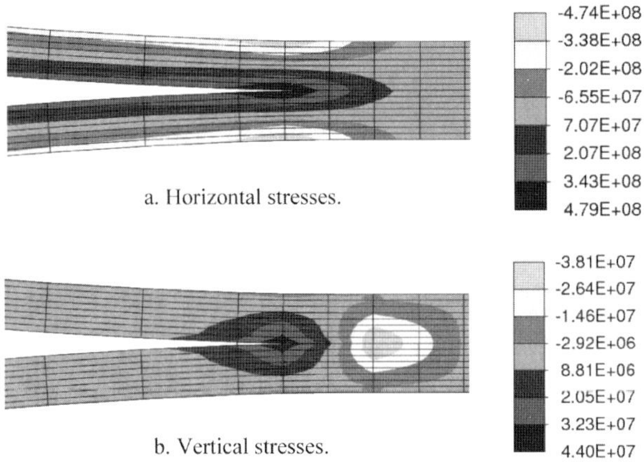

a. Horizontal stresses.

b. Vertical stresses.

Fig. 4. Contour plots of the stresses in the region close to the macrocrack tip.

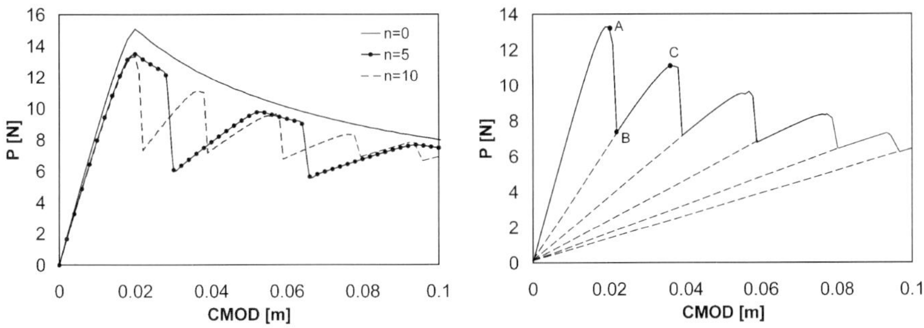

(a) Force vs. CMOD curve for a DCB test with $n = 0$, 5 and 10 defects.

(b) Detail of the unstable response in the case of $n = 10$ defects.

Fig. 5. Unstable mechanical responses due to imperfect bonding conditions.

4 Analogy between Microcracked and Reinforced Zones Ahead of a Macrocrack Tip

The jumps in the reaction force, caused by a discrete crack advancement through the defected zone, are a typical example of crack growth instabilities as those found by Carpinteri and Monetto [39] in case of fracture evolution in multi-cracked solids.

As first noted by Carpinteri [40], it is interesting to point out that the case of a macrocrack with a regular distribution of discrete microcracks ahead of its tip (Figure 6a) is analogous to that of a macrocrack with a regular distribution of discrete reinforcements (Figure 6b) [41]. In both cases the structural response presents a discrete number of instabilities with related peaks and valleys. For the microcracked

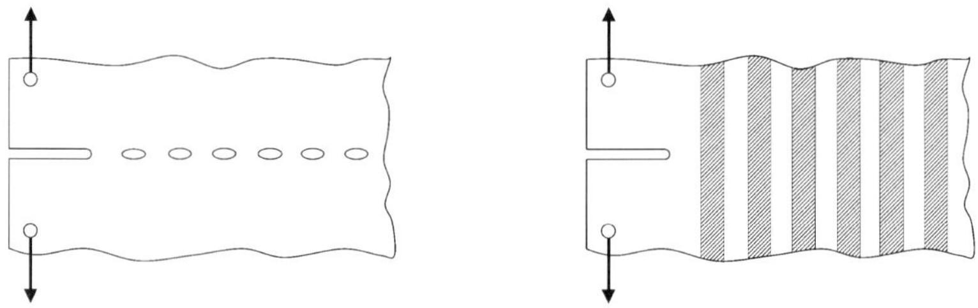

(a) Microcracked zone. (b) Reinforced zone.

Fig. 6. Analogy between microcracked and reinforced zones ahead of a macrocrack tip.

process zone, the valleys represent the stage of approach of the macrocrack tip to a microcrack tip. In this situation the stress state is particularly high and therefore a lower load is sufficient to produce $G_{\mathrm{I}} = G_{\mathrm{IC}}$. On the other hand, the peaks represent the stage subsequent to the macrocrack-microcrack coalescence, when the new macrocrack tip is still far from the next microcrack tip. Thus its stress concentration is still relatively low. In this latter case, a higher load is needed to have $G_{\mathrm{I}} = G_{\mathrm{IC}}$ at the driving tip.

The analogy between reinforced zone and microcracked zone consists therefore in a multiple irregular mechanical response, where the valleys and the peaks represent, respectively, the crack arrest at the fibers or the crack-slip coalescence, and the fiber crossing or the crack growth after coalescence.

5 Conclusion

In this contribution the problem of interface decohesion of laminated beams with non-uniform interface properties has been addressed. To this aim, the implementation of a generalized interface constitutive law in the finite element framework permits to take into account the contact constraint along the interface, which is a condition usually disregarded by the analytical formulations based on a beam on an elastic foundation. Moreover, it has been shown that the proposed methodology permits to model the whole nonlinear behavior of the interface. In particular, cohesive laws more realistic than the tension cut-off relation can be easily considered.

Acknowledgements

The financial support provided by the Italian Ministry of University and Research with the project "PRIN2005 – Modelling and approximation in advanced mechanical problems", and by the European Union with the project "Leonardo – Numerical based medium level training on industrial friction problem (NUFRIC)" is gratefully acknowledged.

References

1. Paggi M (2005) Interface Mechanical Problems in Heterogeneous Materials. Ph.D. thesis Politecnico di Torino, Torino, Italy.
2. Wriggers P, Zavarise G, Zohdi T (1998) *Computational Materials Science*, 12:29–56.
3. Carpinteri A, Lacidogna G, Paggi M (2006) Acoustic emission monitoring and numerical modelling of FRP delamination in RC beams with nonrectangular cross-section. *Materials & Structures*, in press, doi:10.1617/s11527-006-9162-4.
4. Dugdale D (1960) *Journal of the Mechanics and Physics of Solids*, 8:100–104.
5. Barenblatt G (1962) In Dryden H, Karman TV (Eds.), *Advances in Applied Mechanics*, Vol. VII. Academic Press, New York, pp. 55–129.
6. Hillerborg A, Modeer M, Petersson P (1976) *Cement and Concrete Research*, 6:773–782.
7. Carpinteri A (1985) In Shah S (Ed.) *Application of Fracture Mechanics to Cementitious Composites* (Proceedings of a NATO Advanced Research Workshop, Evanston, USA, 1984). Martinus Nijhoff Publishers, Dordrecht, pp. 287–316.
8. Carpinteri A, Tommaso AD, Fanelli M (1986) In Wittmann F (Ed.), *Fracture Toughness and Fracture Energy of Concrete* (Proceedings of an International Conference on Fracture Mechanics of Concrete, Lausanne, Switzerland, 1985). Elsevier, Amsterdam, pp. 117–135.
9. Carpinteri A (1989) *Journal of the Mechanics and Physics of Solids*, 37:567–582.
10. Carpinteri A (1989) *International Journal for Numerical Methods in Engineering*, 28:1521–1537.
11. Carpinteri A, Colombo G (1989) *Computers & Structures*, 31:607–636.
12. Carpinteri A, Paggi M, Zavarise G (2005) *Strength, Fracture and Complexity*, 3:61–72.
13. RILEM (1985) Determination of the fracture energy of mortar and concrete by means of three-point bend tests on notched beams TC 50 Technical Committee on Fracture Mechanics of Concrete, Draft Recommendation. *Materials & Structures*, 18:285–290
14. Cognard J (1986) *Journal of Adhesion*, 20:1–13.
15. Cognard J (1987) *Journal of Adhesion*, 22:97–108.
16. Crosley P, Ripling E (1991) *Journal of Testing and Evaluation*, 19:24–28.
17. ASTM (1993) Standard test method for fracture strength in cleavage of adhesives in bonded joints. Technical Report D3433-93. American Society for Testing and Materials.
18. Massabò R, Cox B (1999) *Journal of the Mechanics and Physics of Solids*, 47:1265–1300.
19. Kanninen M (1973) *International Journal of Fracture*, 9:83–92.
20. Williams J, Hadavinia H (2002) *Journal of the Mechanics and Physics of Solids*, 50:809–825.
21. Qiao P, Wang J, Davalos J (2003) *International Journal of Solids and Structures*, 40:1865–1884.
22. Li S, Wang J, Thouless M (2004) *Journal of the Mechanics and Physics of Solids*, 52:193–214.
23. Cannarozzi A, Tralli A (1984) *International Journal of Solids and Structures*, 20:179–189.

24. Andrews M, Massabò R, Cox B (2004) In *Proceedings of XXI International Congress of Theoretical and Applied Mechanics ICTAM*, Warsaw, Poland.
25. Point N, Sacco E (1996) *International Journal of Fracture*, 79:225–247.
26. Tenchev R, Falzon B (2006) *Finite Elements in Analysis and Design*, 42:698–708.
27. Iannucci L (2006) *Computers & Structures*, 84:1029–1048.
28. Ortiz M, Pandolfi A (1999) *International Journal for Numerical Methods in Engineering*, 44:1267–1282.
29. Bardis J, Kedward K (2002) *Journal of Composites Technology and Research*, 24:30–37.
30. Wriggers P (2002) *Computational Contact Mechanics*, John Wiley & Sons, England.
31. Paggi M, Carpinteri A, Zavarise G (2006) In Wriggers P, Nackenhorst U (Eds.), *Analysis and Simulation of Contact Problems*, Lecture Notes in Applied and Computational Mechanics, Vol. 27. Springer-Verlag, Berlin, pp. 297–304.
32. Tvergaard V (1990) *Material Science and Engineering A*, 107:23–40.
33. Zavarise G, Boso D, Schrefler B (2001) In *Proceedings of CMIS, III Contact Mechanics International Symposium*, Praja de Consolacao, Portugal, pp. 211–218.
34. Zavarise G, Schrefler B, Wriggers P (1992) In Owen D, Onate E, Hinton E (Eds.), *Proc. COMPLASS III, Int. Conf. on Comput. Plasticity*, Pineridge Press, Barcelona, Spain, pp. 349–360.
35. Zavarise G, Borri-Brunetto M, Paggi M (2004) *Wear*, 257:229–245.
36. Zavarise G, Borri-Brunetto M, Paggi M (2007) *Wear*, 262:42–54.
37. Zavarise G, Wriggers P, Stein E, Schrefler B (1992) *International Journal for Numerical Methods in Engineering*, 35:767–785.
38. Paggi M, Carpinteri A, Zavarise G (2006) In *Proceedings of the XVI Convegno Italiano di Meccanica Computazionale (GIMC)*, Bologna, Italy, full paper on CD-ROM.
39. Carpinteri A, Monetto I (1999) *International Journal of Fracture*, 98:225–241.
40. Carpinteri A (Ed.) (1999) *Nonlinear Crack Models for Nonmetallic Materials*. Kluwer Academic Publishers, Dordrecht.
41. Carpinteri A, Massabò R (1996) *International Journal of Fracture*, 81:125–145.

Smooth Frictional Contact between Beams in 3D

Przemysław Litewka

Institute of Structural Engineering, Poznań University of Technology,
ul. Piotrowo 5, 60-965 Poznań, Poland
E-mail: przemyslaw.litewka@ikb.poznan.pl

Abstract. In this paper a smoothing procedure for the 3D beam-to-beam contact is presented. A smooth segment is based on current position vectors of three nodes for two adjacent finite elements. The approximated fragment of a 3D curve modeling a beam axis spans between the centre points of these elements. The curve is described parametrically using three Hermite polynomials. The four boundary conditions used to determine the coefficients for each of these polynomials involve co-ordinates and slopes at the curve ends. The slopes are defined in terms of the element nodal co-ordinates, too, so there is no dependence on nodal rotations and this formulation can be embedded in a beam analysis using any type of beam finite element. This geometric representation of the curve is incorporated into the 3D beam-to-beam frictional contact model with the penalty method used to enforce contact constraints. The residual vector and the corresponding tangent stiffness matrix are determined for the normal part of contact and for the stick or slip state of friction. A numerical example is presented to show the performance of the suggested smoothing procedure in a case of large frictional sliding.

Key words: 3D curve smoothing, Hermite polynomials, frictional contact, consistent linearisation.

1 Introduction

A proper representation of geometry of bodies in contact and sliding is a vital aspect of the FEM modeling. It is necessary to assure the continuity and smoothness. If a too simplified finite element mesh is taken then between adjacent elements a lack of continuity of normal and tangent vectors, which form a base in the contact formulation, takes place. In the consequence if a contact point moves from one element to another convergence of the iterative Newton–Raphson scheme is deteriorated. In some cases, when the number of contact points changing an element is relatively high, then the convergence may not be reached at all. Such a situation is very probable in a beam-to-beam scenario where a point-wise contact is assumed and a number of contact

Peter Wriggers and Udo Nackenhorst (eds.), IUTAM Symposium on Computational Contact Mechanics, 157–176.

spots is low. So the ensuring of smooth curves for contacting beam axes or edges is even more important than in the other cases of contacting bodies.

The bibliography concerning the smoothing procedures for 2D and 3D bodies is quite broad. The mathematical background for these methods is given in [3]. The 2D case was considered in [2] where third order splines were used, [10] with NURBS applied or in [11] with Bezier curves. 3D smoothing requires more sophisticated procedures and more computer power. This problem was solved using surface facets based on Bezier functions [4] or Gregory patches [9].

In most cases the smooth representation of surfaces leads to very complex formulae in contact contributions. It is especially pronounced in 3D cases. Hence, the derivation requires an application of symbolic algebra programs. One of such packages is SMS, based on Mathematica, which was used in [4].

In this paper a simple smoothing procedure for 3D curves is presented. It involves curve segments constructed over two adjacent elements, similarly as in [11] for 2D solids. These segments are approximated using Hermite polynomials. The advantage of this formulation is that it does not include angles of rotations in beam nodes, so it can be easily combined with any, even the simplest, beam element used on the analysis. The details are described in Section 2. Section 3 is devoted to a brief summary of the FEM formulation of a frictional beam-to-beam contact from [5–7, 12, 13] which is discretised using the smooth curve segments. Coulomb model of friction and analogy to plasticity [8] is used therein. In Section 4 a numerical example showing a good performance of the suggested approach is presented. The paper is concluded in Section 5. In the Appendix, Maple V worksheets required to obtain Fortran codes for a generation of some basic matrices used in the contact formulation are given.

2 3D Curve Smoothing Using Hermite Polynomials

The method of 3D curve smoothing presented here is a modification of an approach used to smooth 2D solids described in [11]. The curve consists of several segments. Each of them is constructed on two adjacent beam elements, as shown in Figure 1. C^1-continuity is ensured by four boundary conditions, which force the curve to pass through points C_{12} and C_{23}, being the central points of straight line sections defined by element nodes, 1, 2 and 3, and to be tangent to these sections. The curve is represented parametrically using the local dimensionless coordinate

$$-1 \leq \xi \leq 1. \tag{1}$$

An arbitrary point A lying on the curve is defined by its position vector

$$\mathbf{x} = (x_1, x_2, x_3)^T \tag{2}$$

and each of its three coordinates can be expressed in the form of a Hermite polynomial as a scalar product of vectors

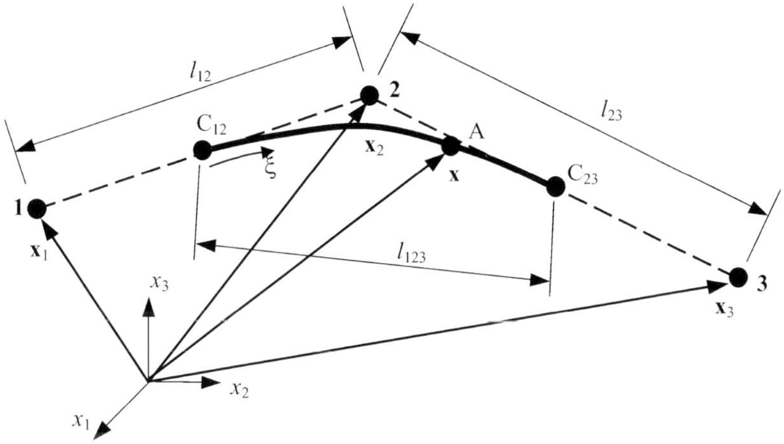

Fig. 1. Construction of a smooth curve segment.

$$x_i = (a_i, b_i, c_i, d_i) \cdot (\xi^3, \xi^2, \xi, 1) = \mathbf{a}_i \cdot \boldsymbol{\xi}. \tag{3}$$

The position vectors of element nodes 1, 2 and 3 are split into their coordinates with the following notation:

$$\begin{aligned} \mathbf{x}_1 &= (x_{11}, x_{12}, x_{13})^T = (x_{1i})^T, \\ \mathbf{x}_2 &= (x_{21}, x_{22}, x_{23})^T = (x_{2i})^T, \\ \mathbf{x}_3 &= (x_{31}, x_{32}, x_{33})^T = (x_{3i})^T. \end{aligned} \tag{4}$$

The boundary conditions for the curve (3), which yield the C^1-continuity forcing the curve to pass through the points C_{12} and C_{23} and to be tangent to the straight lines connecting the nodes, read

$$\begin{aligned} &\xi = -1 \Rightarrow x_i = x_{c12i} = \frac{x_{2i} + x_{1i}}{2}, \\ &\xi = 1 \Rightarrow x_i = x_{c23i} = \frac{x_{3i} + x_{2i}}{2}, \\ &\xi = -1 \Rightarrow \frac{\partial x_i}{\partial \xi} = \varphi_{c12i} = \frac{x_{2i} - x_{1i}}{2l_{12}} l_{123}, \\ &\xi = 1 \Rightarrow \frac{\partial x_i}{\partial \xi} = \varphi_{c23i} = \frac{x_{3i} - x_{2i}}{2l_{23}} l_{123}. \end{aligned} \tag{5}$$

The formulae $(5)_3$ and $(5)_4$ include straight line lengths, which appear due to the relation between the dimensionless coordinate ξ and the global coordinates x_1, x_2

and x_3. These length are shown in Figure 1 and can be calculated from the following relations

$$
\begin{aligned}
l_{12} &= \|\mathbf{x}_2 - \mathbf{x}_1\|, \\
l_{23} &= \|\mathbf{x}_3 - \mathbf{x}_2\|, \\
l_{123} &= \left\| \frac{\mathbf{x}_1 + \mathbf{x}_2}{2} - \frac{\mathbf{x}_2 + \mathbf{x}_3}{2} \right\|.
\end{aligned} \tag{6}
$$

The substitution of coordinates and their derivatives yielding from (3) to the boundary conditions (5) gives a set of four equations for each of three coordinates x_i in (3). Their solution provides three sets of coefficients of Hermite polynomials. They can be expressed in terms of position vectors and their derivatives for the central points C_{12} and C_{23}

$$
\mathbf{a}_i = \frac{1}{4}\begin{bmatrix} 1 & -1 & 1 & 1 \\ 0 & 0 & -1 & 1 \\ -3 & 3 & -1 & -1 \\ 2 & 2 & 1 & -1 \end{bmatrix}\begin{bmatrix} x_{c12i} \\ x_{c23i} \\ \varphi_{c12i} \\ \varphi_{c12i} \end{bmatrix} = \mathbf{A}_1 \begin{bmatrix} x_{c12i} \\ x_{c23i} \\ \varphi_{c12i} \\ \varphi_{c12i} \end{bmatrix} \tag{7}
$$

or in terms of the nodal position vectors (4)

$$
\mathbf{a}_i = \mathbf{A}_1 \frac{1}{2}\begin{bmatrix} 1 & 1 & 0 \\ 0 & 1 & 1 \\ -\frac{l_{123}}{l_{12}} & \frac{l_{123}}{l_{12}} & 0 \\ 0 & -\frac{l_{123}}{l_{23}} & \frac{l_{123}}{l_{23}} \end{bmatrix}\begin{bmatrix} x_{1i} \\ x_{2i} \\ x_{3i} \end{bmatrix} = \mathbf{A}_1\mathbf{A}_2 \begin{bmatrix} x_{1i} \\ x_{2i} \\ x_{3i} \end{bmatrix}. \tag{8}
$$

The vector of coefficients $\mathbf{a}_i$ can now be substituted into (3) to yield the final form of the approximation for each of three coordinates of the position vector (2)

$$
x_i = \boldsymbol{\xi} \cdot [\mathbf{A}_1 \;\; \mathbf{A}_2 \;\; (x_{1i}, x_{2i}, x_{3i})^T]. \tag{9}
$$

It must be noted that this approximation is not linear in the nodal coordinates since the matrix $\mathbf{A}_2$ depends non-linearly on them via lengths (6). It has such a consequence that extra terms in formulation, i.e. linearisation of variation of displacements $\Delta\delta\mathbf{u}$ must be included [7].

On the other hand the approximation (9) is very attractive because it does not involve nodal rotations and therefore can be incorporated easily in any beam analysis program using even the simplest beam elements.

In the further analysis derivatives of the coordinates x_i will be necessary. They can be expressed as

$$
\begin{aligned}
x_{i,\xi} &= \boldsymbol{\xi}' \cdot [\mathbf{A}_1 \;\; \mathbf{A}_2 \;\; (x_{1i}, x_{2i}, x_{3i})^T] \\
x_{i,\xi\xi} &= \boldsymbol{\xi}'' \cdot [\mathbf{A}_1 \;\; \mathbf{A}_2 \;\; (x_{1i}, x_{2i}, x_{3i})^T] \\
x_{i,\xi\xi\xi} &= \boldsymbol{\xi}''' \cdot [\mathbf{A}_1 \;\; \mathbf{A}_2 \;\; (x_{1i}, x_{2i}, x_{3i})^T]
\end{aligned} \tag{10}
$$

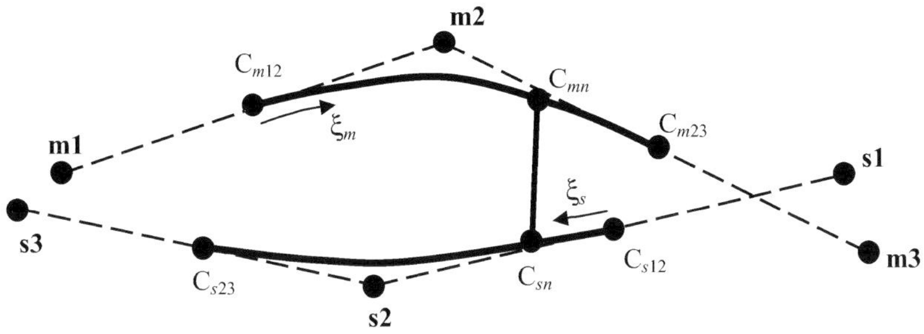

Fig. 2. Smooth beam-to-beam contact element.

and the vectors of derivatives of local coordinate ξ are

$$\begin{aligned}\boldsymbol{\xi}' &= (3\xi^2, 2\xi, 1, 0)^T,\\ \boldsymbol{\xi}'' &= (6\xi, 2, 0, 0)^T,\\ \boldsymbol{\xi}''' &= (6, 0, 0, 0)^T.\end{aligned} \tag{11}$$

3 Frictional Contact Formulation

The formulation of contact used here is the development of what has been derived for beams with circular cross-section [12, 13], rectangular cross-section [5, 6] and modified in [7]. The interested reader may find the details of the derivation in these papers. Here only the final matrix form of contact contributions after the FEM discretisation will be given.

The beam-to-beam contact element based on the 3D curve approximation described in Section 2 is shown in Figure 2. The vector of unknowns for this element includes 18 displacements

$$\begin{aligned}\mathbf{q} &= (u_{m11}, u_{m12}, u_{m13}, u_{m21}, u_{m22}, u_{m23}, u_{m31}, u_{m32}, u_{m33},\\ &\qquad u_{s11}, u_{s12}, u_{s13}, u_{s21}, u_{s22}, u_{s23}, u_{s31}, u_{s32}, u_{s33})^T\\ &= (\mathbf{u}_M^T, \mathbf{u}_S^T)^T,\end{aligned} \tag{12}$$

3 for each of 6 nodes of four beam elements forming the base for smooth segments on each of two contacting beams.

The contact points C_{mn} and C_{sn}, shown in Figure 2, are defined by their position vectors

$$\begin{aligned}\mathbf{x}_{mn} &= (x_{mn1}, x_{mn2}, x_{mn3})^T,\\ \mathbf{x}_{sn} &= (x_{sn1}, x_{sn2}, x_{sn3})^T.\end{aligned} \tag{13}$$

Their difference is denoted by

$$\mathbf{x}_{ms} = \mathbf{x}_{mn} - \mathbf{x}_{sn} = (x_{ms1}, x_{ms2}, x_{ms3})^T \tag{14}$$

and the normal unit vector has the form

$$\mathbf{n} = \frac{\mathbf{x}_{mn} - \mathbf{x}_{sn}}{\|\mathbf{x}_{mn} - \mathbf{x}_{sn}\|} = (n_1, n_2, n_3)^T. \tag{15}$$

The starting point is the FEM representation of variation and linearisation of displacements and their derivatives with respect to local coordinates in the contact points C_{mn} and C_{sn}. These kinematic variables are

$$\begin{aligned}
\Delta\mathbf{u}_{mn} &= \mathbf{G}_{mn}\Delta\mathbf{u}_M, \quad \Delta\mathbf{u}_{sn} = \mathbf{G}_{sn}\Delta\mathbf{u}_S,\\
\delta\mathbf{u}_{mn} &= \mathbf{G}_{mn}\delta\mathbf{u}_M, \quad \delta\mathbf{u}_{sn} = \mathbf{G}_{sn}\delta\mathbf{u}_S,\\
\Delta\mathbf{u}_{mn,m} &= \mathbf{H}_{mn}\Delta\mathbf{u}_M, \quad \Delta\mathbf{u}_{sn,s} = \mathbf{H}_{sn}\Delta\mathbf{u}_S,\\
\delta\mathbf{u}_{mn,m} &= \mathbf{H}_{mn}\delta\mathbf{u}_M, \quad \delta\mathbf{u}_{sn,s} = \mathbf{H}_{sn}\delta\mathbf{u}_S,\\
\Delta\mathbf{u}_{mn,mm} &= \mathbf{M}_{mn}\Delta\mathbf{u}_M, \quad \Delta\mathbf{u}_{sn,sn} = \mathbf{M}_{sn}\Delta\mathbf{u}_S,\\
\delta\mathbf{u}_{mn,mm} &= \mathbf{M}_{mn}\delta\mathbf{u}_M, \quad \delta\mathbf{u}_{sn,sn} = \mathbf{M}_{sn}\delta\mathbf{u}_S.
\end{aligned} \tag{16}$$

Matrices **G**, **H** and **M** with 3×9 dimension can be obtained from the curve approximation (9) and (10) by a partial differentiation with respect to the nodal displacements. This mathematical operations were carried out using the symbolic algebra program Maple 7. The explicit form of matrices is very complicated. For instance, a resulting Fortran code for matrix **G** has over 2300 lines. So, instead of presenting it here, in the Appendix commands of Maple worksheets are given, which lead to the ready-to-use Fortran codes enabling the calculation of these matrices.

Linearisation and variation of local coordinates require definition of a matrix **A**

$$\mathbf{A} = \begin{bmatrix} \mathbf{x}_{mn,m}\cdot\mathbf{x}_{mn,m} + \mathbf{x}_{ms}\cdot\mathbf{x}_{mn,mm} & -\mathbf{x}_{mn,m}\cdot\mathbf{x}_{sn,s} \\ \mathbf{x}_{mn,m}\cdot\mathbf{x}_{sn,s} & -\mathbf{x}_{sn,s}\cdot\mathbf{x}_{sn,s} + \mathbf{x}_{ms,s}\cdot\mathbf{x}_{sn,ss} \end{bmatrix} \tag{17}$$

and are described by the following relations

$$\begin{aligned}
\begin{bmatrix}\Delta\xi_{mn}\\ \Delta\xi_{sn}\end{bmatrix} &= \mathbf{A}^{-1}\begin{bmatrix} -\mathbf{x}^T_{mn,m}\mathbf{G}_{mn} - \mathbf{x}^T_{ms}\mathbf{H}_{mn} & \mathbf{x}^T_{mn,m}\mathbf{G}_{sn} \\ -\mathbf{x}^T_{sn,s}\mathbf{G}_{mn} & (\mathbf{x}^T_{sn,s}\mathbf{G}_{sn} - \mathbf{x}^T_{ms}\mathbf{H}_{sn}) \end{bmatrix}\\
&\times\begin{bmatrix}\Delta\mathbf{u}_M\\ \Delta\mathbf{u}_S\end{bmatrix} = \mathbf{F}\begin{bmatrix}\Delta\mathbf{u}_M\\ \Delta\mathbf{u}_S\end{bmatrix} = \begin{bmatrix}\mathbf{F}_M\\ \mathbf{F}_S\end{bmatrix}\begin{bmatrix}\Delta\mathbf{u}_M\\ \Delta\mathbf{u}_S\end{bmatrix},\\
\begin{bmatrix}\delta\xi_{mn}\\ \delta\xi_{sn}\end{bmatrix} &= \mathbf{F}\begin{bmatrix}\delta\mathbf{u}_M\\ \delta\mathbf{u}_S\end{bmatrix} = \begin{bmatrix}\mathbf{F}_M\\ \mathbf{F}_S\end{bmatrix}\begin{bmatrix}\delta\mathbf{u}_M\\ \delta\mathbf{u}_S\end{bmatrix}.
\end{aligned} \tag{18}$$

Linearisations of local coordinate variations are obtained from

$$\mathbf{A}\begin{bmatrix}\Delta\delta\xi_{mn}\\ \Delta\delta\xi_{sn}\end{bmatrix}+\left[\delta\mathbf{u}_M^T,\delta\mathbf{u}_S^T\right]\begin{bmatrix}\mathbf{R}_m\\ \mathbf{R}_s\end{bmatrix}\begin{bmatrix}\Delta\mathbf{u}_M\\ \Delta\mathbf{u}_S\end{bmatrix}=\begin{bmatrix}0\\0\end{bmatrix} \tag{19}$$

and the involved matrices have the form

$$\begin{aligned}
\mathbf{R}_m &= p_2\cdot\mathbf{F}_s^T\mathbf{F}_s+p_3\cdot(\mathbf{F}_m^T\mathbf{F}_s+\mathbf{F}_s^T\mathbf{F}_m)+p_4\cdot\mathbf{F}_m^T\mathbf{F}_m\\
&\quad+\mathbf{F}_s^T\mathbf{Y}_{11}^T+\mathbf{F}_m^T\mathbf{Y}_{12}^T+\mathbf{Y}_{11}\mathbf{F}_s-\mathbf{Y}_{12}\mathbf{F}_m\\
&\quad+\begin{bmatrix}(\mathbf{H}_{mm}^T\mathbf{G}_{mn}+\mathbf{G}_{mn}^T\mathbf{H}_{mn}) & -\mathbf{H}_{mn}^T\mathbf{G}_{sn}\\ -\mathbf{G}_{sn}^T\mathbf{H}_{mn} & 0\end{bmatrix}\\
&\quad+\begin{bmatrix}\left(\sum_{j=1}^{3}x_{mn,mj}\mathbf{G}_{djmn}+\sum_{j=1}^{3}x_{msj}\mathbf{H}_{djmn}\right) & \mathbf{0}\\ \mathbf{0} & -\sum_{j=1}^{3}x_{mn,mj}\mathbf{G}_{djsn}\end{bmatrix},\\
\mathbf{R}_s &= p_1\cdot\mathbf{F}_s^T\mathbf{F}_s+p_2\cdot(\mathbf{F}_m^T\mathbf{F}_s+\mathbf{F}_s^T\mathbf{F}_m)+p_3\cdot\mathbf{F}_m^T\mathbf{F}_m\\
&\quad+\mathbf{F}_s^T\mathbf{Y}_{21}^T-\mathbf{F}_m^T\mathbf{Y}_{11}^T+\mathbf{Y}_{21}\mathbf{F}_s-\mathbf{Y}_{11}\mathbf{F}_m\\
&\quad+\begin{bmatrix}\mathbf{0} & \mathbf{G}_{mn}^T\mathbf{H}_{sn}\\ \mathbf{H}_{sn}^T\mathbf{G}_{mm} & (-\mathbf{H}_{sn}^T\mathbf{G}_{sn}-\mathbf{G}_{sn}^T\mathbf{H}_{sn})\end{bmatrix}\\
&\quad+\begin{bmatrix}\sum_{j=1}^{3}x_{sn,sj}\mathbf{G}_{djmn} & \mathbf{0}\\ \mathbf{0} & \left(-\sum_{j=1}^{3}x_{sn,sj}\mathbf{G}_{djsn}+\sum_{j=1}^{3}x_{msj}\mathbf{H}_{djsn}\right)\end{bmatrix}.
\end{aligned} \tag{20}$$

The simplifying notation used in (20) is

$$\begin{aligned}
p_1 &= \mathbf{x}_{sn,sss}\cdot\mathbf{x}_{ms}-3\mathbf{x}_{sn,ss}\cdot\mathbf{x}_{sn,s},\\
p_2 &= \mathbf{x}_{sn,ss}\cdot\mathbf{x}_{mn,m},\\
p_3 &= \mathbf{x}_{sn,s}\cdot\mathbf{x}_{mn,mm},\\
p_4 &= \mathbf{x}_{mn,mmm}\cdot\mathbf{x}_{ms}+3\mathbf{x}_{mn,mm}\cdot\mathbf{x}_{mn,m},
\end{aligned} \tag{21}$$

$$\mathbf{Y}_{11}=\begin{bmatrix}-\mathbf{H}_{mn}^T\mathbf{x}_{mn,m}\\ -\mathbf{H}_{sn}^T\mathbf{x}_{sn,n}\end{bmatrix},$$

$$\mathbf{Y}_{12} = \begin{bmatrix} \mathbf{M}_{mn}^T \mathbf{x}_{ms} + 2\mathbf{H}_{mn}^T \mathbf{x}_{mn,m} + \mathbf{G}_{mn}^T \mathbf{x}_{mn,mm} \\ -\mathbf{G}_{sn}^T \mathbf{x}_{mn,mm} \end{bmatrix},$$

$$\mathbf{Y}_{21} = \begin{bmatrix} \mathbf{G}_{mn}^T \mathbf{x}_{sn,ss} \\ \mathbf{M}_{sn}^T \mathbf{x}_{ms} - 2\mathbf{H}_{sn}^T \mathbf{x}_{sn,s} - \mathbf{G}_{sn}^T \mathbf{x}_{sn,ss} \end{bmatrix}, \tag{22}$$

and matrices $\mathbf{G}_{djmn}$, $\mathbf{G}_{djsn}$, $\mathbf{H}_{djmn}$, $\mathbf{H}_{djsn}$ follow from the linearisation of variation of displacements. These operations can be performed using small perturbations and the finite difference scheme to matrices $\mathbf{G}$ and $\mathbf{H}$ and lead to

$$\Delta\delta\mathbf{u}_{mn} = \begin{bmatrix} \delta\mathbf{u}_M^T \mathbf{G}_{d1mn} \Delta\mathbf{u}_M \\ \delta\mathbf{u}_M^T \mathbf{G}_{d2mn} \Delta\mathbf{u}_M \\ \delta\mathbf{u}_M^T \mathbf{G}_{d3mn} \Delta\mathbf{u}_M \end{bmatrix},$$

$$\Delta\delta\mathbf{u}_{sn} = \begin{bmatrix} \delta\mathbf{u}_S^T \mathbf{G}_{d1sn} \Delta\mathbf{u}_S \\ \delta\mathbf{u}_S^T \mathbf{G}_{d2sn} \Delta\mathbf{u}_S \\ \delta\mathbf{u}_S^T \mathbf{G}_{d3sn} \Delta\mathbf{u}_S \end{bmatrix},$$

$$\Delta\delta\mathbf{u}_{mn,m} = \begin{bmatrix} \delta\mathbf{u}_M^T \mathbf{H}_{d1mn} \Delta\mathbf{u}_M \\ \delta\mathbf{u}_M^T \mathbf{H}_{d2mn} \Delta\mathbf{u}_M \\ \delta\mathbf{u}_M^T \mathbf{H}_{d3mn} \Delta\mathbf{u}_M \end{bmatrix},$$

$$\Delta\delta\mathbf{u}_{sn,s} = \begin{bmatrix} \delta\mathbf{u}_S^T \mathbf{H}_{d1sn} \Delta\mathbf{u}_S \\ \delta\mathbf{u}_S^T \mathbf{H}_{d2sn} \Delta\mathbf{u}_S \\ \delta\mathbf{u}_S^T \mathbf{H}_{d3sn} \Delta\mathbf{u}_S \end{bmatrix}. \tag{23}$$

In the friction formulation also some kinematic variables defined in the previous contact points C_{mp} and C_{sp} are necessary. They can be expressed in the following form:

$$\begin{aligned} \Delta\mathbf{u}_{mp} &= \mathbf{G}_{mp} \Delta\mathbf{u}_M, \\ \Delta\mathbf{u}_{sp} &= \mathbf{G}_{sp} \Delta\mathbf{u}_S, \\ \delta\mathbf{u}_{mp} &= \mathbf{G}_{mp} \delta\mathbf{u}_M, \\ \delta\mathbf{u}_{sp} &= \mathbf{G}_{sp} \delta\mathbf{u}_S, \end{aligned} \tag{24}$$

$$\Delta\delta\mathbf{u}_{mp} = \begin{bmatrix} \delta\mathbf{u}_M^T \mathbf{G}_{d1mp} \Delta\mathbf{u}_M \\ \delta\mathbf{u}_M^T \mathbf{G}_{d2mp} \Delta\mathbf{u}_M \\ \delta\mathbf{u}_M^T \mathbf{G}_{d3mp} \Delta\mathbf{u}_M \end{bmatrix},$$

$$\Delta\delta\mathbf{u}_{sp} = \begin{bmatrix} \delta\mathbf{u}_S^T \mathbf{G}_{d1sp} \Delta\mathbf{u}_S \\ \delta\mathbf{u}_S^T \mathbf{G}_{d2sp} \Delta\mathbf{u}_S \\ \delta\mathbf{u}_S^T \mathbf{G}_{d3sp} \Delta\mathbf{u}_S \end{bmatrix}. \tag{25}$$

The matrices in (24) are obtained in the same way as those in (16), while the matrices in (25) can be calculated as those in (23).

The previous contact points are defined by their position vectors

$$\begin{aligned} \mathbf{x}_{mp} &= (x_{mp1}, x_{mp2}, x_{mp3})^T, \\ \mathbf{x}_{sp} &= (x_{sp1}, x_{sp2}, x_{sp3})^T, \end{aligned} \tag{26}$$

and there are also unit tangent vectors introduced

$$\begin{aligned} \mathbf{t}_m &= \frac{\mathbf{x}_{mn} - \mathbf{x}_{mp}}{\|\mathbf{x}_{mn} - \mathbf{x}_{mp}\|} = (t_{m1}, t_{m2}, t_{m3})^T, \\ \mathbf{t}_s &= \frac{\mathbf{x}_{sn} - \mathbf{x}_{sp}}{\|\mathbf{x}_{sn} - \mathbf{x}_{sp}\|} = (t_{s1}, t_{s2}, t_{s3})^T. \end{aligned} \tag{27}$$

To simplify the notation in tangent stiffness matrices and residual vectors a number of another auxiliary matrices are introduced

$$\mathbf{L} = [\mathbf{G}_{mn} \quad -\mathbf{G}_{sn}] + [\mathbf{x}_{mn,m} \quad -\mathbf{x}_{sn,s}]\mathbf{F}, \tag{28}$$

$$\begin{aligned} \mathbf{S}_m &= \mathbf{x}_{mn,m}\mathbf{F}_m + [(\mathbf{G}_{mn} - \mathbf{G}_{mp}) \quad \mathbf{0}], \\ \mathbf{S}_s &= \mathbf{x}_{sn,s}\mathbf{F}_s + [\mathbf{0} \quad (\mathbf{G}_{sn} - \mathbf{G}_{sp})], \end{aligned} \tag{29}$$

$$\begin{aligned} \mathbf{Z}_m &= \mathbf{F}_m^T \mathbf{t}_m^T [\mathbf{H}_{mn} \quad \mathbf{0}], \\ \mathbf{Z}_s &= \mathbf{F}_s^T \mathbf{t}_s^T [\mathbf{0} \quad \mathbf{H}_{sn}], \end{aligned} \tag{30}$$

$$\begin{aligned} \mathbf{W}_m &= \begin{bmatrix} \sum_{j=1}^{3} t_{mj}(\mathbf{G}_{djmn} - \mathbf{G}_{dmjp}) & \mathbf{0} \\ \mathbf{0} & \mathbf{0} \end{bmatrix}, \\ \mathbf{W}_s &= \begin{bmatrix} \mathbf{0} & \mathbf{0} \\ \mathbf{0} & \sum_{j=1}^{3} t_{sj}(\mathbf{G}_{djsn} - \mathbf{G}_{dsjp}) \end{bmatrix}. \end{aligned} \tag{31}$$

The friction formulation involves also linearisation of proportion parameters [7]. To define it the following notation is introduced: additive split of tangential gaps,

g_{Tmn} and g_{Tsn}, into stick parts, g^e_{Tmn} and g^e_{Tsn}, and slip parts, g^p_{Tmn} and g^p_{Tsn}; trial friction force $\mathbf{F}^t_T$ and its magnitude F^t_T and the penalty parameters for friction ε_{Tm} and ε_{Ts}. The proportion parameters are calculated as ratios of friction force components in direction of respective tangent vectors, $\mathbf{t}_m$ and $\mathbf{t}_s$, to the resultant magnitude

$$p_m = \frac{F^t_{Tmn}}{F^t_T},$$

$$p_s = \frac{F^t_{Tsn}}{F^t_T}. \tag{32}$$

With these in hand, the matrices used to define Δp_m and Δp_s can be written as

$$\begin{aligned}
\mathbf{P}_m = \Bigg\{ & \left[\frac{\varepsilon_{Tm}}{F^t_T} - \frac{p_m \varepsilon_{Tm}}{(F^t_T)^2}\mathbf{F}^t_T \cdot \mathbf{t}_m\right]\mathbf{t}^T_m \mathbf{S}_m \\
& - \frac{p_m \varepsilon_{Tm} g^e_{Tmn}}{(F^t_T)^2 \|\mathbf{x}_{mn} - \mathbf{x}_{mp}\|}\mathbf{F}^t_T(\mathbf{1} - \mathbf{t}_m \otimes \mathbf{t}_m)\mathbf{S}_m - \frac{p_m \varepsilon_{Ts}}{(F^t_T)^2}(\mathbf{F}^t_T \cdot \mathbf{t}_s)\mathbf{t}^T_s \mathbf{S}_s \\
& - \frac{p_m \varepsilon_{Ts} g^e_{Tsn}}{(F^t_T)^2 \|\mathbf{x}_{sn} - \mathbf{x}_{sp}\|}\mathbf{F}^t_T(\mathbf{1} - \mathbf{t}_s \otimes \mathbf{t}_s)\mathbf{S}_s \Bigg\}, \\
\mathbf{P}_s = \Bigg\{ & \left[\frac{\varepsilon_{Ts}}{F^t_T} - \frac{p_s \varepsilon_{Ts}}{(F^t_T)^2}\mathbf{F}^t_T \cdot \mathbf{t}_s\right]\mathbf{t}^T_s \mathbf{S}_s \\
& - \frac{p_s \varepsilon_{Tm} g^e_{Tmn}}{(F^t_T)^2 \|\mathbf{x}_{mn} - \mathbf{x}_{mp}\|}\mathbf{F}^t_T(\mathbf{1} - \mathbf{t}_m \otimes \mathbf{t}_m)\mathbf{S}_m - \frac{p_s \varepsilon_{Tm}}{(F^t_T)^2}(\mathbf{F}^t_T \cdot \mathbf{t}_m)\mathbf{t}^T_m \mathbf{S}_m \\
& - \frac{p_s \varepsilon_{Ts} g^e_{Tsn}}{(F^t_T)^2 \|\mathbf{x}_{sn} - \mathbf{x}_{sp}\|}\mathbf{F}^t_T(\mathbf{1} - \mathbf{t}_s \otimes \mathbf{t}_s)\mathbf{S}_s \Bigg\}.
\end{aligned} \tag{33}$$

The normal contact part of the residual vector for the contact element has the form

$$\mathbf{R}^\varepsilon_N = \varepsilon_N g_N \mathbf{R}_1, \tag{34}$$

with the standard notation for normal gap g_N and normal contact penalty parameter ε_N. The vector $\mathbf{R}_1$ is

$$\mathbf{R}_1 = \begin{bmatrix} \mathbf{G}^T_{mn}\mathbf{n} \\ -\mathbf{G}^T_{sn}\mathbf{n} \end{bmatrix}. \tag{35}$$

The friction part of the residual vector for the stick case is

$$\mathbf{R}^{e\varepsilon}_T = \varepsilon_{Tm} g^e_{Tmn} \mathbf{R}_2 + \varepsilon_{Ts} g^e_{Tsn} \mathbf{R}_3 \tag{36}$$

and for the slip case it is

$$\mathbf{R}^{p\varepsilon}_T = \mu \varepsilon_N g_N (p_m s_m \mathbf{R}_2 + p_s s_s \mathbf{R}_3), \tag{37}$$

where μ is the Coulomb friction coefficient, s_m and s_s are sliding direction controls (see [7]). The auxiliary vectors in (36) and (37) are

$$\begin{aligned}\mathbf{R}_2 &= \mathbf{S}_m^T \mathbf{t}_m,\\ \mathbf{R}_3 &= \mathbf{S}_s^T \mathbf{t}_s.\end{aligned} \tag{38}$$

The tangent stiffness matrix for the normal part of contact has the form

$$\mathbf{K}_N^{\varepsilon} = \varepsilon_N \mathbf{K}_1 + \varepsilon_N g_N \mathbf{K}_2, \tag{39}$$

with the submatrices defined as

$$\begin{aligned}\mathbf{K}_1 &= \mathbf{R}_1 \otimes \mathbf{R}_2,\\ \mathbf{K}_2 &= \begin{bmatrix} \mathbf{H}_{mn}^T \mathbf{n} & \mathbf{0} \\ 0 & -\mathbf{H}_{sn}^T \mathbf{n} \end{bmatrix} \mathbf{F} + \mathbf{F}^T \begin{bmatrix} \mathbf{n}^T \mathbf{H}_{mn} & \mathbf{0} \\ \mathbf{0} & -\mathbf{n}^T \mathbf{H}_{sn} \end{bmatrix}\\ &+ \begin{bmatrix} \mathbf{G}_{nm} & \mathbf{0} \\ \mathbf{0} & -\mathbf{G}_{ns} \end{bmatrix} + \mathbf{F}^T \begin{bmatrix} \mathbf{n}^T \mathbf{x}_{mn,mm} & \mathbf{0} \\ \mathbf{0} & -\mathbf{n}^T \mathbf{x}_{sn,ss} \end{bmatrix} \mathbf{F}\\ &+ \frac{1}{g_N} \mathbf{L}^T (\mathbf{1} - \mathbf{n} \cdot \mathbf{n}^T) \mathbf{L}.\end{aligned} \tag{40}$$

The friction contribution to the tangent stiffness matrix in the stick case is

$$\mathbf{K}_T^{e\varepsilon} = \varepsilon_{Tm} \mathbf{R}_2 \otimes \mathbf{R}_2 + \varepsilon_{Tm} g_{Tmn}^e \mathbf{K}_3 + \varepsilon_{Ts} \mathbf{R}_3 \otimes \mathbf{R}_3 + \varepsilon_{Ts} g_{Tsn}^e \mathbf{K}_4, \tag{41}$$

and in the slip case

$$\begin{aligned}\mathbf{K}_T^{p\varepsilon} &= \mu \varepsilon_N p_m s_m \mathbf{R}_2 \otimes \mathbf{R}_1 + \mu \varepsilon_N g_N p_m s_m \mathbf{K}_3 + \mu \varepsilon_N g_N s_m \mathbf{R}_2 \otimes \mathbf{P}_m\\ &+ \mu \varepsilon_N p_s s_s \mathbf{R}_3 \otimes \mathbf{R}_1 + \mu \varepsilon_N g_N p_s s_s \mathbf{K}_4 + \mu \varepsilon_N g_N s_s \mathbf{R}_3 \otimes \mathbf{P}_s\end{aligned} \tag{42}$$

with the submatrices

$$\begin{aligned}\mathbf{K}_3 &= \frac{1}{d_{gTmn}} \mathbf{S}_m^T (\mathbf{1} - \mathbf{t}_m \otimes \mathbf{t}_m) \mathbf{S}_m + \mathbf{t}_m^T \mathbf{x}_{mn,mm} \mathbf{F}_m^T \mathbf{F}_m + \mathbf{Z}_m + \mathbf{Z}_m^T\\ &- \mathbf{t}_m^T \mathbf{x}_{mn,m} (a_{mm} \mathbf{R}_m + a_{ms} \mathbf{R}_s) + \mathbf{W}_m,\\ \mathbf{K}_4 &= \frac{1}{d_{gTsn}} \mathbf{S}_s^T (\mathbf{1} - \mathbf{t}_s \otimes \mathbf{t}_s) \mathbf{S}_s + \mathbf{t}_s^T \mathbf{x}_{sn,ss} \mathbf{F}_s^T \mathbf{F}_s + \mathbf{Z}_s + \mathbf{Z}_s^T\\ &- \mathbf{t}_s^T \mathbf{x}_{sn,s} (a_{sm} \mathbf{R}_m + a_{ss} \mathbf{R}_s) + \mathbf{W}_s\end{aligned} \tag{43}$$

and s_m and s_s are parameters of sliding direction control.

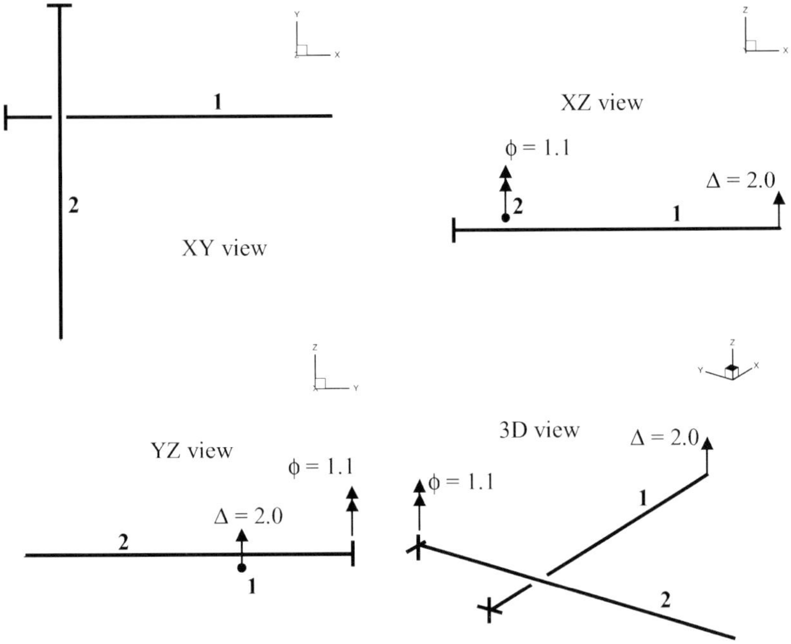

Fig. 3. Example – initial configuration of beam axes.

4 Numerical Example

The purpose of the presented example is to check the performance of the derived smooth beam-to-beam contact element in the case of large sliding between fragments of curves representing axes of contacting beams. The case of beams with circular cross-section is considered. The beams are divided into 10 identical co-rotational beam finite elements [1].

The beams are subjected to forced nodal displacements applied in 30 increments. The process of deformation is described by the dimensionless parameter T varying from 0 to 1.

Iterations of the Newton–Raphson scheme were stopped when the relative energy calculated with respect to the first iteration in each increment dropped below 10^{-20}. The energy itself was calculated as the scalar product of a residual vector and a vector of displacement increments.

Two cantilever beams, whose axes in initial configuration are shown in Figure 3, were considered. Beam **1** tip was moved in Z direction by $\Delta = 2.0$ and the fixed end of beam **2** was rotated about axis Z by the angle $\phi = 1.1$. The following data, identical for both beams were used: Young's modulus $205 \cdot 10^5$, Poisson's ratio 0.3,

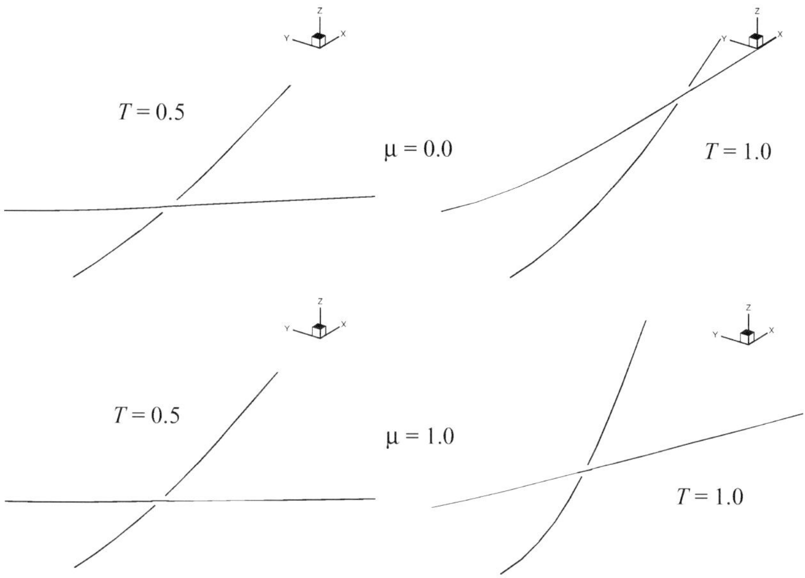

Fig. 4. Example – deformed configurations of beam axes.

radius of the cross-section 0.1, length 6.0. Initial gap was 0.001, friction coefficient $\mu = 0.0$ or 1.0, penalty parameters $\varepsilon_N = 10000$, $\varepsilon_{Tm} = \varepsilon_{Ts} = 1000$.

Deformed configurations of beam axes for two selected stages of deformation process and both considered cases of friction are presented in Figure 4. The influence of friction coefficient is clearly visible. Though both cases are characterised by the slip state, in the friction case relative sliding is limited what is reflected in the bending deformation of beam **1**.

The performance of the smooth contact element is visible if one considers the graphs in Figure 5, which show the variation of stick parts of tangential gaps during the deformation process and sliding of the contact point between elements on both beams. These values are used to calculate the friction force in contact and one can see that they are smooth during the entire process. This proves the excellent behaviour of the element in the case of large sliding.

In Figure 6 the slip parts of the tangential gaps are shown.

5 Concluding Remarks

A smoothing procedure for 3D beams using Hermite polynomials was presented in this paper. The construction of a 3D curve representing a beam axis is very attractive because it bases only on displacements of nodes of beam finite elements without any dependence on angles of rotation.

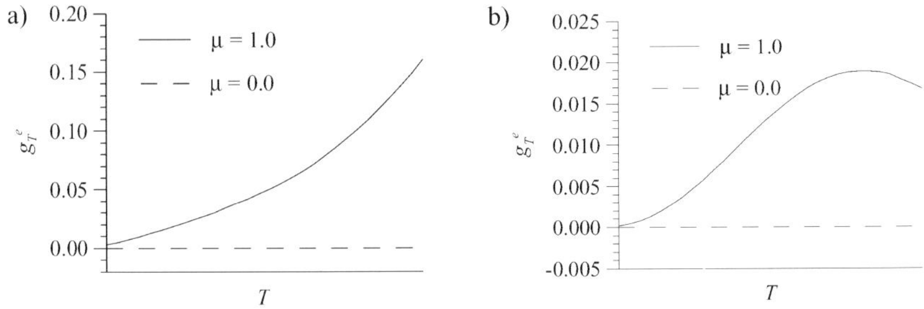

Fig. 5. Example – stick parts of the tangential gaps: (a) on beam **1**, (b) on beam **2**.

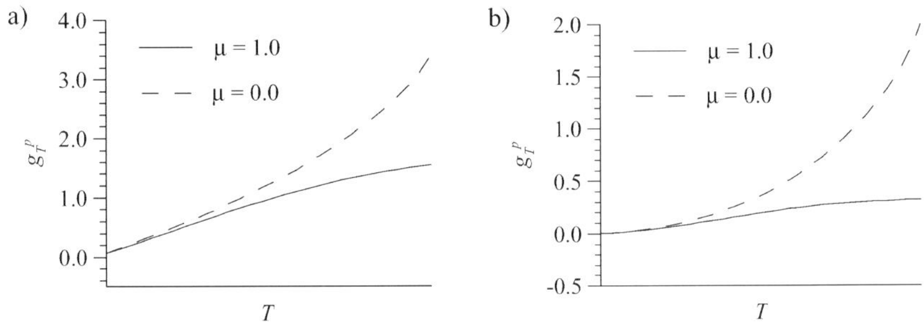

Fig. 6. Example – slip parts of the tangential gaps: (a) on beam **1**, (b) on beam **2**.

The application of Hermite polynomials is another aspect contributing to the simplicity of the formulation. Though not effective for solids, leading to knots instead of smooth surfaces [3, 4, 11], they prove to be robust and reliable in the case of beam-to-beam contact.

The numerical example presented in the paper confirms that the suggested formulation can be used effectively in the case of large sliding from one to another facet.

Appendix

A1. Elements of Matrix G

The symbolic algebra program Maple 7 was used to calculate matrices **G**, present in the formulation of contact $(16)_{1,2}$. Here the commands of Maple worksheet leading to the ready-to-use optimised Fortran code are presented.

1. Coefficients of three Hermite polynomials (3): a1, b1, c1, d1; a2, b2, c2, d2; a3, b3, c3, d3, expressed in terms of coordinates of nodal position vectors xij, where i is the node number and j – number of coordinate. These relations follow from the matrix formula (8).

```
>restart;
>a1:=(1/8-1/4/wl12)*x11+(1/4/wl12-1/4/wl23)*x21+(-
 1/8+1/4/wl23)*x31;
>a2:=(1/8-1/4/wl12)*x12+(1/4/wl12-1/4/wl23)*x22+(-
 1/8+1/4/wl23)*x32;
>a3:=(1/8-1/4/wl12)*x13+(1/4/wl12-1/4/wl23)*x23+(-
 1/8+1/4/wl23)*x33;
>b1:=1/4/wl12*x11+(-1/4/wl12-1/4/wl23)*x21+1/4/wl23*x31;
>b2:=1/4/wl12*x12+(-1/4/wl12-1/4/wl23)*x22+1/4/wl23*x32;
>b3:=1/4/wl12*x13+(-1/4/wl12-1/4/wl23)*x23+1/4/wl23*x33;
>c1:=(-3/8+1/4/wl12)*x11+(-1/4/wl12+1/4/wl23)*x21+(3/8-
 1/4/wl23)*x31;
>c2:=(-3/8+1/4/wl12)*x12+(-1/4/wl12+1/4/wl23)*x22+(3/8-
 1/4/wl23)*x32;
>c3:=(-3/8+1/4/wl12)*x13+(-1/4/wl12+1/4/wl23)*x23+(3/8-
 1/4/wl23)*x33;
>d1:=(1/4-1/4/wl12)*x11+(1/2+1/4/wl12+1/4/wl23)*x21+(1/4-
 1/4/wl23)*x31;
>d2:=(1/4-1/4/wl12)*x12+(1/2+1/4/wl12+1/4/wl23)*x22+(1/4-
 1/4/wl23)*x32;
>d3:=(1/4-1/4/wl12)*x13+(1/2+1/4/wl12+1/4/wl23)*x23+(1/4-
 1/4/wl23)*x33;
```

2. Straight line lengths (6).

```
>wl:=.5*sqrt((x31-x11)^2+(x32-x12)^2+(x33-x13)^2);
>wl12:=2*sqrt((x21-x11)^2+(x22-x12)^2+(x23-x13)^2)/wl;
>wl23:=2*sqrt((x31-x21)^2+(x32-x22)^2+(x33-x23)^2)/wl;
```

3. Elements of matrix **G** calculated as partial derivatives of (9) with respect to the coordinates of nodal position vectors.

```
>g11:=xsi^3*(diff(a1,x11))+xsi^2*(diff(b1,x11))+xsi*(diff(c1,
 x11))+diff(d1,x11);
>g12:=xsi^3*(diff(a1,x12))+xsi^2*(diff(b1,x12))+xsi*(diff(c1,
 x12))+diff(d1,x12);
>g13:=xsi^3*(diff(a1,x13))+xsi^2*(diff(b1,x13))+xsi*(diff(c1,
 x13))+diff(d1,x13);
>g14:=xsi^3*(diff(a1,x21))+xsi^2*(diff(b1,x21))+xsi*(diff(c1,
 x21))+diff(d1,x21);
>g15:=xsi^3*(diff(a1,x22))+xsi^2*(diff(b1,x22))+xsi*(diff(c1,
 x22))+diff(d1,x22);
>g16:=xsi^3*(diff(a1,x23))+xsi^2*(diff(b1,x23))+xsi*(diff(c1,
 x23))+diff(d1,x23);
>g17:=xsi^3*(diff(a1,x31))+xsi^2*(diff(b1,x31))+xsi*(diff(c1,
```

```
 x31))+diff(d1,x31);
>g18:=xsi^3*(diff(a1,x32))+xsi^2*(diff(b1,x32))+xsi*(diff(c1,
 x32))+diff(d1,x32);
>g19:=xsi^3*(diff(a1,x33))+xsi^2*(diff(b1,x33))+xsi*(diff(c1,
 x33))+diff(d1,x33);
>g21:=xsi^3*(diff(a2,x11))+xsi^2*(diff(b2,x11))+xsi*(diff(c2,
 x11))+diff(d2,x11);
>g22:=xsi^3*(diff(a2,x12))+xsi^2*(diff(b2,x12))+xsi*(diff(c2,
 x12))+diff(d2,x12);
>g23:=xsi^3*(diff(a2,x13))+xsi^2*(diff(b2,x13))+xsi*(diff(c2,
 x13))+diff(d2,x13);
>g24:=xsi^3*(diff(a2,x21))+xsi^2*(diff(b2,x21))+xsi*(diff(c2,
 x21))+diff(d2,x21);
>g25:=xsi^3*(diff(a2,x22))+xsi^2*(diff(b2,x22))+xsi*(diff(c2,
 x22))+diff(d2,x22);
>g26:=xsi^3*(diff(a2,x23))+xsi^2*(diff(b2,x23))+xsi*(diff(c2,
 x23))+diff(d2,x23);
>g27:=xsi^3*(diff(a2,x31))+xsi^2*(diff(b2,x31))+xsi*(diff(c2,
 x31))+diff(d2,x31);
>g28:=xsi^3*(diff(a2,x32))+xsi^2*(diff(b2,x32))+xsi*(diff(c2,
 x32))+diff(d2,x32);
>g29:=xsi^3*(diff(a2,x33))+xsi^2*(diff(b2,x33))+xsi*(diff(c2,
 x33))+diff(d2,x33);
>g31:=xsi^3*(diff(a3,x11))+xsi^2*(diff(b3,x11))+xsi*(diff(c3,
 x11))+diff(d3,x11);
>g32:=xsi^3*(diff(a3,x12))+xsi^2*(diff(b3,x12))+xsi*(diff(c3,
 x12))+diff(d3,x12);
>g33:=xsi^3*(diff(a3,x13))+xsi^2*(diff(b3,x13))+xsi*(diff(c3,
 x13))+diff(d3,x13);
>g34:=xsi^3*(diff(a3,x21))+xsi^2*(diff(b3,x21))+xsi*(diff(c3,
 x21))+diff(d3,x21);
>g35:=xsi^3*(diff(a3,x22))+xsi^2*(diff(b3,x22))+xsi*(diff(c3,
 x22))+diff(d3,x22);
>g36:=xsi^3*(diff(a3,x23))+xsi^2*(diff(b3,x23))+xsi*(diff(c3,
 x23))+diff(d3,x23);
>g37:=xsi^3*(diff(a3,x31))+xsi^2*(diff(b3,x31))+xsi*(diff(c3,
 x31))+diff(d3,x31);
>g38:=xsi^3*(diff(a3,x32))+xsi^2*(diff(b3,x32))+xsi*(diff(c3,
 x32))+diff(d3,x32);
>g39:=xsi^3*(diff(a3,x33))+xsi^2*(diff(b3,x33))+xsi*(diff(c3,
 x33))+diff(d3,x33);
```

4. Generation of Fortran optimised code

```
>with(codegen,fortran);
>fortran(g11,optimized,mode=double,precision=double);
>fortran(g12,optimized,mode=double,precision=double);
>fortran(g13,optimized,mode=double,precision=double);
>fortran(g14,optimized,mode=double,precision=double);
>fortran(g15,optimized,mode=double,precision=double);
>fortran(g16,optimized,mode=double,precision=double);
>fortran(g17,optimized,mode=double,precision=double);
>fortran(g18,optimized,mode=double,precision=double);
>fortran(g19,optimized,mode=double,precision=double);
>fortran(g21,optimized,mode=double,precision=double);
>fortran(g22,optimized,mode=double,precision=double);
>fortran(g23,optimized,mode=double,precision=double);
>fortran(g24,optimized,mode=double,precision=double);
>fortran(g25,optimized,mode=double,precision=double);
>fortran(g26,optimized,mode=double,precision=double);
>fortran(g27,optimized,mode=double,precision=double);
>fortran(g28,optimized,mode=double,precision=double);
>fortran(g29,optimized,mode=double,precision=double);
>fortran(g31,optimized,mode=double,precision=double);
>fortran(g32,optimized,mode=double,precision=double);
>fortran(g33,optimized,mode=double,precision=double);
>fortran(g34,optimized,mode=double,precision=double);
>fortran(g35,optimized,mode=double,precision=double);
>fortran(g36,optimized,mode=double,precision=double);
>fortran(g37,optimized,mode=double,precision=double);
>fortran(g38,optimized,mode=double,precision=double);
>fortran(g39,optimized,mode=double,precision=double);
```

A2. Elements of Matrix H

The symbolic algebra program Maple 7 was used to calculate matrices **H**, present in the formulation of contact $(16)_{3,4}$. Here the commands of Maple worksheet leading to the ready-to-use optimised Fortran code are presented.

Parts 1 and 2 identical as in Appendix A1 for matrix **G**.

3. Elements of matrix **H** calculated as partial derivatives of $(10)_1$ with respect to the coordinates of nodal position vectors.

```
>h11:=3*xsi^2*(diff(a1,x11))+2*xsi*(diff(b1,x11))+(diff(c1,x1
 1));
>h12:=3*xsi^2*(diff(a1,x12))+2*xsi*(diff(b1,x12))+(diff(c1,x1
 2));
>h13:=3*xsi^2*(diff(a1,x13))+2*xsi*(diff(b1,x13))+(diff(c1,x1
 3));
>h14:=3*xsi^2*(diff(a1,x21))+2*xsi*(diff(b1,x21))+(diff(c1,x2
 1));
>h15:=3*xsi^2*(diff(a1,x22))+2*xsi*(diff(b1,x22))+(diff(c1,x2
 2));
>h16:=3*xsi^2*(diff(a1,x23))+2*xsi*(diff(b1,x23))+(diff(c1,x2
 3));
>h17:=3*xsi^2*(diff(a1,x31))+2*xsi*(diff(b1,x31))+(diff(c1,x3
 1));
>h18:=3*xsi^2*(diff(a1,x32))+2*xsi*(diff(b1,x32))+(diff(c1,x3
 2));
>h19:=3*xsi^2*(diff(a1,x33))+2*xsi*(diff(b1,x33))+(diff(c1,x3
 3));
>h21:=3*xsi^2*(diff(a2,x11))+2*xsi*(diff(b2,x11))+(diff(c2,x1
 1));
>h22:=3*xsi^2*(diff(a2,x12))+2*xsi*(diff(b2,x12))+(diff(c2,x1
 2));
>h23:=3*xsi^2*(diff(a2,x13))+2*xsi*(diff(b2,x13))+(diff(c2,x1
 3));
>h24:=3*xsi^2*(diff(a2,x21))+2*xsi*(diff(b2,x21))+(diff(c2,x2
 1));
>h25:=3*xsi^2*(diff(a2,x22))+2*xsi*(diff(b2,x22))+(diff(c2,x2
 2));
>h26:=3*xsi^2*(diff(a2,x23))+2*xsi*(diff(b2,x23))+(diff(c2,x2
 3));
>h27:=3*xsi^2*(diff(a2,x31))+2*xsi*(diff(b2,x31))+(diff(c2,x3
 1));
>h28:=3*xsi^2*(diff(a2,x32))+2*xsi*(diff(b2,x32))+(diff(c2,x3
 2));
>h29:=3*xsi^2*(diff(a2,x33))+2*xsi*(diff(b2,x33))+(diff(c2,x3
 3));
>h31:=3*xsi^2*(diff(a3,x11))+2*xsi*(diff(b3,x11))+(diff(c3,x1
 1));
>h32:=3*xsi^2*(diff(a3,x12))+2*xsi*(diff(b3,x12))+(diff(c3,x1
 2));
>h33:=3*xsi^2*(diff(a3,x13))+2*xsi*(diff(b3,x13))+(diff(c3,x1
 3));
```

```
>h34:=3*xsi^2*(diff(a3,x21))+2*xsi*(diff(b3,x21))+(diff(c3,x2
 1));
>h35:=3*xsi^2*(diff(a3,x22))+2*xsi*(diff(b3,x22))+(diff(c3,x2
 2));
>h36:=3*xsi^2*(diff(a3,x23))+2*xsi*(diff(b3,x23))+(diff(c3,x2
 3));
>h37:=3*xsi^2*(diff(a3,x31))+2*xsi*(diff(b3,x31))+(diff(c3,x3
 1));
>h38:=3*xsi^2*(diff(a3,x32))+2*xsi*(diff(b3,x32))+(diff(c3,x3
 2));
>h39:=3*xsi^2*(diff(a3,x33))+2*xsi*(diff(b3,x33))+(diff(c3,x3
 3));
```

4. As in Appendix A1 for matrix **G**, but the entries `gij` must be replaced by `hij`.

A3. Elements of matrix M

The symbolic algebra program Maple 7 was used to calculate matrices **M**, present in the formulation of contact $(16)_{5,6}$. Here the commands of Maple worksheet leading to the ready-to-use optimised Fortran code are presented.

Parts 1 and 2 identical as in Appendix A1 for matrix **G**.

3. Elements of matrix M calculated as partial derivatives of $(10)_2$ with respect to the coordinates of nodal position vectors.

```
>m11:=6*xsi*(diff(a1,x11))+2*(diff(b1,x11));
>m12:=6*xsi*(diff(a1,x12))+2*(diff(b1,x12));
>m13:=6*xsi*(diff(a1,x13))+2*(diff(b1,x13));
>m14:=6*xsi*(diff(a1,x21))+2*(diff(b1,x21));
>m15:=6*xsi*(diff(a1,x22))+2*(diff(b1,x22));
>m16:=6*xsi*(diff(a1,x23))+2*(diff(b1,x23));
>m17:=6*xsi*(diff(a1,x31))+2*(diff(b1,x31));
>m18:=6*xsi*(diff(a1,x32))+2*(diff(b1,x32));
>m19:=6*xsi*(diff(a1,x33))+2*(diff(b1,x33));
>m21:=6*xsi*(diff(a2,x11))+2*(diff(b2,x11));
>m22:=6*xsi*(diff(a2,x12))+2*(diff(b2,x12));
>m23:=6*xsi*(diff(a2,x13))+2*(diff(b2,x13));
>m24:=6*xsi*(diff(a2,x21))+2*(diff(b2,x21));
>m25:=6*xsi*(diff(a2,x22))+2*(diff(b2,x22));
>m26:=6*xsi*(diff(a2,x23))+2*(diff(b2,x23));
>m27:=6*xsi*(diff(a2,x31))+2*(diff(b2,x31));
>m28:=6*xsi*(diff(a2,x32))+2*(diff(b2,x32));
```

```
>m29:=6*xsi*(diff(a2,x33))+2*(diff(b2,x33));
>m31:=6*xsi*(diff(a3,x11))+2*(diff(b3,x11));
>m32:=6*xsi*(diff(a3,x12))+2*(diff(b3,x12));
>m33:=6*xsi*(diff(a3,x13))+2*(diff(b3,x13));
>m34:=6*xsi*(diff(a3,x21))+2*(diff(b3,x21));
>m35:=6*xsi*(diff(a3,x22))+2*(diff(b3,x22));
>m36:=6*xsi*(diff(a3,x23))+2*(diff(b3,x23));
>m37:=6*xsi*(diff(a3,x31))+2*(diff(b3,x31));
>m38:=6*xsi*(diff(a3,x32))+2*(diff(b3,x32));
>m39:=6*xsi*(diff(a3,x33))+2*(diff(b3,x33));
```

4. As in Appendix A1 for matrix **G**, but the entries `gij` must be replaced by `mij`.

References

1. Crisfield MA (1990) A consistent co-rotational formulation for non-linear, three-dimensional beam-elements. *Comp Meth Appl Mech Eng*, 81:131–150.
2. El-Abbasi N, Meguid SA, Czekanski A (2001) On the modelling of smooth contact surfaces using cubic splines. *Int J Num Meth Eng*, 50:953–967.
3. Farin G (1993) *Curves and Surfaces for Computer Aided Geometric Design*, 3rd edn. Academic Press/Harcourt Brace Jovanovich Publishers, Boston.
4. Krstulović-Opara L, Wriggers P, Korelc J (2002) A C^1-continuous formulation for 3D finite deformation frictional contact. *Comput Mech*, 29:27–42.
5. Litewka P, Wriggers P (2002) Contact between 3D beams with rectangular cross sections. *Int J Num Meth Eng*, 53:2019–2041.
6. Litewka P, Wriggers P (2002) Frictional contact between 3D beams. *Comput Mech*, 28:26–39.
7. Litewka P (2006) Hermite polynomial smoothing in beam-to-beam frictional contact problem. *Comput Mech*, published online 30.11.2006, DOI 10.1007/s00466-006-0143-9.
8. Michałowski R, Mróz Z (1978) Associated and non-associated sliding rules in contact friction problems. *Arch Mech*, 30:259–276.
9. Puso MA, Laursen TA (2002) A 3D contact smoothing method using Gregory patches. *Int J Num Meth Eng*, 54:1161–1194.
10. Stadler M, Holzapfel GA, Korelc J (2003) C^n continuous modelling of smooth contact surfaces using NURBS and application to 2D problems. *Int J Num Meth Eng*, 57:2177–2203.
11. Wriggers P, Krstulović-Opara L (1999) Development of 2D smooth polynomial frictional contact elements based on symbolic approach. In: W Wunderlich (Ed.), *Proceedings of ECCM'99*, München, Germany, 31 July–3 August, 1999.
12. Wriggers P, Zavarise G (1997) On contact between three-dimensional beams undergoing large deflections. *Comm Num Meth Eng*, 13:429–438.
13. Zavarise G, Wriggers P. (2000) Contact with friction between beams in 3-D space. *Int J Num Meth Eng*, 49:977–1006.

Nonlinear Boundary Problems for a Cylinder

Alexander Shermenev

Wave Research Center, Russian Academy of Sciences, Moscow 117942, Russia
E-mail: sher@orc.ru, alex.sher@hotmail.com

Abstract. Nonlinear boundary problems for in a cylinder is studied within quadratic approximation. When the cylinder coordinates are used, the usual perturbation techniques in separation of variables method inevitably lead to a series of overdetermined systems of linear algebraic equations for the unknown coefficients (in contrast with the Cartesian coordinates). However, if we formally introduce a new function $W(r)$ satisfying the first system of this series, all these overdetermined systems become compatible (remaining overdetermined). Using the new function and the quadratic polynomials of the Bessel functions of radius, we explicitly express the solutions of the nonlinear boundary problem.

Key words: Bessel functions, nonlinear interaction, separation of variables, wave equation.

1 Introduction

The three-dimensional nonlinear wave equation for a potential $\phi(x_1, x_2, x_3, t)$

$$-\kappa\phi_{tt} + \Delta\phi + \frac{1}{2}\alpha(\nabla\phi \cdot \nabla\phi)_t + \frac{1}{2}\beta(\phi_t^2)_t = 0 \tag{1}$$

describes the pressure elastic waves, rotational elastic waves, waves in an isentropic gas flow for the non-dissipative case, and numerous other physical processes. In two-dimensional case, Equation (1) is the shallow water equation describing the long surface water waves.

The aim of this paper is to derive a general approach for solving Equation (1) in the cylinder coordinates (θ, r, z).

Given in the cylinder coordinates, Equation (1) for $\varphi(r, \theta, z, t) = \phi(x_1, x_2, x_3, t)$ can be written as follows:

$$-\kappa\varphi_{tt} + \frac{1}{r}\varphi_r + \varphi_{rr} + \frac{1}{r^2}\varphi_{\theta\theta} + \varphi_{zz} + \frac{1}{2}\alpha\left(\varphi_r^2 + \frac{1}{r^2}\varphi_\theta^2 + \varphi_z^2\right)_t + \frac{1}{2}\beta(\varphi_t^2)_t = 0. \tag{1'}$$

Peter Wriggers and Udo Nackenhorst (eds.), IUTAM Symposium on Computational Contact Mechanics, 177–187.

Applying the separation of variables method to the linear version of Equation (1′), we obtain a solution of the following form:

$$\varphi_{lin}(r,\theta,z,t) = \sum_{i=1}^{n} [S^i \sin(K_i z + \omega_i t) + C^i \cos(K_i z + \omega_i t)],$$

where coefficients S^i and C^i are functions depending on (r,θ).

The functions $S^i(r,\theta)$ and $C^i(r,\theta)$ satisfy the Helmholtz equations

$$Z_{rr} + \frac{1}{r} Z_r + \frac{1}{r^2} Z_{\theta\theta} + k_i^2 Z = 0, \tag{2}$$

where $k_i = \sqrt{\kappa\omega_i^2 - K_i^2}$ and, consequently, their Fourier expansions with respect to θ have the following forms:

$$\begin{aligned} S^i &= a_0^i J_0^i + (a_1^i \sin\theta + b_1^i \cos\theta) J_1^i + \cdots + \\ &\quad + (a_m^i \sin m\theta + b_m^i \cos m\theta) J_m^i + \cdots, \end{aligned} \tag{3}$$

$$\begin{aligned} S^j &= c_0^j J_0^j + (c_1^j \sin\theta + d_1^j \cos\theta) J_1^j + \cdots + \\ &\quad + (c_m^j \sin m\theta + d_m^j \cos m\theta) J_m^j + \cdots, \end{aligned} \tag{4}$$

where $J_m^i \overset{\text{def}}{=} J_m(k_i r)$ and $J_m(*)$ are Bessel functions. I assume that $k_i \neq 0$ and consider only the first N terms of Fourier expansions (6).

In the quadratic approximation, the potential $\varphi(r,\theta,z,t)$ (regular at $r = 0$) can be written as follows:

$$\begin{aligned} \varphi(r,\theta,z,t) &= \varepsilon \sum_{i=1}^{n} (S^i \sin(K_i z + \omega_i t) + C^i \cos(K_i z + \omega_i t)) + \\ &\quad + \varepsilon^2 \left[\begin{array}{l} \sum_{i,j=1}^{n} Sp^{ij} \sin(K_i z + \omega_i t + K_j z + \omega_j t) \\ \qquad + Cp^{ij} \cos(K_i z + \omega_i t + K_j z + \omega_j t) + \\ \sum_{i,j=1}^{n} Sm^{ij} \sin(K_i z + \omega_i t - K_j z - \omega_j t) \\ \qquad + Cm^{ij} \cos(K_i z + \omega_i t - K_j z - \omega_j t) + \end{array} \right], \end{aligned} \tag{5}$$

where coefficients of trigonometric functions: Sp^{ij}, Cp^{ij}, Sm^{ij}, and Cm^{ij} are functions depending on (r,θ). The aim of this article is to give explicit expressions for the functions Sp^{ij}, Cp^{ij}, Sm^{ij}, and Cm^{ij}. The obtained formulas can be used for solving the bondary problems involving the circular cylinders. We concentrate on the finding Sp^{ij} because the problems with Cp^{ij}, Sm^{ij}, and Cm^{ij} are quite analogous.

It is convenient to use the indexes A and B in Sp^{ij} instead of i and j which are used further in numerous summations.

Thus

$$S^A = a_0^A J_0^A + (a_1^A \sin\theta + b_1^A \cos\theta) J_1^A + \cdots + \\ + (a_m^A \sin m\theta + b_m^A \cos m\theta) J_m^A + \cdots, \tag{6}$$

$$S^B = a_0^B J_0^B + (a_1^B \sin\theta + b_1^B \cos\theta) J_1^B + \cdots + \\ + (a_m^B \sin m\theta + b_m^B \cos m\theta) J_m^B + \cdots, \tag{7}$$

where $J_m^A \overset{\text{def}}{=} J_m(k_A r)$, $J_m^B \overset{\text{def}}{=} J_m(k_B r)$, and $J_m(*)$ are Bessel functions (see, for example [2]).

Assume that $k_A \neq 0$, $k_B \neq 0$ and consider only the first N terms of Fourier expansions (6) and (7).

The function Sp^{AB} can also be expanded in the Fourier series with respect to θ:

$$Sp^{AB}(r,\theta) = M_0^{AB}(r) + \sum_{m=1}^{\infty} (M_m^{AB}(r) \sin m\theta + N_m^{AB}(r) \cos m\theta). \tag{8}$$

The problem of finding the functions $Sp^{AA}(r,\theta)$ arises when we try to describe a motion with a given period. This problem was investigated in [7, 10]. It was proved that they can be found as quadratic polynomials of J_0^A, J_1^A, (or J_0^B, J_1^B):

$$R_{00}(J_0^A)^2 + R_{01} J_0^A J_1^A + R_{11}(J_1^A)^2, \tag{9}$$

where R_{00}, R_{01}, and R_{11} are polynomials of r^{-1} and r with unknown constant coefficients and of unknown degree n of the form

$$\sum_{k=-n}^{n} C_k r^k. \tag{10}$$

The unknown constant coefficients of (10) are found by solving a series (for each $\sin j\theta$ and $\cos j\theta$) of *overdetermined* (but *compatible*) systems of linear algebraic equations.

In this article we find expression for Sp^{AB} $(A \neq B)$ or $M_m^{AB}(r)$ and $N_m^{AB}(r)$.

First we try to find $M_j^{AB}(r)$ $(j > 0)$, $N_j^{AB}(r)$ in the form

$$R_{00} J_0^A J_0^B + R_{01} J_0^A J_1^B + R_{10} J_1^A J_0^B + R_{11} J_1^A J_1^B, \tag{11}$$

but the corresponding overdetermined systems for coefficients of (11) are *incompatible*.

However, if we introduce a new function $W(r) = M_0^{AB}(r)$ (a solution of the first problem of the series), all remaining involved functions: $M_j^{AB}(r)$ $(j > 0)$, $N_j^{AB}(r)$ can be found in the following form:

$$R_{00} J_0^A J_0^B + R_{01} J_0^A J_1^B + R_{10} J_1^A J_0^B + R_{11} J_1^A J_1^B + R_0 W(r) + R_1 W'(r), \quad (12)$$

where R_{ij} and R_i are of the form (10). The constant coefficients are found as solutions of series (for each $\sin j\theta$ and $\cos j\theta$) of systems of linear algebraic equations, which are again overdetermined but now compatible.

The function $W(r)$ is defined as a particular solution (regular at $r = 0$) of the following equation:

$$Z_{rr} + \frac{1}{r} Z_r + k^2 Z = J_0^A J_0^B. \quad (13)$$

It arises in a series of problems of mathematical physics (see [11]) and merits detailed consideration. Its properties and graph are sketched in Section 3.

The construction allows us to find the solutions of Equation (1′) in the form (5), (8), (12) and can be considered as a generalization of the separation of variables method for the nonlinear case. The expressions of type (11) and (12) play role of trigonometric expressions in the case of Carteian coordinates.

The natural area of application are the problems of wave propagation, scattering, or diffraction involving bodies which can be written in the cylinder coordinates (cylinders, wedges). Then the coefficients a_m and b_m can usually be found from some boundary conditions on the surfaces $r = \text{const}$ (cylinders) or $\theta = \text{const}$ (planes, wedges).

In the two-dimensional case, the derived solution gives a nonlinear correction to the classical linear solution describing the wave motion in a circular basin (see [1]). Two simple numerical examples are presented in Section 4.

2 Periodic Problem, First Reduction

Substitute (2) in (1′) and consider coefficients of $\sin(K_A z + \omega_A t + K_B z + \omega_B t)$ and $\cos(K_A z + \omega_A t + K_B z + \omega_B t)$. Collecting the terms with $\sin j\theta$ and $\cos j\theta$, we obtain the following equations in the order ε^2:

$$\begin{aligned} & S_{rr}^{A+B} + \frac{1}{r} S_r^{A+B} + \frac{1}{r^2} S_{\theta\theta}^{A+B} + k^2 S^{A+B} \\ & = \frac{1}{2}(\alpha K_A K_B + \beta \omega_A \omega_B)(\omega_A + \omega_B)(C^A C^B - S^A S^B) - \\ & - \frac{1}{2}\alpha(\omega_A + \omega_B) r^{-2} (C_\theta^A C_\theta^B - S_\theta^A S_\theta^B) - \\ & - \frac{1}{2}\alpha(\omega_A + \omega_B)(C_r^A C_r^B - S_r^A S_r^B), \end{aligned} \quad (14)$$

$$C_{rr}^{A+B} + \frac{1}{r}C_r^{A+B} + \frac{1}{r^2}C_{\theta\theta}^{A+B} + k^2 C^{A+B}$$

$$= -\frac{1}{2}(\alpha K_A K_B + \beta\omega_A\omega_B)(\omega_A + \omega_B)(C^A S^B + C^B S^A) +$$

$$+ \frac{1}{2}\alpha(\omega_A + \omega_B)r^{-2}(C_\theta^A S_\theta^B + C_\theta^B S_\theta^A) +$$

$$+ \frac{1}{2}\alpha(\omega_A + \omega_B)(C_r^A S_r^B + C_r^B S_r^A), \tag{15}$$

The right-hand sides of (14) and (15) can be represented as sums: $R_{\sin}^1 + R_{\sin}^2$ and $R_{\cos}^1 + R_{\cos}^2$, where

$$R_{\sin}^1 \equiv \frac{\alpha}{2}(K_A K_B - \kappa\omega_A\omega_B)(\omega_A + \omega_B)(C^A C^B - S^A S^B) -$$

$$- \frac{\alpha}{2}(\omega_A + \omega_B)r^{-2}(C_\theta^A C_\theta^B - S_\theta^A S_\theta^B) + \frac{\alpha}{2}(\omega_A + \omega_B)(C_r^A C_r^B - S_r^A S_r^B),$$

$$R_{\sin}^2 \equiv \frac{1}{2}(\beta + \alpha\kappa)\omega_A\omega_B(\omega_A + \omega_B)(C^A C^B - S^A S^B)$$

$$R_{\cos}^1 \equiv -\frac{\alpha}{2}(K_A K_B - \kappa\omega_A\omega_B)(\omega_A + \omega_B)(C^A S^B + C^B S^A) +$$

$$+ \frac{\alpha}{2}(\omega_A + \omega_B)r^{-2}(C_\theta^A S_\theta^B + C_\theta^B S_\theta^A) + \frac{\alpha}{2}(\omega_A + \omega_B)(C_r^A S_r^B + C_r^B S_r^A),$$

$$R_{\cos}^2 \equiv -\frac{1}{2}(\beta + \alpha\kappa)\omega_A\omega_B(\omega_A + \omega_B)(C^A S^B + C^B S^A)$$

It can be easily verified that

$$\frac{\alpha}{4}(\omega_A + \omega_B)(C^A C^B - S^A S^B)$$

is a particular solution to the equation:

$$S_{rr}^{A+B} + \frac{1}{r}S_r^{A+B} + \frac{1}{r^2}S_{\theta\theta}^{A+B} + k^2 S^{A+B} = R_{\sin}^1 \tag{16}$$

and

$$\frac{\alpha}{4}(\omega_A + \omega_B)(C^A S^B + C^B S^A)$$

is a particular solution to the equation:

$$C_{rr}^{A+B} + \frac{1}{r}C_r^{A+B} + \frac{1}{r^2}C_{\theta\theta}^{A+B} + k^2 C^{A+B} = R_{\cos}^1.$$

So the problem is reduced to finding particular solutions to the equations

$$S_{rr}^{A+B} + \frac{1}{r}S_r^{A+B} + \frac{1}{r^2}S_{\theta\theta}^{A+B} + k^2 S^{A+B} = R_{\sin}^2, \tag{17}$$

$$C_{rr}^{A+B} + \frac{1}{r}C_r^{A+B} + \frac{1}{r^2}C_{\theta\theta}^{A+B} + k^2 C^{A+B} = R_{\cos}^2. \tag{18}$$

or to the equations

$$S_{rr}^{A+B} + \frac{1}{r}S_r^{A+B} + \frac{1}{r^2}S_{\theta\theta}^{A+B} + k^2 S^{A+B} = C^A C^B - S^A S^B, \tag{19}$$

$$C_{rr}^{A+B} + \frac{1}{r}C_r^{A+B} + \frac{1}{r^2}C_{\theta\theta}^{A+B} + k^2 C^{A+B} = C^A S^B + C^B S^A. \tag{20}$$

which is a subject of the following sections.

The general solutions to (14) and to (15) can be represented as linear combinations of the particular solutions and the general solution to the homogeneous equation

$$Z_{rr} + \frac{1}{r}Z_r + \frac{1}{r^2}Z_{\theta\theta} + k^2 Z = 0, \tag{21}$$

which are the Bessel functions too.

3 Main Results

First we sketch a number of properties of Bessel functions, which can be verified by calculations.

Denote by $\mathbf{B}_m$ the differential operator acting on the functions depending on r

$$\frac{d^2}{dr^2} + \frac{1}{r}\frac{d}{dr} + \left(k^2 - \frac{m^2}{r^2}\right),$$

Lemma 1. Assume that $f(r)$ is a twice differentiable function of r and $\mathbf{B}_m(f) = h$. Then

$$\mathbf{B}_{m+1}\left(f' - \frac{m}{r}f\right) = h' - \frac{m}{r}h.$$

Lemma 2. Assume that $m = i + j$ and $f = J_i^A J_j^B$. Then

$$\mathbf{B}_{m+1}\left(f' - \frac{m}{r}f\right) = -k_A J_{i+1}^A J_j^B - k_B J_i^A J_{j+1}^B.$$

Lemma 3. Assume that $m = i + j$, $i > 0$, and $j > 0$. Then the following identity is valid:

$$\mathbf{B}_m(J_i^A J_j^B) = -k_A k_B J_{i-1}^A J_{j+1}^B + (k^2 - k_A^2 - k_B^2) J_i^A J_j^B - k_A k_B J_{i+1}^A J_{j-1}^B.$$

The main result is the following theorem.

Theorem 1. Assume that i and j are natural numbers. Then the equation:

$$\mathbf{B}_{i+j}(Z) = J_i^A(r)J_j^B(r)$$

have a particular solution of the form

$$R_{00}J_0^A J_0^B + R_{01}J_0^A J_1^B + R_{10}J_1^A J_0^B + R_{11}J_1^A J_1^B + R_0 W + R_1 W', \tag{22}$$

where R_{ij} and R_i are polynomials of r^{-1} and r of the form:

$$\sum_{k=-n}^{m} C_k^{ij} r^k. \tag{23}$$

The proof is performed by induction with respect to m.

For $m = 0$, the statement is true by definition of function $W(r) = M_0^{A+B}(r)$.

Suppose that the statement is proved for m and let us to prove it for $m + 1$.

Assume that i is a positive natural number and $i \leq m$. Consider a three-dimensional linear space $\mathbf{K}_i$ generated by functions $J_{i-1}^A(r)J_{m-i+2}^B(r)$, $J_i^A(r)J_{m-i+1}^B(r)$, $J_{i+1}^A(r)J_{m-i}^B(r)$. According to Lemma 2, there exist functions $f_{i,m+1}$ and $f_{i+1,m+1}$ such that

$$\mathbf{B}_{m+1}(f_{i,m+1}) = -k_A J_{i-1}^A J_{m-i+2}^B - k_B J_i^A J_{m-i+1}^B, \tag{24}$$

$$\mathbf{B}_{m+1}(f_{i+1,m+1}) = -k_A J_i^A J_{m-i+1}^B - k_B J_{i+1}^A J_{m-i}^B. \tag{25}$$

In addition, according to Lemma 3, we have

$$\begin{aligned}\mathbf{B}_{m+1}(J_i^A J_{m-i+1}^B) = {} & -k_A k_B J_{i-1}^A J_{m-i+1}^B + \\ & + (k^2 - k_A^2 - k_B^2)J_i^A J_{m-i+1}^B - k_A k_B J_{i+1}^A J_m^B.\end{aligned} \tag{26}$$

The determinant of (26)–(28) is equal to $k^2 k_A k_B$. Therefore $\mathbf{B}_{m+1}(f_{i,m+1})$, $\mathbf{B}_{m+1}(f_{i+1,m+1})$, $\mathbf{B}_{m+1}(J_i^A J_{m-i+1}^B)$ are linearly independent and, consequently, generate the space $\mathbf{K}_i$ that proves the theorem.

* * *

The function $W(r)$ is a regular solution of Equation (13). It can be presented in the following form:

$$\begin{aligned}W(r) = {} & Y_0(kr)\int_1^r \frac{\pi}{2} q J_0(kq)J_0(k_A q)J_0(k_B q)dq - \\ & - J_0(kr)\int_1^r \frac{\pi}{2} q J_0(k_B q)J_0(k_A q)Y_0(kq)dq - \mu Y_0(kr),\end{aligned}$$

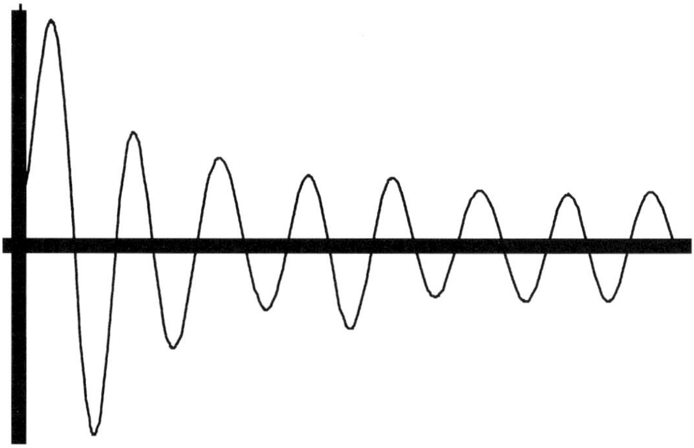

Fig. 1. Function W for $k_A = 3.83171$, $k_B = 7.01559$, $k = 7.99377$.

where

$$\mu = \frac{\pi}{2}\int_1^0 q J_0(k_B q) J_0(k_A q) Y_0(kq) dq$$

The graph of function W is shown in Figure 1.

The function $W = W(r, k, k_A, k_B)$ formally depends on four variables but can be reduced to the function of three variables using a suitable scaling. It can be approximated and stored for solving various problems.

4 Examples

Consider a horizontal motion of gas ($K_A = 0$, $K_B = 0$) in a cylinder with radius 1 and assume (without loss in generality) that $\kappa = 1$. Then $\omega_A = k_A$ and $\omega_B = k_B$ and, in linear approximation, we have

$$\varphi(r, \theta, z, t) = \sum_{m=0,\, n=0}^{\infty,\infty} [(a_n^m \sin \omega_n^m t + b_n^m \cos \omega_n^m t) \sin n\theta + \\ + (c_n^m \sin \omega_n^m t + d_n^m \cos \omega_n^m t) \cos n\theta] J_n(\omega_n^m r),$$

where ω_n^m are roots of equations

$$J_n'(\omega_n^m) = 0, \quad \text{(boundary condition)}.$$

The coefficients a_n^m, b_n^m, c_n^m, d_n^m are determined from the other boundary (or initial) conditions. Assume additively that $\alpha = 1$, $\beta = 1$ and search for quadratic correction of the linear solution.

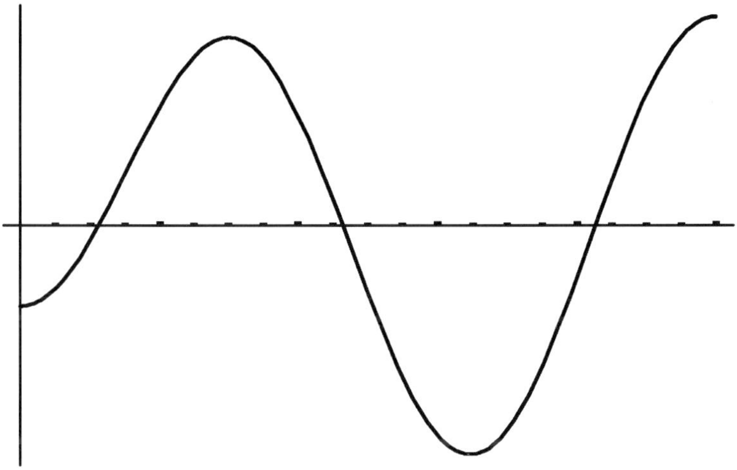

Fig. 2. Dependence $f_1(r)$ on r.

Example 1.

Consider a motion of gas depending only on radius r. Assume that ω_A , ω_B are the first two roots of equation $J_0'(x) = 0$ and

$$\varphi_{lin}(r, \theta, z, t) = \varepsilon[J_0(k_A r) \sin \omega_A t + J_0(k_B r) \sin \omega_B t],$$

where

$$k_A = \omega_A = 3.83171 \quad \text{and} \quad k_B = \omega_B = 7.01559.$$

Then a part of the quadratic correction interesting to us is equal to

$$\varphi_{\text{nonlin}}(r, \theta, z, t) = \varepsilon^2 f_1(r) \sin(\omega_A + \omega_B)t,$$

where

$$f_1(r) = \frac{\omega_A + \omega_B}{4} J_0(k_A r) J_0(k_B r) + \frac{\omega_A + \omega_B}{4} (k^2 - k_A^2 - k_B^2 + 2\omega_A \omega_B) W(r).$$

The graphic of the coefficient $f_1(r)$ is shown in Figure 2.

Example 2.

Assume that

$$\varphi_{\text{lin}}(r, \theta, z, t) = \varepsilon[J_0(k_A r) \sin \omega_A t + J_1(k_B r) \sin \theta \sin \omega_B t]$$

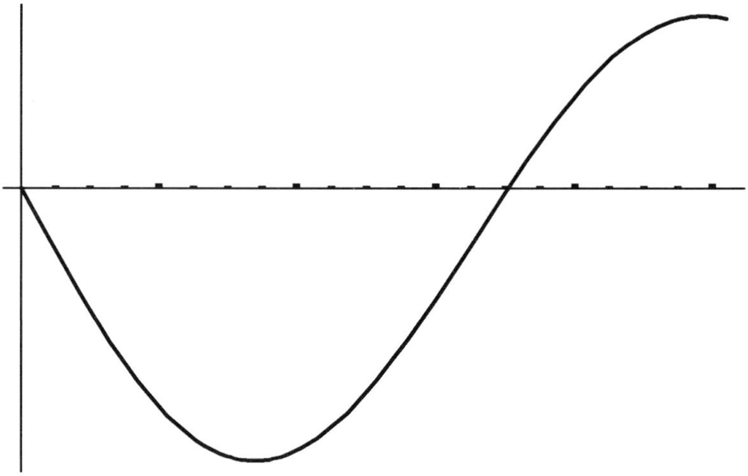

Fig. 3. Dependence $f_2(r)$ on r.

and the boundary condition in the first approximation gives

$$k_A = \omega_A = 3.83171 \quad \text{and} \quad k_B = \omega_B = 1.84118.$$

Then

$$\varphi_{\text{nonlin}}(r, \theta, z, t) = \varepsilon^2 f_2(r) \sin\theta \sin(\omega_A + \omega_B)t$$

$$f_2(r) = -\frac{\omega_A + \omega_B}{4} J_0(k_A r) J_1(k_B r) - \frac{\omega_A + \omega_B}{8k^2 k_B}(k^2 - k_A^2 - k_B^2 + 2\omega_A\omega_B)$$
$$\times [k_A J_1(k_A r) J_0(k_B r) - k_B J_0(k_A r) J_1(k_B r) - (k^2 - k_A^2 - k_B^2) W(r)].$$

The graph of the coefficient $f_2(r)$ of this harmonic is shown in Figure 3.

5 Conclusions

A new function $W(r)$ is introduced. Using this function, nonlinear solutions: (5), (8), (12) to nonlinear wave equation are found with the accuracy of ε^2 in the cylinder coordinates. The solutions can be checked by substitution into Equation (1′). They describe the nonlinear interaction of two waves.

The linear versions of these problems when the terms only of the first order in ε are retained were studied in a number of classical books of (see [1, 2]).

The derived formulas are obtained by a method of unknown coefficients as solutions of some *overdetermined* systems of algebraic linear equations. The reason for their solvability remains obscure at the moment.

Using this approach, a general boundary problem for a cylinder can be solved. In the first stage, we solve a linear problem. In the second stage, we calculate its nonlinear correction. In the third stage, we add the solution of the homogeneous equation (21) for satisfying the boundary conditions in second order with respect to ε.

A similar approach was used in [5, 6] for describing the long periodic water waves on a slope in the high-order shallow water approximation.

References

1. Lamb, H. (1932) *Hydrodynamics*, sixth edition. Cambridge University Press.
2. Rayleigh, W. (1929) *Theory of Sound*. McMillan, London.
3. Skudrzyk, E. (1971) *The Foundation of Acoustics*. Springer-Verlag, Wien, New York.
4. Mei, C.C. (1983) *The Applied Dynamics of Ocean Surface Waves*. Wiley.
5. Shermenev, A. and Shermeneva, M. (2000) Long periodic waves on an even beach. *Physical Review E*, 5:6000–6002.
6. Shermenev, A. (2001) Nonlinear periodic waves on a beach. *Geophysical and Astrophysical Fluid Dynamics*, 1–2:1–14.
7. Shermenev, A. (2003) Nonlinear acoustic waves in tubes. *Acta Acustica*, 89:426–429.
8. Shermenev, A. (2003) *Nonlinear Periodic Waves in Shallow Water*, LNCS, Vol. 2630. Springer-Verlag, pp. 375–386.
9. Shermenev, A. (2004) Separation of variables for the nonlinear wave equation in polar coordinates. *Journal of Physics, A*, 37:1–9.
10. Shermenev, A. (2005) Separation of variables for the nonlinear wave equation in cylindrical coordinates. *Physica D: Nonlinear Phenomena*, 212(3–4):205–215.
11. Shermenev, A. (2005) *Nonlinear Waves in a Rod*, LNCS, Vol. 3718. Springer-Verlag, pp. 379–386.

Evolution from the Hertzian Contact Model to Non-Hertzian Conditions for Fast Dynamic Simulations

Hugues Chollet, Michel Sebes and Jean Bernard Ayasse

INRETS, 2 av. Malleret-Joinville, F-94114 Arcueil Cedex, France
E-mail: {chollet, sebes, ayasse@inrets.fr}

Abstract. In the domain of railway dynamics, specific models have been developed based on Hertz and Kalker theory to calculate the normal and tangential stresses in a semi-analytic way, and inserted into multibody codes, in order to ensure fast and reliable simulation of the contact stresses in the presence of dry friction. Recently several advanced contact models have been developed which are able to manage non-Hertzian simulations online. The particular developments proposed by the authors are presented.

Key words: Hertz theory, Kalker theory, Winkler foundation, Coulomb friction, rolling contact, friction, railway, wheel, rail, stress.

This paper is dedicated to Joost Jacques Kalker, who left us in the beginning of 2006.

1 Introduction

Despite its long history, the railway transport system is still a research domain. The present work has been initiated by the need to improve the control on the fatigue and wear of modern rails.

In the domain of railway dynamic simulations, the use of contact models is constrained by the great number of contact states which are necessary to describe the wheelset dynamics, at a frequency currently over 1 kHz, for simulations covering several minutes, for example by simulating a passage in a curve.

2 The Railway Context

The most specific component of the railway system is the guiding and rolling mechanism of the rigid wheelset, used for more than 150 years.

Peter Wriggers and Udo Nackenhorst (eds.), IUTAM Symposium on Computational Contact Mechanics, 189–205.

2.1 Historical Background

The early development of railways has been mainly empirical, with the stabilisation of the rail and wheel shapes in the beginning of the 19th century. Then came the commercial development of the networks, starting from Europe and rapidly growing everywhere in the world, at a time where the energy was not so abundant than now. The low energy losses between the steel wheel and the rail, and the possibility of hauling heavy loads, were due to the contact qualities of the steel on steel elastic contact.

At the end of the 19th century, it became possible to model the normal contact forces between the wheelset and the track on the basis of the Hertzian theory (1885). Later, works from Carter on the stability of electric locomotives proposed a first model of the rolling-sliding contact (1926) to calculate the tangential forces up to saturation. In the sixties, Kalker [1] proposed an accurate description of the tangential creep forces generated by the "rolling contact with slip and spin".

These tangential force models give access to a reliable calculation of the critical speed of the vehicles on a given track, and it became possible [15] to take into account the shear stresses in the evaluation of the fatigue load of the rails.

2.2 Rolling Contact Fatigue

The first step before the evaluation of initiation and propagation of cracks is the evaluation of the loading conditions. This can be done partly by experiments: measurement of the forces, but it does not give access to the local position of the forces on the rail surface. The contact location must be described by models, where the dynamic variations of the load is given by a vehicle multibody model including itself a contact model.

Thirty years ago, the rail users were interested in the initiation and propagation of cracks from metallurgical defects inside the rail, an the Hertzian description of the contact without friction was sufficient, the maximal Von Mises stress being located under the contact at a depth of $0.4a$ (a being the longitudinal semi-axis of the contact ellipse), where a lot of crack initiation was observed. This type of cracks has been reduced by a better control of the steel inclusions. The residual problems occurs today with cracks named "squats" and "head checks". These defects are initiated in the neighbourhood of the surface; it has become useful to improve the description of the contact pressure and shear at the interface.

2.3 Multiple Contact Situations

Modern dynamic simulation models used in railway dynamics are based on Hertz' and Kalker's complementary theories, with some adaptations to use them in the non-strictly Hertzian situations.

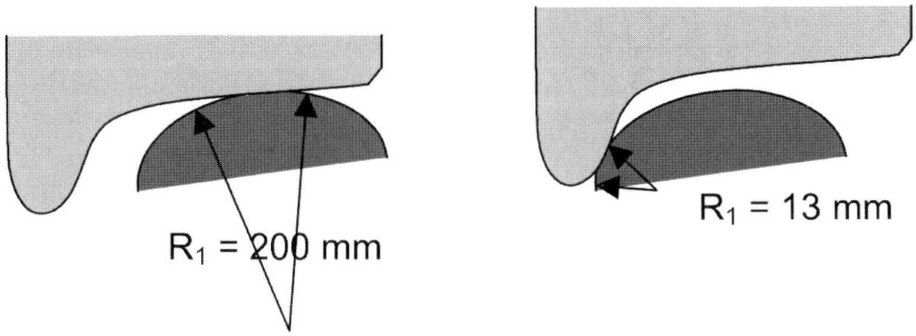

Fig. 1. Rail contact radius for two positions of the wheel.

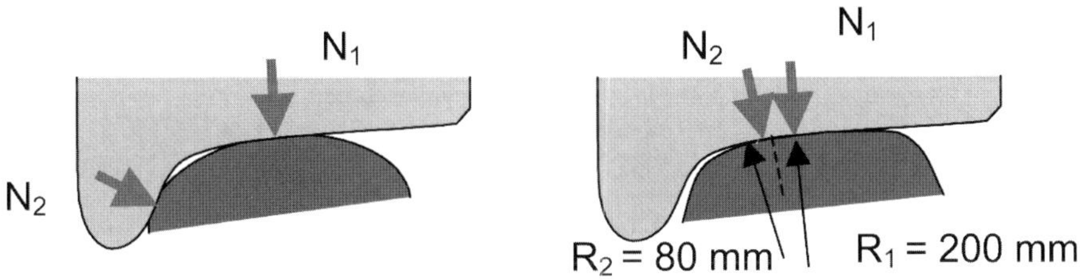

Fig. 2. Double contact in two different cases.

Between the two flange contacts, the wheelset can be shifted in any position in the "track play" up to derailment, as in Figure 1 right.

A wheel and a rail are not exactly two half space solids, and their transversal curvatures in different parts of the wheel and rail are not exactly constant. The standard rail has a radius of 200–300 mm on the tread, where the load is applied the most part of the time, and 13 mm at the flange. When it is clear that only one contact point occur, the curvatures are taken in the so called "rigid contact point" corresponding to the minimum distance between the two rigid profiles.

Basically, the shape of the wheel equipped with a flange can lead in curves to a double contact between the tread and the flange (Figure 2 left); in this case it can be considered as two separate Hertzian contacts.

Multi-Hertzian models [3, 7, 8] have been proposed to describe these situations, with particular considerations to dispatch the dynamic load into several Hertzian areas, considered independently.

When the two contact are far one from the other, it is possible to consider them independently, and to determine (Figure 3) the overlap Δty between the two loads as a function of the lateral shift ty.

An analytical formula is proposed by Ayasse and Chollet [14, p. 113], based on the Hertzian elasticity and the relative local deformations δ_i, different for each ellipse.

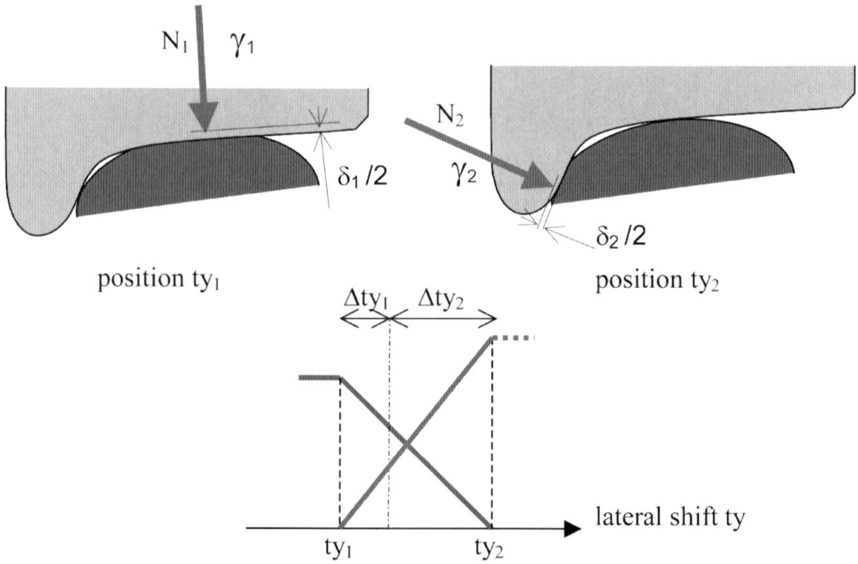

Fig. 3. Determination of the contact overlap from the local deformations.

$$\Delta t y_1 = \frac{\delta_1}{2} \frac{\cos \gamma_2}{\sin |\gamma_2 - \gamma_1|} . \tag{1}$$

The adequate "interpenetration" of the rigid profiles $h_o = f(\delta)$ will be discussed in Section 3. A simplified choice can be $\delta/2$ (Figure 3).

On the tread too, the variable curvature can lead again to non-Hertzian conditions. In some cases, for example with the European standard wheel and rail profiles, the discontinuous curvatures on the rail can lead to place the two contact very close one to the other (Figure 2, right). The analytical formula can deal with this situation, by considering the limits ty_1 and ty_2 where there are two single contacts. However the curvatures are no more constant in the contact vicinity, Hertz assumption are weak hypothesis between these two borders.

The improvement of the calculation power of micro computers led the authors to a more general method, suitable for these limit cases: "conformal" contacts, and in the case of variable profiles along the track like in a switch.

The load transfer determination is based on the elastic interaction between the wheel and the rail in order to determine a repartition of several normal loads between them, on a reduced set of contact points. This give the idea to extend it up to a sort of "Winkler foundation", using the complete "virtual interpenetration" or "indentation" of the two bodies [3, 4] (Figure 4).

However, it was necessary to find expressions of the normal and tangent stresses in continuity to the previous Hertzian model.

2.4 From Multiple Hertzian Contacts to Semi-Hertzian Models

The Winkler foundation method [4–6] can be used to describe a non-Hertzian pressure more directly, where the lateral repartition is proportional to the normal "interpenetration" of the transversal rigid profiles.

Particularly adapted to rolling contact, the method presented in this paper and named STRIPES is called "semi-Hertzian" because it considers a constant curvature in the longitudinal direction along contact strips.

3 The Normal Problem

The presentation of the new method starts from a review of the Hertzian theory.

3.1 Hertzian Solution to the Contact Determination

In Hertzian conditions, the vertical relative distance between two bodies 1 and 2 before being pressed can be written with a polynomial combination of the curvatures A and B:

$$z_1 + z_2 = d(x, y) = Ax^2 + By^2. \tag{2}$$

After being pressed, the classical Hertzian contact shape is an ellipse, whose semi-axes b and a are determined on the basis of the curvature ratio λ in the centre of the contact, or "rigid" contact point $\lambda = A/B$ [9].

In the following sections some adaptations of the Hertzian calculation method are presented, which are useful to further develop the semi-Hertzian model proposed by the authors.

3.2 Adaptation of the Hertzian Calculation Method

A/B and b/a ratios vary in the same way: if $A > B$, then $b > a$. The equality ($B/A = 1$) leads to a circular contact patch. The traditional calculation is based on the determination of the ellipticity $g < 1$ function of $\lambda = B/A$ by using an intermediate parameter, the angle θ defined (see for example [9, 14]). m, n and r are non-dimensional coefficients tabulated as a function of the ratio $g = n/m$ or the angle θ.

Because of symmetry properties, the traditional table is limited to $0 < g < 1$. This leads to a switch between $g = b/a$ or $g = a/b$, while at the same time $g = n/m$ becomes $g = m/n$. It is proposed to avoid this ambiguous situation (Table 1) with a description of the traditional tables extended to the domain $0 < g < \infty$. Then, in all the cases, $g = a/b = m/n$.

The same principle will be adopted further for Kalker tables, used in the tangential force domain.

Table 1. Hertz coefficients from 0 to 180°.

$\theta°$	0	5	10	30	60	90	120	150	170	175	180
$\lambda = A/B$	0	0.0019	0.0077	0.0717	0.3333	1	3.0	13.93	130.6	524.6	∞
n/m	0	0.0212	0.0470	0.1806	0.4826	1	2.0720	5.5380	21.26	47.20	∞
m	∞	11.238	6.612	2.731	1.486	1	0.7171	0.4931	0.311	0.2381	0
r	0	0.2969	0.4280	0.7263	0.9376	1	0.9376	0.7263	0.4280	0.2969	0

Practically, in the railway dynamic codes, the parameters are calculated with a polynomial fit of these extended tables, the solution of which is faster than the interpolation.

3.3 Determination of the Hertzian Ellipse by Indentation

As an alternative to the Hertzian tabulated method, it is possible to interpenetrate virtually the two surfaces and to determine the contour of the interpenetration region. It is also an ellipse. However its size and patch is not the same than Hertz solution.

The first considerations of the method are applied to a Hertzian contact, where the parameters: curvatures, interpenetration, etc., are considered in the centre of the contact.

The last Hertzian parameter of Table 1, r, is used to calculate the parameter δ, sometimes called erroneously "interpenetration". It is in fact the value of the elastic approach between the two bodies, the undeformed references being taken far from the contact: at the curvature centres by example.

Kalker and Pascal and Sauvage [3] observed that this δ value led to a larger contact area than the Hertzian ellipse, and avoided to use it. Vohla [5] and Ayasse and Chollet [9] observed that an interpenetration value close to $\delta/2$ gives a good approximation of the contact area with the contour of the interpenetration region.

These authors mentioned the discrepancy between a basic geometrical interpenetration method and Hertz' b/a ratio for slender ellipses. In the following section, an exact analytical expression (4) of the interpenetration is established after the treatment of a curvature (3) proposed to solve the last problem.

3.3.1 Compensation of the Curvature to Obtain the Elliptic Ratio

With the interpenetration method, the geometrical semi-axis related to the normal "interpenetration" h_0 (Figure 4) gives a ratio different from the Hertzian one, b/a except for a circular patch.

To obtain the correct ratio, it is proposed to compensate the local value of one or the two curvatures A and B. The simplest expression is obtained by the compensation of the A curvature only, which corresponds to the rolling direction, using only the m and n parameters:

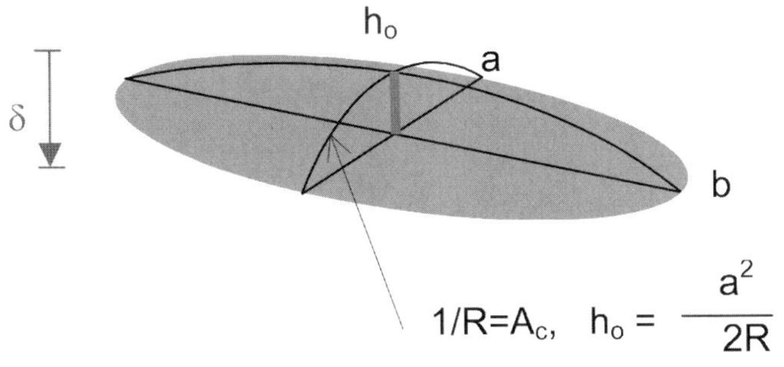

Fig. 4. Geometrical correspondence between the "indentation" h_0 and (a, b).

$$\frac{A_c}{A} = \frac{(b/a)^2}{\lambda} = \frac{(n/m)^2}{\lambda}. \tag{3}$$

In the following part of the paper this compensated curvature will be used only for the calculation of the length a of the ellipse.

The compensation of both A and B curvatures has been proposed in [9]. However, for slender ellipses, a distortion appears again, so the choice has been made of the compensation of the A curvature sum only in the railway case. In the case of a very large contact, $b \gg a$, Cloupet [11] proposed a different compensation.

3.3.2 Analytical Expression of the Interpenetration

Due to the correction of one of the curvatures, it is possible to express the interpenetration h_0, both based on B or on A_c with the classical expression of the "bow and arrow" where the arrow is the interpenetration h_0:

$$h_0 = a^2 A_c = b^2 B = \frac{n^2}{r} \frac{\delta}{1+\lambda}. \tag{4}$$

This expression is restricted to the Hertzian case (Figure 4).

3.3.3 Interpenetration in the Case of Non-Constant Curvatures

This is the general case. A simplified representation is given in Figure 5.

The displacement h_0 is applied to the rigid profiles. The length a_i of each strip is determined by the previous formula following Figure 4.

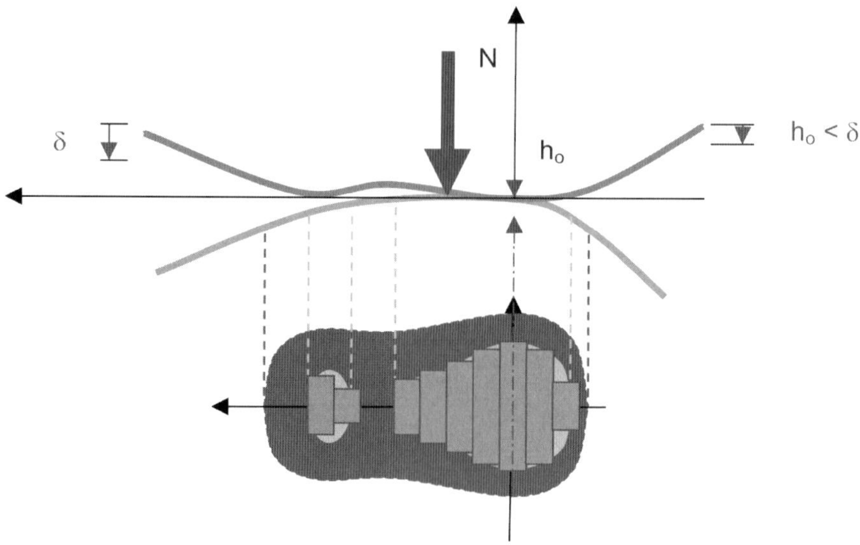

Fig. 5. Semi-Hertzian model of the STRIPES algorithm.

4 The Tangential Contact

4.1 Creepages and Forces Applied on a Wheelset

In the case of the railway contact, the normal and the tangential problem are considered independently because of the use of comparable steels.

4.1.1 Creepages

The tangential contact forces are calculated from the "creepages": relative speed at the level of the contact point. Using a multibody approach, the local speeds are calculated from the rigid bodies velocities.

The longitudinal creepage generating the longitudinal steering forces can be deduced from the difference of rolling radius $\gamma t_y / r_0$, generated by the lateral shift t_y, while the lateral creepage generating the lateral guiding forces is mainly dependent of the yaw angle α (Figure 6). See for example [15] for a complete definition.

The spin creepage has a minor influence on the wheelset dynamics when the contacts occur on the tread, because the two couples generated by the two wheels are opposed. However when flanging, it must be considered, and the local effect on the distribution of contact stresses cannot be neglected.

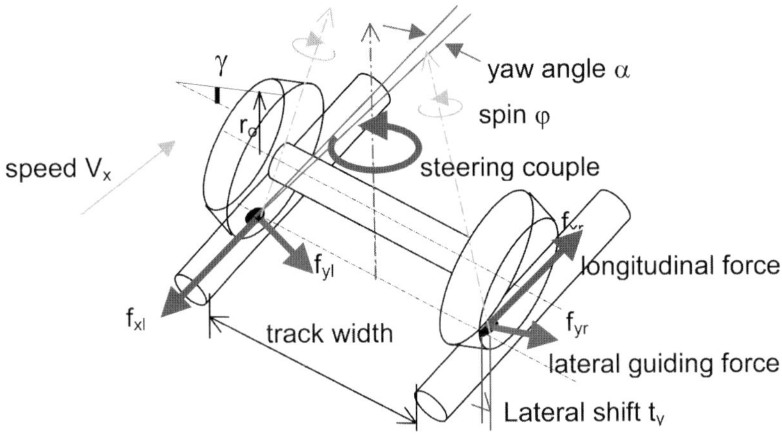

Fig. 6. Forces and creepages in the railway wheelset case.

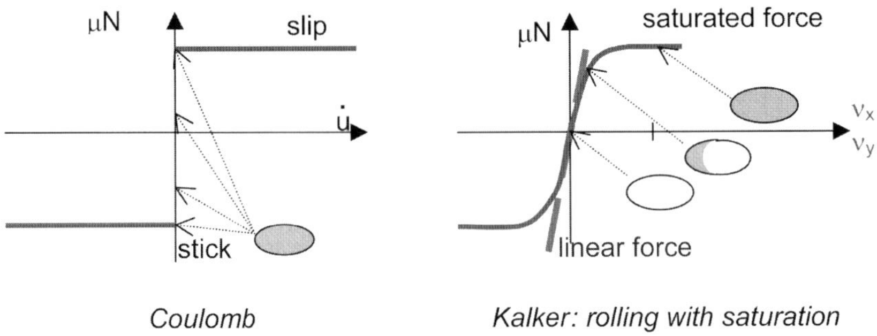

Fig. 7. Coulomb's and rolling contact model.

4.1.2 Saturation of the Tangential Forces

The basic saturation expressions proposed by Carter, Vermeulen and Johnson [20] or Kalker [1], are in continuity with the Coulomb friction model (Figure 7).

Kalker, in the simplified case of Hertzian contacts, proposed the calculation of a compliance in series with the coulomb model, the slope c_{ij} being calculated from the local ellipticity b/a. These stiffness coefficients have been tabulated by Kalker [1] for several Poisson's coefficients. A table for the case of steel is given in [15].

4.2 Kalker's Non-Hertzian Contact Model

In the general case where the contact is not Hertzian and where there is spin, this model cannot be used directly.

The rolling contact general model proposed by Kalker used a variational approach to solve both the normal and the tangential problem, under the software name CONTACT. Such a method is limited to the elastic domain inside the two contacting bodies, but it is able to manage non-Hertzian pressures. In the railway domain, this code is only used for comparison because it needs several seconds to calculate one contact case.

Instead of accurate iterative models, like FEM or Boussinesq models, simplified and faster methods are necessary in order to remain close to real time simulations for engineering calculation.[1] This led for example Kalker to propose on the one hand a complete theory for rolling contact in its PhD dissertation in 1967 [1] and on the other hand a simplified theory restricted to Hertzian rolling contact for engineering applications [2].

4.3 Kalker's Simplified Theory and the FASTSIM Algorithm

In 1987 Kalker published a complete algorithm to deal with the rolling contact stresses in the case of ellipses with spin. The contact is discretised into a matrix, where elements are defined along longitudinal strips.

The name "simplified theory" comes from the shear stress repartition strategy. Pre-established variations are proposed by Kalker in order to accelerate the numerical method.

Figure 9 presents two different cases; in the general case they are combined on each strip and saturated by the same friction coefficient, in the sense of Coulomb.

4.3.1 Elliptic Pressure and Parabolic Traction Bound

The normal pressure n_i is assumed elliptic in the longitudinal direction, along a strip. However, the traction bound μn_i remains parabolic.

This was proposed originally by Kalker to obtain the leading edge saturation when the creepages reach a certain level, which is impossible with an elliptic traction bound. Beyond this level, the Coulomb friction must occur.

The difference between the elliptic repartition and the parabolic one is an approximation on the shear stress description and/or on the local friction coefficient value μ_i, however it must be compared to the other approximations and particularly the linear gradient of this simplified theory.

4.4 Adaptation of the Tangential Method to a Semi-Hertzian Repartition

In the general case, the contact parameters will not be considered in the centre of the potential contact surface but in the centres of all the longitudinal strips on the contact surface.

[1] The calculation of a complete high speed train can lead to calculate 60 wheel rail contact at each time step, at a frequency target of 10 kHz.

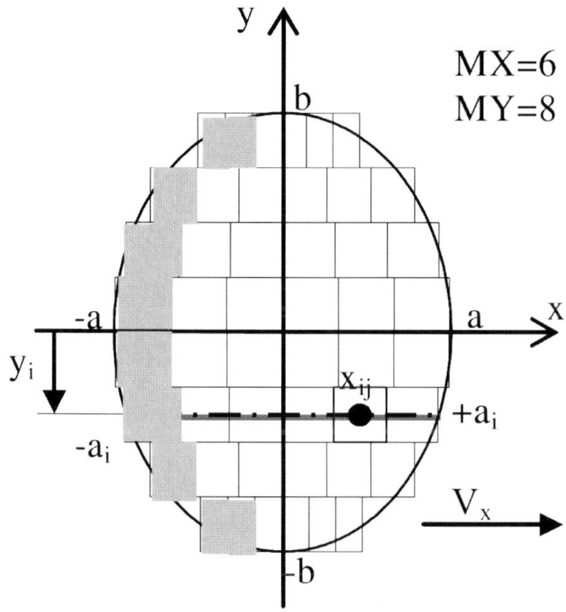

Fig. 8. FASTSIM: discretisation in strips and elements.

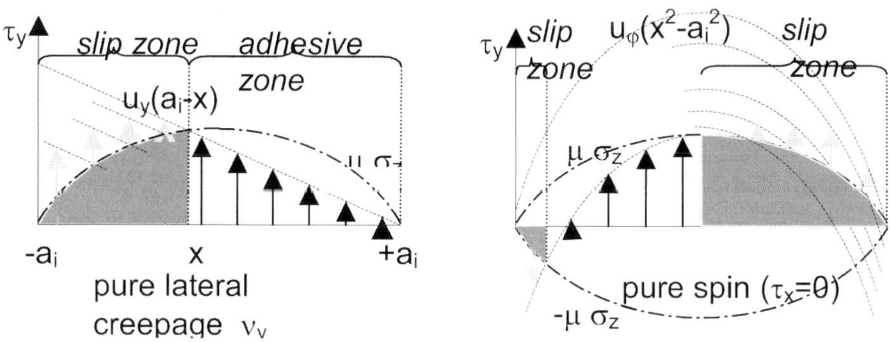

Fig. 9. FASTSIM: step-by-step variations from the leading (a_i) to the trailing edge ($-a_i$), example for two pure cases.

The rolling contact of a wheel with a regular rolling radius on a straight rail led to the choice of longitudinal strips on the rail, where the pressure shape is considered elliptic between the limits $+a_i$ and $-a_i$. The normal pressure distribution in the lateral direction will no more remain elliptic, as the lateral curvatures are not constant in the contact patch. For these reasons the contact model is named semi-Hertzian.

4.4.1 Local Stresses on the Strips

The local stresses are calculated from the local parameters only. The local radial interpenetration h_i determines the local contact bound but not the pressure repartition.

The shear stress gradient expression in Kalker's theory is respected in the expression of the longitudinal compliance:

$$L_1 = \frac{8a}{3c_{11}G} = \frac{8a_1}{3c_{11}G(a_i/a)} = \frac{8a_i}{3c_{11}Gk_i}. \tag{5}$$

Note that the lateral compliance L_2 is unchanged and the spin corresponding one L_3 is not used with the present method (see [9, §3.3]).

As it can be seen from (5), the parameter k_i is expressed as a function of local parameters a_i, k_i, h_i, and no more from the ellipse value a, with the help of:

$$k_i^2 = \frac{8a}{3c_{11}G} = \frac{8a_i}{3c_{11}G(a_i/a)} = \frac{8a_i}{3c_{11}Gk_i}. \tag{6}$$

Thus, the shear stresses and traction bound expressions from Kalker becomes, in the case where A only is compensated:

$$\left.\begin{aligned} \sigma_{xi}(x, y_i) &= \frac{3}{8}Gc_{11}\nu_{xi}\left(1 - \frac{x}{a_i}\right)k_i, \\ \sigma_{yi}(x, y_i) &= \left(\frac{3}{8}Gc_{22}\nu_y\frac{2}{\pi}\sqrt{\frac{n}{m}}Gc_{23}\varphi_i(a_i + x)\right)\left(1 - \frac{x}{a_i}\right)k_i, \\ \mu\sigma_{zi}(x, y_i) &= \frac{4}{3\pi}\mu\frac{1+\lambda}{n^3}\frac{E}{1-\nu^2}\frac{h_i}{a_i}\left(1 - \left(\frac{x}{a_i}\right)^2\right)k_i. \end{aligned}\right\} \tag{7}$$

The only difference between the local expressions (7) and those from Kalker (9–11) is a term of spin in the longitudinal stress, which disappears in the semi-Hertzian method and is replaced by the variation of the longitudinal creepage ν_{xi} [9, figure 3].

In some cases, when there is a jump between two close patches, the area between them can give locally a negative value to B_i. This area must be considered as a conformal contact when the sign of the local virtual interpenetrations h_i does not change.

The corresponding B_i values are modified to obtain a minimal numerical value in accordance with the limit of coefficients given by Hertz and Kalker. For the corresponding strips, the limit of the k_i value is taken to be 1. On the example of Figure 10 it is possible to locate the modified values of B in the considered region of contact.

It has been observed that the present method led generally to intermediate conformal contacts between the main areas, while FEM models does not present this continuity; this can be attributed to the Winkler foundation approximation, for which there is no local interference between the strips.

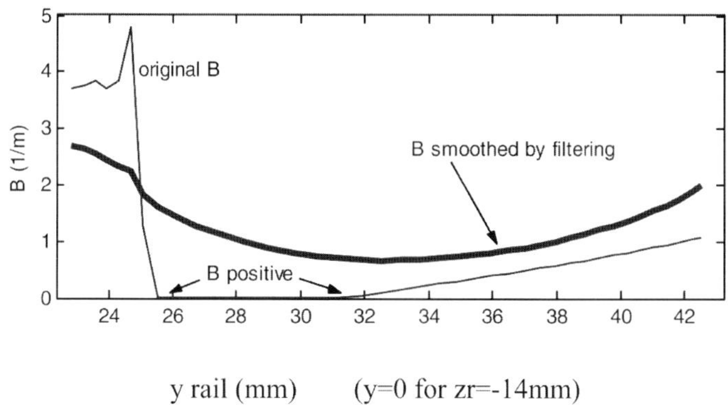

Fig. 10. Smoothing of the curvature sum B, case S1002/UIC60 centered.

4.4.2 Smoothing of the Combined Curvature B

The rude method gives sharp variations where there are discontinuities of the curvatures along the profile. In order to take into account the shear phenomenon which is neglected by the basic strip method adopted by Kalker with FASTSIM, the semi-Hertzian method is completed by a smoothing process of the function $B(s)$ into $B_f(s)$. The filter proposed in [9] is based on the Boussinesq approach: the characteristic width depends of the curvature value and is $\sqrt{ab}$ for a virtual circular contact $(mn = 1)$.

This last step of the method is unfortunately not based on analytical considerations. The smoothing process has been adjusted and relatively validated by comparison to contact patches found with FEM tools [13]. It is not possible to obtain a so good correspondence between this direct method and the FEM, but this lack of accuracy is counterbalanced by the possibility to calculate faster.

This method, for which the name STRIPES has been chosen, has been implemented online by the authors in their own dynamic MBS code, VOCOLIN. It is possible also to use it off line, in open loop, starting from the output of a MBS with a multi-Hertzian contact model, as Enblom and Berg did [17] with the MBS code GENSYS.

5 Application and Extension of the Method

Practical application have shown a more reliable calculation of the flanging situation, particularly when the contact appears between worn profiles, and a simplification of the simulation in turnouts.

The STRIPES method has been applied to the simulation of the wheel rail contact when passing a turnout at high speed [18]. The industrial question was the sensibility

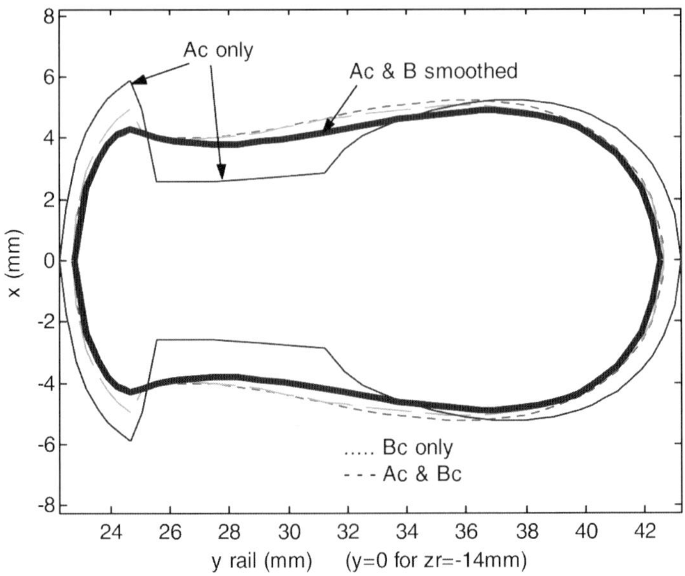

Fig. 11. Effect of correction and smoothing on patch in the case of theoretical profiles, S1002/UIC60 centered.

of the vehicle dynamics to the grinding of some parts of the switch which were between the main contact areas. The classical multi-Hertzian models, considering the Hertzian parameters in the centre of each contact, were unable to see a difference, even with grinded zones situated at a few millimetres near a contact centre.

The advantage of the new method is to take into account the whole potential surfaces in contact, without specific considerations on the contact jumps and on separate contact areas between the wheel and the rails (Figure 12).

The main improvement of this method is a better expression of the contact stresses for the determination of the surface fatigue and abrasive wear behaviour of rails and wheels. Note that due to the Saint-Venant principle, this improvement has no influence on the internal fatigue.

This research has opened the field to more accurate wear [9, 10, 17] and rolling contact fatigue simulations in the railway domain, but also in the ball bearing domain in the case of dry friction.

An extension of the method is in preparation for the description of the rubber tyre contact with the road in multibody simulations.

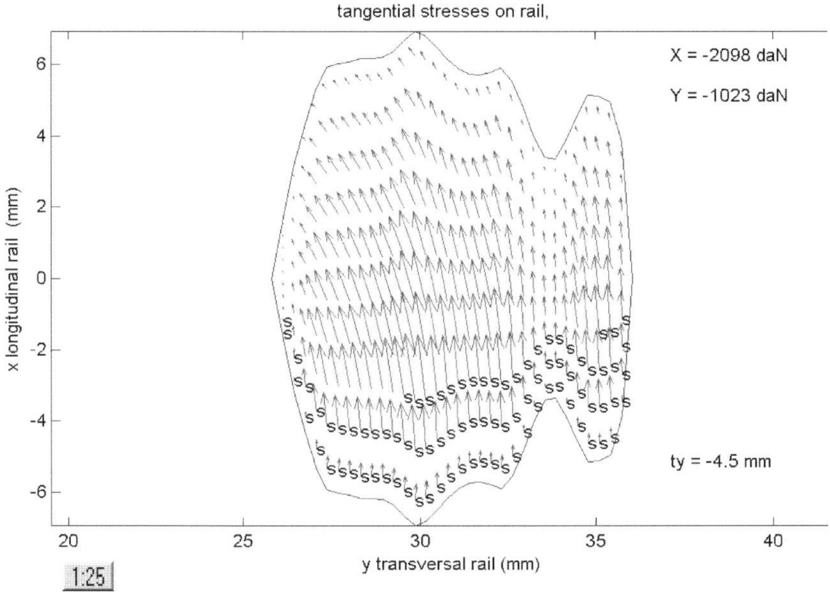

Fig. 12. Contact patch in the case of worn profiles, example.

6 Conclusion

This research is dedicated to the fast resolution of the rolling contact with slip and spin, following the way opened by Professor J.J. Kalker.

In Hertzian cases, it led to the improvement of the so called "interpenetration method" to obtain more accurately the Hertzian patch with a compensation of the transversal curvatures A. With this modification it has been shown that, in the Hertzian case, the interpenetration value has an analytical formulation depending only of the known Hertzian parameters. A formulation of the Hertzian tables function of A/B must be preferred to prepare the parameters used in the semi-Hertzian method.

In a non-Hertzian case, in comparison to more complete methods, the new one leads to a similar contact patch if a smoothing operation on the longitudinal B curvatures is done with a "Boussinesq filter".

For the tangential stresses when rolling, expressions (13) compatible with those from Kalker have been established with minor simplifications.

The industrial success of this research led to a good compromise between the accuracy and the speed computing performance of the new algorithm comparatively to FEM models and to the previous methods, giving access to real time simulations when simulating railway vehicles at 10 kHz step time.

Acknowledgements

This research was supported by the railway operators RATP, SNCF, and the rail manufacturer CORUS, in the framework of the IDR2 research group.

References

1. Kalker, J.J., On the rolling contact of two elastic bodies in the presence of dry friction, PhD dissertation, Department of Mechanical Engineering, Delft, 1967.
2. Kalker, J.J., A fast algorithm for the simplified theory of rolling contact (FASTSIM program). *Vehicle Systems Dynamics*, 11:1–13, 1982.
3. Pascal, J.P. and Sauvage, G., The available methods to calculate the wheel rail forces in non-Hertzian contact and rail damaging. *Vehicle Systems Dynamics*, 22:263–275, 1993.
4. Kik, W. and Piotrowski, J., A fast, approximate method to calculate normal load at contact between wheel and rail and creep forces during rolling. In: *Proceedings of 2nd Mini-Conference on Contact Mechanics and Wear of Rail/Wheel Systems*, Budapest, July 29–31, 1996.
5. Vohla, G., Werkzeuge zur realitatsnahen Simulation der Laufdynamik von Schienenfahrzeugen. Diss. ETH Nr 11300, VDI Verlag Düsseldorf, 1996.
6. Linder, C., Verschleiss von Eisenbahnradern mit Unrundheiten. Diss, ETH Nr 12342, Zurich, 1997.
7. Ayasse, J.B., Chollet, H. and Maupu, J.L. Paramètres caractéristiques du contact roue-rail. Rapport de Recherche INRETS No. 225, ISSN 0768-9756, January 2000 [in French].
8. Chevalier, L. and Chollet, H., Endommagement des pistes de roulement, *Mec. Ind.*, 1:77–103, 2000 [in French].
9. Ayasse, J.B. and Chollet, H., Determination of the wheel rail contact patch in semi-Hertzian conditions, *Vehicle Systems Dynamics*, 43(3):161–172, 2005.
10. Chollet, H. and Piotrowski, J., Wheel rail contact models for vehicle system dynamics including multi-point contact, *IAVSD State of the Art* (Special Issue), *Vehicle Systems Dynamics*, 43(6–7):455–483, 2005.
11. Cloupet, S., Computational of superficial wear by creepage of rolling contact between roller and cam: Probabilistic approach of experimental dispersions, PhD Dissertation, ENS-Cachan, 2006 [in French].
12. Quost, X., Sensibilité des motrices TGV au vent traversier, PhD Thesis, ECL, Lyon, 2005.
13. Quost, X., Sebes, M., Ayasse, J-B., Chollet, H., Gautier, P-E. and Thouverez, F., Assessment of a semi Hertzian method for determination of wheel-rail contact patch, *Vehicle Systems Dynamics*, 44(10): 789–814, 2006.
14. Ayasse, J.B. and Chollet, H., Wheel-rail contact, in *Handbook of Railway Vehicle Dynamics*, S. Iwnicki (Ed.), Taylor & Francis, 2006, pp. 85–120.
15. Piotrowski, J., Contact loading of a high rail in curves. Physical simulation method to investigate shelling. *Vehicle Systems Dynamics*, 17:57–79, 1998.
16. Kalker, J.J., *Three Dimensional Elastic Bodies in Rolling Contact*, Kluwer Academic Publishers, Dordrecht, 1990.
17. Enblom, R. and Berg, M., Impact of non-elliptic contact modelling in wheel wear simulation, in *Proceedings Contact Mechanics and Wear of Rail/Wheel Systems Conference*, Brisbane, September 2006.

18. Sebes, M., Ayasse, J.B., Pouligny, P. and Pirat, B., Application of a Semi-Hertzian method to the simulation of vehicles in high speed switches, in *Proceedings of XIX IAVSD Symposium*, Milano, 2005.
19. Johnson, K.L., *Contact Mechanics*, Cambridge University Press, 2001.
20. Vermeulen, P.J. and Johnson, K.L., Contact of non spherical elasic bodies transmitting tangential forces, *Trans. ASME, Journal of Applied Mechanics*, 338–340, June 1964.

A Fully Implicit Approach for the Numerical Treatment of Tractive Rolling

M. Ziefle and U. Nackenhorst

Institute of Mechanics and Computational Mechanics, Leibniz University of Hannover, Appelstr. 9A, 30167 Hannover, Germany
E-mail: {ziefle, nackenhorst}@ibnm.uni-hannover.de

Abstract. For the finite element analysis of rolling contact problems Arbitrary Lagrangian Eulerian (ALE) methods are well established. These techniques enable a time independent formulation for elastic bodies under stationary rolling conditions and for local mesh refinement concentrated to the contact region. A drawback is on the computation of history dependent material properties because the path of material points is not traced inherently. This affects inelastic constitutive behavior as well as frictional contact. In this contribution a mathematically sound approach for the treatment of frictional rolling within the ALE description based on a spatially fixed finite element mesh is suggested. By this novel and fully implicit algorithm the slip velocities are integrated along their path-lines which enables for the treatment of frictional contact as in a material picture. Quadratic convergence behavior and physical reliability will be demonstrated as well as the computability of large scaled finite element tire models.

Key words: Rolling contact, tire, Arbitrary Lagrangian Eulerian (ALE) method, frictional contact.

1 Introduction

In many industrial applications contact occurs between rolling bodies. A famous example are car-tires rolling on roads, where the local reactions in the finite contact patch even under high speed conditions are essential for safety, comfort etc. Thus, reliable and efficient computational methods for rolling contact including frictional processes are needed to give essential advices in construction and improvement of tires, brakes or electronic driving safety systems.

Analytical solutions for rolling contact problems, e.g. Carter [3], are based on the theory of elastic half-spaces and therefore limited to linear elastic constitutive behavior and small displacement kinematics. First finite element approaches for the solution of steady state rolling have been introduced by Oden and Lin [15] and

Peter Wriggers and Udo Nackenhorst (eds.), IUTAM Symposium on Computational Contact Mechanics, 207–223.

Padovan [16]. These early theories based on plane kinematics were developed rapidly, e.g. extended to three-dimensional motions and applied to large scale tire models [4]. The relationship to the more general Arbitrary Lagrangian Eulerian (ALE) methods was mentioned first in [11]. Nowadays the ALE approach which provides a more general kinematical framework is well established for rolling contact analysis in tire industries. However, in contrast to the rapid development of Arbitrary Lagrangian Eulerian methods in other fields of engineering application, like fluid-structure interaction or metal forming processes for example, compare [5], the improvements in rolling contact mechanics have been moderately. An example is the treatment of inelastic material properties, where usually engineering approaches are used to integrate the history variables along predefined rings of integration points, e.g. [4, 14, 15]. This approach seems to be established in the rolling contact community although problems with unstructured meshes are obvious and a sound mathematical basis is missing. A first step in this direction has been presented by Le Tallec and Rahier [10] who applied methods established in fluid mechanics for a continuous transport of viscoelastic history variables within the spatially fixed ALE mesh. These authors suggested finite difference upwind-schemes for the solution of the underlying advection problem. In contrast to this only conditional stable approach an unconditional stable algorithm based on a Time Discontinuous Galerkin (TDG) schema has been proposed more recently [13, 20]. These methods enable the treatment of arbitrary inelastic material properties within a general thermodynamic consistent constitutive theory. In general, these approaches end up in overall explicit algorithms for tracing the history of material particles and their feedback onto the equations of motion.

For the frictional contact similar problems arise within the ALE framework, the history of particles which initially got into contact has to be traced to evaluate the friction law. This rather delicate problem has not been discussed extensively in literature so far. A common procedure is the penalization of the slip velocities which are computed directly within the relative kinematic formulation, cf. e.g. [6, 9] or the commercial software ABAQUS (Version 6.6) [1]. However, in [12] it has been shown that a simple penalization of the slip velocities is not sufficient to obtain physically reliable results. In addition the impulse flux over the boundaries of the contact region has to be considered for the enforcement of the stick conditions. A further complication arises from the fact that by this approach the tangential contact traction is not computed directly from a constitutive law. Additional intermediate projection steps are needed for the evaluation of the friction law which ends up in an overall explicit schema. In [12] a staggered schema for the treatment of frictional rolling contact has been suggested which suffers under poor convergence behavior.

In this contribution a novel fully implicit algorithm for the treatment of the frictional rolling contact is suggested. These developments are based on the experiences made for the treatment of inelastic material properties by tracing the path of the material points. This enables for the application of efficient algorithms well developed within the Lagrangian picture, e.g. return mapping schemes for frictional contact. The physical reliability and computational behavior of the presented approach is discussed in comparison with analytical solutions. The applicability for industrial problems is demonstrated by computations on a rather detailed tire-model.

2 Theoretical Framework

2.1 ALE Formulation of Rolling Bodies

The rolling contact is described in an Arbitrary Lagrangian Eulerian (ALE) formulation outlined just briefly in this section, for details the reader is referred to [12]. Introducing an arbitrary moved reference configuration between the initial and the current configuration leads to a decomposition of motion into a rigid body rotation and a pure deformation. Whereas the rigid body motion is described in spatial (Eulerian) coordinates, the generally large deformations are formulated in a material (Lagrangian) manner, measured relative to the reference configuration. Mathematically this decomposition is formulated by a multiplicative split of the deformation gradient,

$$\boldsymbol{F} = \hat{\boldsymbol{F}} \cdot \boldsymbol{R}, \tag{1}$$

where $\hat{\boldsymbol{F}}$ and $\boldsymbol{R}$ describe the material deformation and the rigid body motion respectively. From this relative kinematics a spatially fixed reference configuration $\chi(B)$ is concluded.

The material velocity splits into a relative and a convective part,

$$\boldsymbol{v} = \left.\frac{\mathrm{d}\boldsymbol{\varphi}}{\mathrm{d}t}\right|_X = \left.\frac{\partial \boldsymbol{\varphi}}{\partial t}\right|_\chi + \operatorname{Grad}\boldsymbol{\varphi} \cdot \boldsymbol{w} = \hat{\boldsymbol{v}} + \boldsymbol{c}, \tag{2}$$

where the gradient is computed with respect to the coordinates of the reference configuration, $\boldsymbol{\varphi}$ is the spatial location which is actually passed by an arbitrary material point and $\boldsymbol{w}$ is the guiding velocity due to the rigid body motion of the wheel. The symbols $\hat{\boldsymbol{v}}$ and $\boldsymbol{c}$ describe the relative and the convective velocity. Note that for stationary motions the relative velocity vanishes.

The weak form of the equation of motion written with respect to coordinates in the reference configuration reads

$$\int_{\chi(B)} \hat{\varrho}\, \dot{\boldsymbol{v}} \cdot \delta\boldsymbol{\varphi}\, \mathrm{d}\hat{V} = - \int_{\chi(B)} \hat{\boldsymbol{P}} \cdot\cdot \operatorname{Grad}\delta\boldsymbol{\varphi}\, \mathrm{d}\hat{V} + \int_{\chi(B)} \hat{\varrho}\, \boldsymbol{b} \cdot \delta\boldsymbol{\varphi}\, \mathrm{d}\hat{V} + \int_{\partial_t \chi(B)} \boldsymbol{t} \cdot \delta\boldsymbol{\varphi}\, \mathrm{d}\hat{A}, \tag{3}$$

with the virtual displacement $\delta\boldsymbol{\varphi}$. $\hat{\boldsymbol{P}}$ represents the first Piola–Kirchhoff stress tensor and $\boldsymbol{b}$ and $\boldsymbol{t}$ denote the external volume and surface loads. For stationary motions the inertia term is reformulated as follows

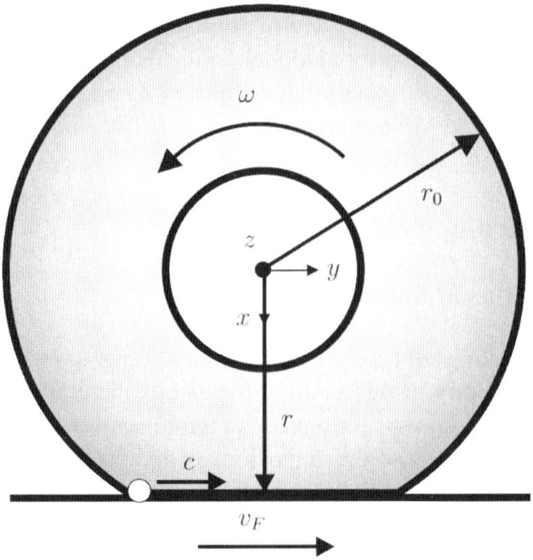

Fig. 1. Kinematics of rolling.

$$\int_{\chi(B)} \hat{\varrho}\, \dot{\boldsymbol{v}} \cdot \delta\boldsymbol{\varphi}\, \mathrm{d}\hat{V} = - \int_{\chi(B)} \hat{\varrho}\, \boldsymbol{c} \cdot (\mathrm{Grad}\, \delta\boldsymbol{\varphi} \cdot \boldsymbol{w})\, \mathrm{d}\hat{V} + \int_{\partial\chi(B)} \hat{\varrho}\, \delta\boldsymbol{\varphi} \cdot \boldsymbol{c}\, \boldsymbol{w} \cdot \hat{\boldsymbol{n}}\, \mathrm{d}\hat{A}, \quad (4)$$

where the first part describes the centrifugal forces due to rotation and the surface integral expresses the impulse flux over the boundary of the domain. It is argued that this part vanishes on natural boundaries of a rolling wheel, because the outward unit normal $\hat{\boldsymbol{n}}$ is always perpendicular to the guiding velocity. The linearized and discretized form of this nonlinear equation (3) results in the incremental finite element equation for the motion of stationary rolling bodies,

$$\left[{}^{t}\boldsymbol{K} - \boldsymbol{W}\right]\left[\Delta\,\boldsymbol{\varphi}\right] = \left[{}^{t+\Delta t}\boldsymbol{f}_e + {}^{t}\boldsymbol{f}_i - {}^{t}\boldsymbol{f}_\sigma\right]. \quad (5)$$

Herein ${}^{t}\boldsymbol{K}$ describes the tangential stiffness matrix and $\boldsymbol{W}$ is the ALE inertia matrix. On the right hand side ${}^{t+\Delta t}\boldsymbol{f}_e$ describes the externally applied forces, ${}^{t}\boldsymbol{f}_i$ are the equivalent nodal forces caused from inertia effects and ${}^{t}\boldsymbol{f}_\sigma$ describes the internal forces due to the divergence of the stress state.

2.2 Treatment of ALE Rolling Contact Problems

In Figure 1 the kinematics of the stationary rolling wheel is sketched. This model can be associated with a deformable wheel that rolls on a rigid drum with a radius $R_d \to \infty$. The wheel with initial radius r_0 rotates with a constant angular velocity ω. Because of the axis load the wheel is flattened and the rolling radius changes to r. The material particles of the rigid drum move with the velocity $\boldsymbol{v}_F$ in positive y-direction. This velocity is interpreted as ground velocity. For any material particle of the wheel the guiding velocity $\boldsymbol{w}$ and the convective velocity $\boldsymbol{c}$ can be easily computed from (2).

The normal contact constraints can be treated as spatial quantities. For the enforcement of the rolling contact conditions in that direction standard approaches well established for the material description of contact problems, e.g. [8, 19], can be applied and will not be discussed in detail in this paper. In tangential direction the rolling contact kinematics is deduced in a different way. The classical way within a material picture is to compute the relative slip distance s_α of two contacting particles. The time derivative $\dot{s}_\alpha$ of the slip is called slip velocity. Both, s_α and $\dot{s}_\alpha$ with $\alpha = 1, 2$, are computed at the contact surface with the tangential vectors $\bar{\boldsymbol{a}}_1, \bar{\boldsymbol{a}}_2$. The sticking condition then reads

$$s_\alpha = 0 \quad \text{respectively} \quad \dot{s}_\alpha = 0. \tag{6}$$

In consequence of the Eulerian description of the rigid body motion the relative slip distance that is accumulated by a material particle since its entry into the contact zone is not determined directly. One possibility is the integration of the known slip velocities. In the stationary ALE formulation of rolling the integration in time is substituted by the spatial integration along the path-line of a material particle which moves through the contact area with the convective velocity $c_\alpha = \boldsymbol{c} \cdot \bar{\boldsymbol{a}}_\alpha$. In a componentwise notation the integration along the path coordinate ζ then follows to

$$s_\alpha(\zeta) = s_{0\alpha} + \int \dot{s}_\alpha \, \mathrm{d}t = s_{0\alpha} + \int\limits_{\zeta_0}^{\zeta} \frac{\dot{s}_\alpha(\bar{\zeta})}{c_\alpha(\bar{\zeta})} \, \mathrm{d}\bar{\zeta} \quad \text{with} \quad \mathrm{d}\bar{\zeta} = c_\alpha(\zeta) \, \mathrm{d}t. \tag{7}$$

In Figure 2 the distribution of the local slip velocity and the computed distribution of the local slip distance are shown. Obviously, the right curve represents the spatial integral of the left curve.

In the special case of elastic-rigid rolling contact the slip velocities are computed as difference between the tangential part of the convective velocity $\boldsymbol{c}$ and the prescribed ground velocity $\boldsymbol{v}_F$,

$$\dot{s}_\alpha = (\boldsymbol{c} - \boldsymbol{v}_F) \cdot \bar{\boldsymbol{a}}_\alpha = (\operatorname{Grad} \boldsymbol{\varphi} \cdot \boldsymbol{w} - \boldsymbol{v}_F) \cdot \bar{\boldsymbol{a}}_\alpha. \tag{8}$$

The local convective velocities, defined in (2), depend on the displacement gradient and are element-wise constant quantities within a C^0-smooth finite element approximation. In the case of sticking the local velocities of the particles of the rolling body are equal to the ground velocity. For the efficient numerical evaluation of the described problem (7) is not solved directly. Recalling the material time derivative (2) within this special kind of ALE description, under steady state conditions the slip velocity can be expressed as

$$\dot{s}_\alpha = \operatorname{Grad} s_\alpha \cdot \boldsymbol{w}. \tag{9}$$

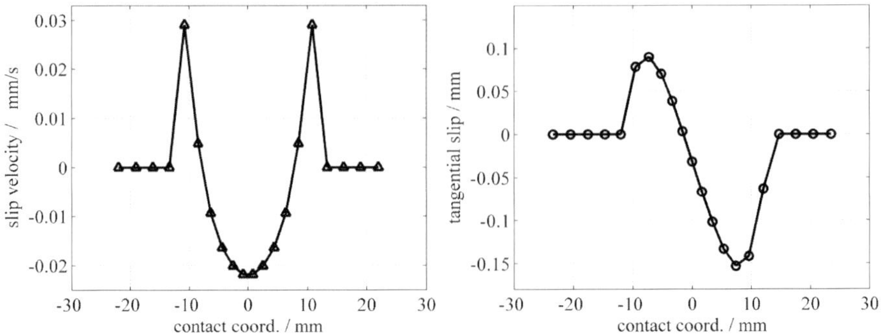

Fig. 2. Computation of the local slip distribution (right) via spatial integration of the local slip velocities (left).

This proposed ansatz requires a C^0-smooth distribution of the nodal slip variables s_α which are introduced as additional variables in the active part of the contact area. The combination of (8) and (9) leads to

$$\operatorname{Grad} s_\alpha \cdot \boldsymbol{w} = (\operatorname{Grad} \boldsymbol{\varphi} \cdot \boldsymbol{w} - \boldsymbol{v}_F) \cdot \bar{\boldsymbol{a}}_\alpha . \tag{10}$$

For a proper physical description of stationary rolling the entry condition of the material particles has to be included in that equation. Here, for the stationary rolling conditions the assumption is stated that the particles enter the contact zone smoothly without any tangential impact. Thus, homogeneous Dirichlet conditions are applied to the nodes at the leading edge of the contact area,

$$s_\alpha = 0 \quad \text{at} \quad \Gamma_0 \subset \partial_c \Phi(B), \tag{11}$$

where $\partial_c \Phi(B)$ denotes the contact area in the current configuration. At the trailing edge the particles leave the contact zone without any restraints. In this part of the contact zone the so called *Snap-Out*-effect often is observed, when the particles are accelerated back to their initial path velocity.

For the computation of the slip variables s_α we start from the weak representation of (10),

$$\int_{\partial_c \Phi(B)} (\operatorname{Grad} s_\alpha \cdot \boldsymbol{w})\, \delta \dot{s}^\alpha \, \mathrm{d}a = \int_{\partial_c \Phi(B)} \left[(\operatorname{Grad} \boldsymbol{\varphi} \cdot \boldsymbol{w} - \boldsymbol{v}_F) \cdot \bar{\boldsymbol{a}}_\alpha \right] \delta \dot{s}^\alpha \, \mathrm{d}a . \tag{12}$$

Herein, $\delta \dot{s}_\alpha$ represents the material variation of the slip velocity $\dot{s}_\alpha$ in the tangential plane. For the solution of the nonlinear problem within a standard incremental iterative schema (12) has to be linearized with respect to both, the nodal slip variables s_α and the displacement variables $\boldsymbol{\varphi}$,

$$
\begin{aligned}
&- \int_{\partial_c \Phi(B)} \left[(\text{Grad}\, \Delta\boldsymbol{\varphi} \cdot \boldsymbol{w}) \cdot \bar{\boldsymbol{a}}_\alpha \right] \delta \dot{s}^\alpha \; da + \int_{\partial_c \Phi(B)} (\text{Grad}\, \Delta s_\alpha \cdot \boldsymbol{w})\, \delta \dot{s}^\alpha \; da \\
&\quad = \int_{\partial_c \Phi(B)} \left[\left(\text{Grad}^{\,t} \boldsymbol{\varphi} \cdot \boldsymbol{w} - \boldsymbol{v}_F \right) \cdot \bar{\boldsymbol{a}}_\alpha - \text{Grad}^{\,t} s_\alpha \cdot \boldsymbol{w} \right] \delta \dot{s}^\alpha \; da.
\end{aligned} \tag{13}
$$

By solving this equation the spatial slip distribution in the contact area can be computed directly. Once obtained this slip distribution the standard concepts for the treatment of frictional contact problems can be used, see e.g. [18]. The determination of the tangential contact stress $\boldsymbol{\tau}$ requires the distinction between the case of sticking and sliding. Assuming Coulomb's law for simplicity, the frictional function reads

$$
f = \|\boldsymbol{\tau}\| - \tau_{\max} \le 0 \quad \text{with} \quad \tau_{\max} = \mu\, p, \tag{14}
$$

where μ is the constant friction coefficient. The stick-slip distinction follows to

$$
\begin{aligned}
&\text{Stick:} \qquad f < 0 \qquad \text{and} \qquad s_\alpha = 0, \\
&\text{Slip:} \qquad f = 0 \qquad \text{and} \qquad s_\alpha > 0.
\end{aligned} \tag{15}
$$

According to this, in the sticking case no relative displacement s_α between the contacting particles is allowed. However, in the regularized form of the friction law a very small reversible slip is tolerated which is interpreted as an elastic micro deformation. The total slip distance splits into a very small elastic (sticking) part and an irreversible (sliding) part,

$$
s_\alpha = s_\alpha^{\text{el}} + s_\alpha^{\text{irr}} \quad \text{and} \quad \dot{s}_\alpha = \dot{s}_\alpha^{\text{el}} + \dot{s}_\alpha^{\text{irr}}. \tag{16}
$$

For the elastic part s_α^{el} a linear constitutive dependency

$$
s_\alpha^{\text{el}} = -\frac{1}{\epsilon_t}\, \tau_\alpha \quad \Leftrightarrow \quad \tau_\alpha = -\epsilon_t\, s_\alpha^{\text{el}} \tag{17}
$$

is assumed which might be interpreted as a penalty regularization of the stick conditions too.

In case of sliding, i.e. the friction law is violated by the initial assumption of sticking, the irreversible part s_α^{irr} of the total slip is computed from the evaluation of the non-associative friction rule

$$
\dot{s}_\alpha^{\text{irr}} = \dot{\lambda}\, \frac{\partial f}{\partial \tau_\alpha} = \dot{\lambda}\, \frac{\tau_\alpha}{\|\boldsymbol{\tau}\|} = \dot{\lambda}\, n_{t\,\alpha}, \tag{18}
$$

which can be integrated directly by the use of radial return mapping schemes.

The frictional rolling contact conditions are included into the weak form of the equation of motion (3) by adding the term of the virtual work of the tangential contact forces,

$$\mathfrak{C} = -\int\limits_{\partial_c \Phi(B)} \tau_\alpha\, \delta s^\alpha \,\mathrm{da} \quad \text{with} \quad \delta s_\alpha = \delta\boldsymbol{\varphi}\cdot\bar{\boldsymbol{a}}_\alpha\,. \tag{19}$$

In contrast to the material variation of the slip velocity in (12) δs_α is a spatial variation of the current location of the particle. The linearization of (19) yields

$$\begin{aligned}\int\limits_{\partial_c \Phi(B)} {}^{t+\Delta t}\tau_\alpha\, \delta s^\alpha \,\mathrm{da} = &\int\limits_{\partial_c \Phi(B)} {}^{t}\tau_\alpha\, \delta s^\alpha \,\mathrm{da} + \int\limits_{\partial_c \Phi(B)} {}^{t}\tau_\alpha\, \Delta\delta s^\alpha \,\mathrm{da}\\ &+ \int\limits_{\partial_c \Phi(B)} \Delta\tau_\alpha\, \delta s^\alpha \,\mathrm{da} + \int\limits_{\partial_c \Phi(B)} {}^{t}\tau_\alpha\, \delta s^\alpha\, \Delta\mathrm{da},\end{aligned} \tag{20}$$

with $\Delta\delta s_\alpha = 0$ for the contact on a rigid flat surface. The non-smoothness in the tangential contact law is treated within a predictor-corrector schema. In the predictor step a trial state is computed under the assumption of sticking

$${}^{t+\Delta t}\tau_\alpha^{\text{trial}} = {}^{t}\tau_\alpha - \epsilon_t\, \Delta s_\alpha = -\epsilon_t\,({}^{t+\Delta t}s_\alpha - {}^{t}s_\alpha^{\text{irr}}). \tag{21}$$

Depending on the distinction of stick ($f(\boldsymbol{\tau}^{\text{trial}}) < 0$) or slip ($f(\boldsymbol{\tau}^{\text{trial}}) > 0$) a corrector step is carried out

$${}^{t+\Delta t}\tau_\alpha = {}^{t+\Delta t}\tau_\alpha^{\text{trial}} - \epsilon_t\, \Delta\lambda\; {}^{t+\Delta t}n_{t\,\alpha} = {}^{t+\Delta t}\tau_\alpha^{\text{trial}} - f(\boldsymbol{\tau}^{\text{trial}})\; {}^{t+\Delta t}n_{t\,\alpha}. \tag{22}$$

The incremental tangential contact stress then yields to

$$\Delta\boldsymbol{\tau} = -\epsilon_t\, \Delta\boldsymbol{s} \tag{23}$$

in the case of local stick, and

$$\Delta\boldsymbol{\tau} = \epsilon_n \mu\, \Delta d_n\, \boldsymbol{n}_t + \underbrace{\epsilon_t\, \frac{\mu p}{\|\boldsymbol{\tau}^{\text{trial}}\|}\, (\mathbf{1} - \boldsymbol{n}_t \otimes \boldsymbol{n}_t)}_{\boldsymbol{\epsilon}_t^{\text{fric}}}\, \Delta\boldsymbol{s} \tag{24}$$

in the case of slip. Herein Δd_n describes the incremental rate on normal penetration.

In the contact area the surface of the rolling body is discretized with four-node contact-elements. Because of the restriction to the deformable-rigid contact only the *slave*-body is discretized. Within a *segment-to-surface* strategy the contact stresses are computed at the integration points. In the contact patch the global displacement field $\boldsymbol{\varphi}$ and the spatial slip variables s_α are discretized with bilinear shape functions $H^i(\xi,\eta) = \frac{1}{4}\,(1+\xi\xi_i)\,(1+\eta\eta_i)$,

$$\boldsymbol{\varphi}(\xi,\eta) = \sum_{i=1}^{4} H^i(\xi,\eta)\,\tilde{\boldsymbol{\varphi}}_i = \mathbf{H}_\varphi \tilde{\boldsymbol{\varphi}}, \quad \mathbf{s}(\xi,\eta) = \sum_{i=1}^{4} H^i(\xi,\eta)\,\tilde{\mathbf{s}}_i = \mathbf{H}_s \tilde{\mathbf{s}}, \tag{25}$$

with the local coordinates ξ_i, η_i of the isoparametric reference element. In addition to the three nodal degrees of freedom for the displacements there are two nodal degrees of freedom for the slip variables s_α if the contact is active. The related element matrices are labeled by the indices φ and s.

The finite element discretization of the global displacement field $\boldsymbol{\varphi}$ and the spatial slip variables s_α yields to a coupled system for the linearized ALE formulation of the rolling contact problem ,

$$\left[\begin{array}{c|c} {}^t\mathbf{K} - \mathbf{W} + \mathbf{K}_{c_{\varphi\varphi}} & \mathbf{K}_{c_{\varphi s}} \\ \hline \mathbf{K}_{c_{s\varphi}} & \mathbf{K}_{c_{ss}} \end{array}\right] \left[\begin{array}{c} \Delta\tilde{\boldsymbol{\varphi}} \\ \hline \Delta\tilde{\mathbf{s}} \end{array}\right] = \left[\begin{array}{c} {}^{t+\Delta t}\mathbf{f}_{ext} + {}^t\mathbf{f}_i - {}^t\mathbf{f}_\sigma - {}^t\mathbf{f}_{c_\varphi} \\ \hline - {}^t\mathbf{f}_{c_s} \end{array}\right] . \tag{26}$$

Whereas the frictionless rolling contact just depends on the incremental displacements $\Delta\boldsymbol{\varphi}$, the solution of the frictional problem is coupled in the displacement variable $\boldsymbol{\varphi}$ and the nodal slip variable $\boldsymbol{s}$. A four-node contact-element then has twelve degrees of freedom for the displacements and additionally eight degrees of freedom for the nodal slip. In the following the matrices corresponding to the weak form of the tangential contact forces (19) and the computation of the nodal slip (12) are derived. These matrices are labeled by the index c in eqn. (26). The subindices φ and s are used for the corresponding equations and degrees of freedom.

At first equation (13) for the computation of the total nodal slip is discretized,

$$\begin{aligned} \delta\tilde{\mathbf{s}}^T \left({}^t\mathbf{K}_{c_{ss}}\, \Delta\tilde{\mathbf{s}} + {}^t\mathbf{K}_{c_{s\varphi}}\, \Delta\tilde{\boldsymbol{\varphi}} + {}^t\mathbf{f}_{c_s}\right) = \delta\tilde{\mathbf{s}}^T \Bigg(& \int_{\partial_c\Phi(B)} \mathbf{A}_s^T \mathbf{A}_s \,\mathrm{d}a\, \Delta\tilde{\mathbf{s}} \\ & - \int_{\partial_c\Phi(B)} \mathbf{A}_s^T \mathbf{a}_t^T \mathbf{A}\,\mathrm{d}a\, \Delta\tilde{\boldsymbol{\varphi}} + \int_{\partial_c\Phi(B)} \mathbf{A}_s^T \left[\left({}^t\mathbf{c} - \mathbf{v}_F\right) \mathbf{a}_t - \mathbf{A}_s\, {}^t\tilde{\mathbf{s}}\right] \mathrm{d}a \Bigg) . \end{aligned} \tag{27}$$

In this equation the following matrix representations are used for the description of the velocities and the material variation $\delta\dot{s}$ of the sliding velocity,

$$\mathbf{P} = \operatorname{Grad} \boldsymbol{s} \cdot \boldsymbol{w} = \mathbf{A}_s\, \tilde{\mathbf{s}}, \quad \delta\dot{s} = \mathbf{A}_s\, \delta\tilde{\mathbf{s}}, \quad \boldsymbol{c} = \operatorname{Grad} \boldsymbol{\varphi} \cdot \boldsymbol{w} = \mathbf{A}\, \tilde{\boldsymbol{\varphi}}. \tag{28}$$

The only difference between the matrices $\mathbf{A}_s$ and $\mathbf{A}$ is the mapping onto two, respectively three degrees of freedom per element node. The matrix $\mathbf{a}_t$ represents the two tangential vectors $\bar{\boldsymbol{a}}_1$, $\bar{\boldsymbol{a}}_2$. In the case of local stick the discretized form of the linearized virtual work of the tangential contact forces (20) reads

$$\delta\tilde{\boldsymbol{\varphi}}^T\left({}^t\mathbf{K}_{c_{\varphi s}}\Delta\tilde{s} + {}^t\mathbf{K}_{c_{\varphi\varphi}}\Delta\tilde{\boldsymbol{\varphi}} + {}^t\mathbf{f}_{c_\varphi}\right) = \delta\tilde{\boldsymbol{\varphi}}^T\left(\int\limits_{\partial_c\Phi(B)} \epsilon_t \mathbf{H}^T\mathbf{a}_t\mathbf{H}_s\,da\,\Delta\tilde{\mathbf{s}} - \int\limits_{\partial_c\Phi(B)}\left(\mathbf{H}^T\mathbf{a}_t\,{}^t\boldsymbol{\tau}\mathbf{H}_{,\chi}\right)\mathrm{d}a\,\Delta\tilde{\boldsymbol{\varphi}}^e - \int\limits_{\partial_c\Phi(B)}\mathbf{H}^T\mathbf{a}_t\,{}^t\boldsymbol{\tau}\,\mathrm{da}\right). \tag{29}$$

Herein the spatial variation of the sliding distance is given by

$$\delta\mathbf{s}^T = \delta\boldsymbol{\varphi}^T\mathbf{H}^T\,\mathbf{a}_t\ . \tag{30}$$

In the case of local slip the tangential stress depends on the chosen friction law. Thus, for Coulomb's law the specific algorithmic consistent tangent operator $\boldsymbol{\epsilon}_t^{\text{fric}}$ of (24) has to be used,

$$\delta\tilde{\boldsymbol{\varphi}}^T\left({}^t\mathbf{K}^e_{c_{\varphi s}}\Delta\tilde{s} + {}^t\mathbf{K}^e_{c_{\varphi\varphi}}\Delta\tilde{\boldsymbol{\varphi}} + {}^t\mathbf{f}_{c_\varphi}\right) = \delta\tilde{\boldsymbol{\varphi}}^T\left(\int\limits_{\partial_c\Phi(B)} \mathbf{H}^T\mathbf{a}_t\boldsymbol{\epsilon}_t^{\text{fric}}\,\mathbf{H}_s\,da\,\Delta\tilde{\mathbf{s}} + \int\limits_{\partial_c\Phi(B)}\left(\epsilon_n\mu\,\mathbf{n}_t\mathbf{H}^T\mathbf{a}_t\mathbf{a}_n^T\mathbf{H} - \mathbf{H}^T\mathbf{a}_t\,{}^t\boldsymbol{\tau}\mathbf{H}_{,\chi}\right)da\,\Delta\tilde{\boldsymbol{\varphi}}^e - \int\limits_{\partial_c\Phi(B)}\mathbf{H}^T\,\mathbf{a}_t\,{}^t\boldsymbol{\tau}\,\mathrm{da}\right). \tag{31}$$

The assemblage of the element matrices leads to the coupled system of equations (26). This system is non-symmetric because $\mathbf{K}_{c_{\varphi s}} \neq \mathbf{K}^T_{c_{s\varphi}}$ and its size depends on the active-set because of the additional nodal slip variables. Concerning an efficient solution of the non-symmetric system of coupled equations (26) the possibility of a staggered solution neglecting the strong coupling has been evaluated, but only poor or none convergence of the iterative schema has been obtained. In contrast, the fully coupled solution of the system yields quadratic convergence rates. The typical convergence behavior of the computation of a tractive rolling state is shown in Figure 3 which will be discussed in the following section.

3 Numerical Examples

3.1 Rolling Disc

For testing the proposed algorithmic approach a simple disc rolling on a rigid flat surface as depicted in Figure 4a has been studied. The wheel has an initial outer radius $r_o = 40$ mm, an inner radius $r_i = 17.5$ mm, a thickness $t = 18$ mm, and it rotates with the constant angular velocity $\omega = 10$ rad/s. Dirichlet boundary conditions are

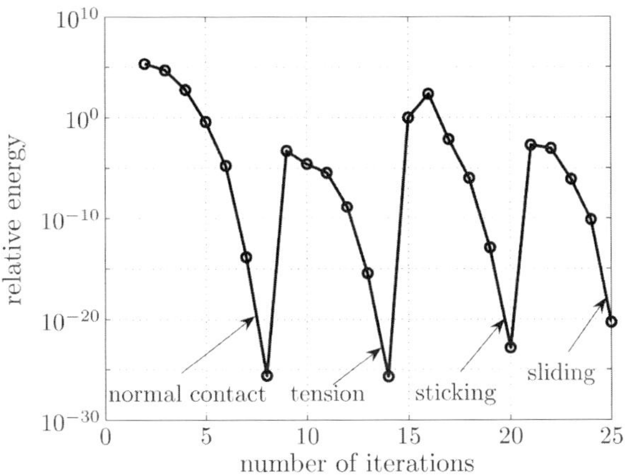

Fig. 3. Convergence behavior.

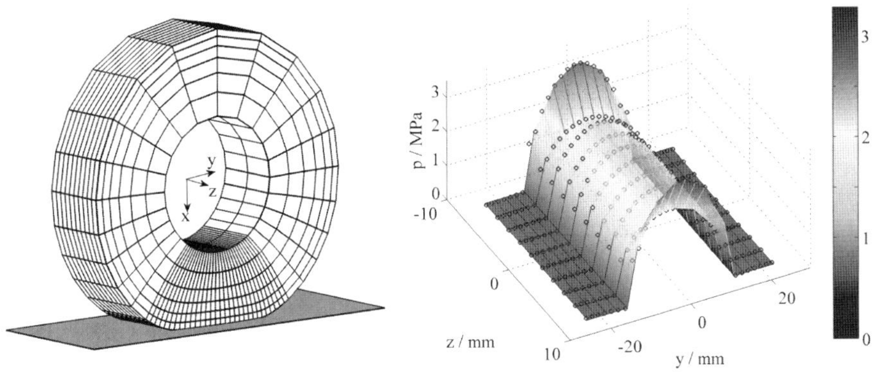

Fig. 4. (a) Rolling disc and (b) contact pressure distribution for traction free rolling.

prescribed at the interior hole for modeling a rigid rim. The material behavior is described by a hyperelastic Mooney–Rivlin Model with the coefficients $c_{10} = 1$ MPa, $c_{01} = 0.67$ MPa and a bulk modulus $\kappa = 16.7$ MPa. The penalty-stiffness for the contact has been chosen to $\epsilon_n = 10^6$ MPa/mm for the normal and $\epsilon_t = 10^4$ MPa/mm in tangential direction. The friction coefficient for the Coulomb law has been chosen as $\mu = 0.5$.

In a first step the disc has been pressed displacement controlled in one incremental step of $\Delta u = 4$ mm against the rigid plane contact surface. In this step only the deformation state for the frictionless normal contact problem has been computed. Due to tensile contact stresses computed after the first equilibrium iteration the active set has been changed and the equilibrium has been re-iterated which can be observed as second arc in Figure 3. In a second step the tangential contact conditions are

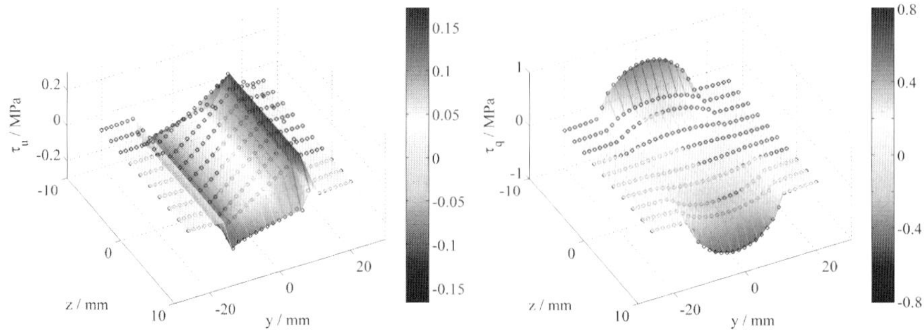

Fig. 5. Contact shear stresses for traction free rolling: (a) longitudinal and (b) lateral.

enforced by the newly introduced approach. The computation of both, the case of local stick and the case of local slip gives a quadratic rate of convergence as depicted in Figure 3.

The state of traction free rolling has been computed for a given ground velocity of $v_F = 397.2$ mm/s which in this kinematical driven process has been found iteratively. The obtained contact pressure distribution is shown in Figure 4b. In this example the homogeneity of typical parabolic shape is disturbed by high edge pressures which are the result of the restricted lateral motion of these particles due to finite friction. The corresponding tangential contact stresses are depicted in Figure 5. In general the longitudinal contact shear stress distribution appears S-shaped, as expected from analytical results. However, due to local effects observed at the lateral edges already discussed at the contact pressure distribution, the longitudinal shear changes the sign at the edges. From the comparison of magnitudes between longitudinal and lateral shear it is obvious that in this example the edge effects are dominated from the lateral contact shear traction depicted in the right of Figure 5.

Starting from the state of traction free rolling the ground velocity is increased to simulate different states of braking. A globally measurable kinematic quantity called global slip is defined as

$$S = \frac{v_F - \omega r_o}{v_F}. \tag{32}$$

to quantify these stages. In Figure 6a the longitudinal shear stress distribution is plotted for a tractive rolling state with braking slip of $S = 4\%$. In Figure 6b the active contact elements are separated into sticking (red) and sliding (blue) elements, whereas the arrows indicate the slip velocities. Obviously, the sliding area is located at the trailing edge of the contact zone whereas the leading edge sticks. These results are in clear correspondence with analytical solutions. Increasing the ground velocity, i.e. the global slip, leads to a rising tangential braking force going ahead with an increasing slip area until each element is in frictional contact. The development of this process is depicted in Figure 7a for braking slip values of 4%, 10% and 30%. In this

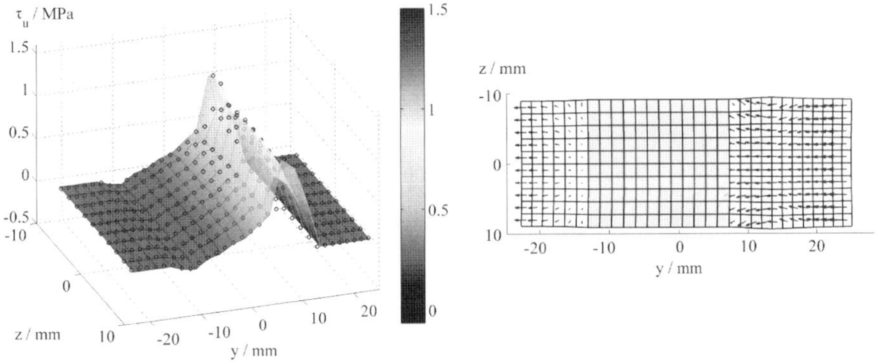

Fig. 6. (a) Longitudinal contact stress for tractive rolling ($S = 4\%$) and (b) local status of contact elements.

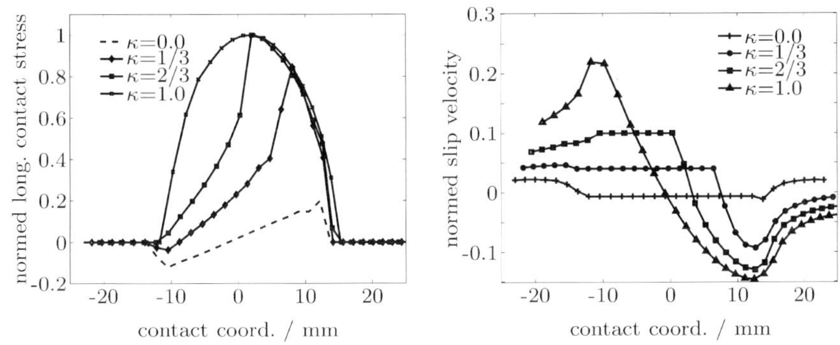

Fig. 7. Increasing braking slip: (a) normalized longitudinal contact shear stress and (b) normalized slip velocity.

plot the normalized longitudinal stress $\tau_u/(\mu p_0)$ is evaluated along the element ring at $z \approx 2$ mm. For Coulomb friction the traction coefficient $\kappa_t = T/\mu N$ is defined as the ratio of the resulting tangential force T and the maximal transmittable tangential force $T_{\max} = \mu N$. In the case of 30% braking slip each element slides and the traction coefficient yields $\kappa_t = 1$. From the distribution of the corresponding normalized slip velocities along that element ring depicted in Figure 7b it is concluded that the stick condition (6) is enforced correctly and that a smooth entry of the material particles is obtained at the leading edge.

3.2 Rolling Tire

In a second example a detailed model of a radial tire with three circumferential grooves has been analyzed to demonstrate the computability of the developed

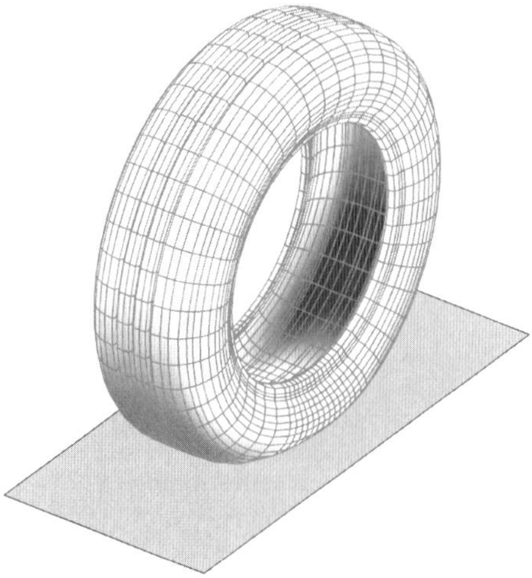

Fig. 8. Discretized model of a radial tire.

approach for industrial problems. The finite element model depicted in Figure 8 consists of about 45,000 degrees of freedom. The cross-section is discretized with 15 distinct material groups. The tire model has been inflated with an internal pressure of 0.2 MPa and set into rotation corresponding to a rolling speed of 80 km/h. Coulomb's friction coefficient has been chosen to $\mu = 1.0$.

The contact stresses computed for a traction free rolling state are depicted in Figures 9 and 10. High pressure concentrations are observed at the lateral edges of the footprint which is typical for virgin radial tires as far as no stress softening of the rubber material has taken place. In longitudinal direction the contact shear stress distribution appears S-shaped while no resulting force is transmitted. An interesting effect can be observed in lateral direction. Due to the grooves in the tread the lateral contact stresses appear oscillating because at the edges of each rib the sticking particles are constrained for lateral spreading. This causes high local traction with changing directions in each rib. This behavior is in good qualitative correspondence with experimental results, see [2, 7].

The proposed algorithm has been coded partly in MATLAB- and partly in FORTRAN-language. The direct solver PARDISO-library [17] has been used for the solution of the non-symmetric linear system within the Newton–Raphson iterations. The overall computation time for the free rolling state of the tire model discussed before took about 15 minutes on a notebook computer including an INTEL T2400 Duo-processor and 2 Gb RAM.

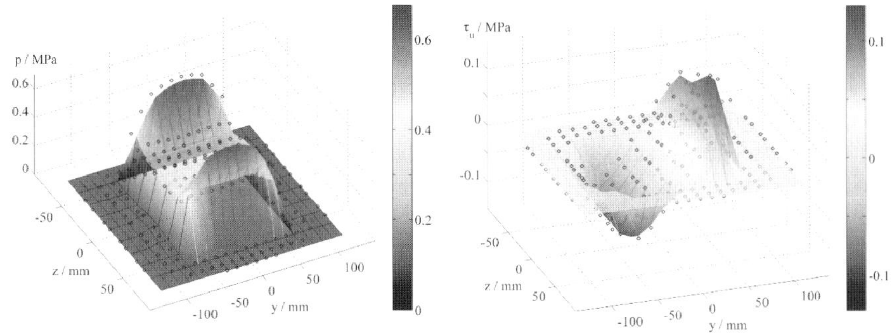

Fig. 9. Traction free rolling: (a) contact pressure, (b) longitudinal contact stress.

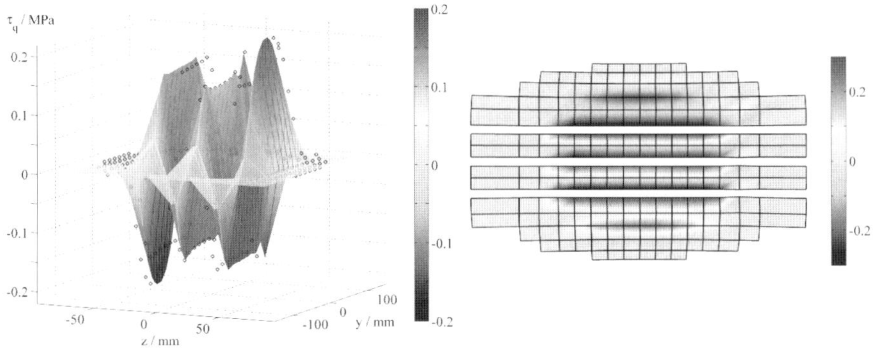

Fig. 10. Traction free rolling: lateral contact stress.

4 Conclusions

A novel and computational efficient approach for the treatment of tractive rolling contact within an ALE framework for steady state rolling has been presented in this contribution. The tangential slip distribution in the contact patch has been recovered from the slip velocity distribution by a mathematically consistent approach which enables for the enforcement of stick conditions and evolution of frictional slip within a material framework. From a weak form for the stick condition a coupled system has been derived which is solved for the unknown displacement-field and the spatial slip distance distribution simultaneously, on the cost of two additional unknowns for the active contact nodes and a non-symmetrical coupling.

The physical reliability and computational efficiency has been demonstrated on the example of an elastic disc rolling on a rigid plane surface. While computed results are qualitatively in good agreement with analytical solutions deviations due to three dimensional edge-effects could have been shown by these fully three-dimensional analysis. The optimal quadratic convergence behavior of the presented approach has been demonstrated for traction free rolling as well as tractive rolling including local frictional slip.

The computability of large scaled finite element models of industrial relevance has been shown by the analysis of a detailed tire model. The contact of a tire with circumferential grooves has been analyzed. The resolution of local effects, e.g. in the lateral shear traction distribution which are in good agreement with high-resolution measurements underline the physical correctness and reliability of the proposed approach.

Acknowledgement

The authors express their gratitude for the financial support by the German Research Foundation (DFG) within the framework of the DFG-Research Unit 492.

References

1. Abaqus. *ABAQUS/Standard Theory Manual, Version 6.6.* 2006.
2. R. Blab and J. T. Harvey. Modeling measured 3d tire contact stresses in a viscoelastic FE-pavement model. *The International Journal of Geomechanics*, 2(3):271–290, 2002.
3. F. W. Carter. On the action of a locomotive driving wheel. *Proceedings of the Royal Society London A*, 122:151–157, 1926.
4. L. O. Faria, J. T. Oden, B. Yavari, W. Tworzydlo, J. M. Bass, and E. B. Becker. Tire modeling by finite elements. *Tire Science & Technology*, 20:33–56, 1992.
5. M. S. Gadala. Recent trends in ALE-formulation and its applications in solid mechanics. *Computer Methods in Applied Mechanical Engineering*, 193:4247–4275, 2004.
6. G. Hu and P. Wriggers. On the adaptive finite element method of steady state rolling contact for hyperelasticity in finite deformations. *Computer Methods in Applied Mechanical Engineering*, 191:1333–1348, 2002.
7. S. H. Koehne, B. Matute, and R. Mundl. Evaluation of tire tread and body interactions in the contact patch. *Tire Science & Technology*, 31(3):159–172, 2003.
8. T. A. Laursen. *Computational Contact and Impact Mechanics*. Springer Verlag, 2002.
9. T. A. Laursen and I. Stanciulescu. An algorithm for incorporation of frictional sliding conditions within a steady state rolling framework. *Communications in Numerical Methods in Engineering*, 22:301–318, 2006.
10. P. Le Tallec and C. Rahier. Numerical models of steady rolling for non-linear viscoelastic structures in finite deformations. *International Journal for Numerical Methods in Engineering*, 37:1159–1186, 1994.
11. U. Nackenhorst. On the finite element analysis of steady state rolling contact. In M. H. Aliabadi and C. A. Brebbia (Eds.), *Contact Mechanics – Computational Techniques*. Computational Mechanics Publications, 1993.
12. U. Nackenhorst. The ALE-formulation of bodies in rolling contact – Theoretical foundations and finite element approach. *Computer Methods in Applied Mechanical Engineering*, 193(39-41):4299–4322, 2004.

13. U. Nackenhorst, M. Brinkmeier, and M. Ziefle. Dynamics of rolling wheels. *Machine Dynamics Problems*, 30(1):45–60, 2006.
14. L. Nasdala, M. Kaliske, A. Becker, and H. Rothert. An efficient viscoelastic formulation for steady-state rolling. *Computational Mechanics*, 22:395–403, 1998.
15. J. T. Oden and T. L. Lin. On the general rolling contact problem for finite deformations of a viscoelastic cylinder. *Computer Methods in Applied Mechanical Engineering*, 57:297–367, 1986.
16. J. Padovan. Finite element analysis of steady and transient moving/rolling nonlinear viscoelastic structure – 1. Theory. *Computers & Structures*, 27:249–257, 1987.
17. O. Schenk and K. Gärtner. Solving unsymmetric sparse systems of linear equations with pardiso. *Journal of Future Generation Computer Systems*, 20(3):475–487, 2004.
18. P. Wriggers. Finite element algorithms for contact problems. *Archives of Computer Methods in Engineering*, 2:1–49, 1995.
19. P. Wriggers. *Computational Contact Mechanics*, 2nd edition. Springer Verlag, 2006.
20. M. Ziefle and U. Nackenhorst. A new update procedure for internal variables in an ALE-description of rolling contact. *Proceedings in Applied Mathematics and Mechanics (PAMM)*, 5:71–74, 2005.

A Study on 3D Contact of Rough Surfaces Considering Finite Elastoplasticity

Alex Alves Bandeira[1], Paulo M. Pimenta[2] and Peter Wriggers[3]

[1] *Department of Civil Engineering at Presbyterian Mackenzie University, São Paulo, Brazil*
E-mail: alex_bandeira@terra.com.br
[2] *Department of Structural and Geotechnical Engineering, Polytechnic School at University of São Paulo, São Paulo, Brazil*
E-mail: ppimenta@usp.br
[3] *Institute for Structural and Computational Mechanics, University of Hannover, Hannover, Germany*
E-mail: wriggers@ibnm.uni-hannover.de

Abstract. In this work the non-penetration condition and the interface models for contact taking into account the surface microstructure are investigated in detail. It is done using a homogenization procedure presented by Bandeira et al. [2–5] in order to obtain by numerical simulation the interface behavior for the normal and tangential contact pressures based on statistical surface models. The contact surfaces of both bodies are rough. This paper can be regarded as a complementary study to that presented by Bandeira et al. [5]. Here the plasticity of the asperities is taken into account by assuming a constitutive equation based on an associated von Mises yield function formulated in principal axes, as shown by Pimenta [22]. The plastic zones in the microstructure are shown to study in detail the contact interface. Numerical examples are selected in order to show the ability of the algorithm to represent interface law for rough surfaces considering elastoplastic behavior of the asperities.

Key words: Contact mechanics, contact surface, interface constitutive equation, elastoplasticity.

1 Introduction

Several formulations concerning the treatment of the contact interface have been presented in the literature, especially when the contact interface of two contacting bodies is rough. Constitutive equations for the normal contact have been developed by investigating micromechanical behavior within the contact surface. Associated models have been developed based on experiments, like Greenwood and Williamson [12], Evseev et al. [9], Zavarise et al. [33], Yovanovich [32] and Kragelsky et al. [15].

Peter Wriggers and Udo Nackenhorst (eds.), IUTAM Symposium on Computational Contact Mechanics, 225–238.

In general, the micromechanical behavior depends on material parameters like hardness and on geometrical parameters like surface roughness. It should be noted that the real micromechanical phenomena are extremely complex due to extremely high local pressure at the asperities. The model used in this paper, attempts only to capture the most important phenomena and assume either elastic or plastic deformation of the asperities having real contact in the interface.

This article concentrates on the behavior of the contact interface. The idea is to study the interface behavior by modeling the contact surfaces using a finite element discretization to take into account the geometrical properties of the asperities. The probabilistic theory is applied based on a statistical model of the micro-geometry like in the microscopic contact mechanics developed by Greenwood and Williamson [12] and Wriggers et al. [30]. Finally a simple homogenization leads to a contact interface law.

A three-dimensional eight-node brick element is used for the treatment of finite elastic-plastic deformation of the contacting surfaces. An augmented Lagrangian method is applied to solve the frictional contact problems because high-pressures occur which cannot be treated adequately by standard penalty procedures, see [6, 7, 11, 13, 16–19, 28–31]. The technique used to solve three-dimensional contact problems with friction, see [8, 26], in finite deformations was already developed and described in [1, 2–5, 20, 24, 27–31].

The finite element program is based on a C++ code developed by Bandeira [5]. All numerical examples given are based on three-dimensional calculations. In the numerical examples high-density meshes are used to represent the geometrical irregularity on the surfaces more precisely.

2 Constitutive Equation for Contact Interface

Different approaches have been proposed to represent microscopic contact mechanisms. The available formulations are based either on curve fitting of experimental results or on theoretical analyses of microscopically rough surface, see [9, 12, 15, 33].

In this paper, the current normal approach g_N is investigated in detail. It can be defined by

$$g_N = \xi - d, \tag{1}$$

where ξ denotes the maximum initial asperities height and d the current mean plane distance. The graphical interpretation of Equation (1) is illustrated in Figure 1.

The normal contact force F_N is obtained as a product of the apparent pressure by the apparent contact area A, as follows

$$F_N = A c_N (g_N)^2, \tag{2}$$

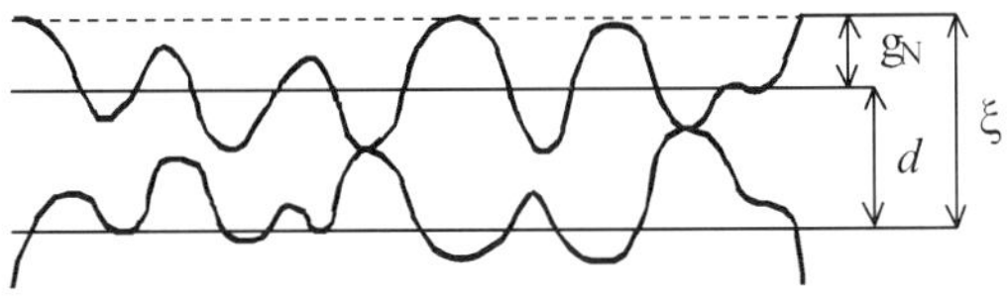

Fig. 1. Physical approach on the contact interface (cross section).

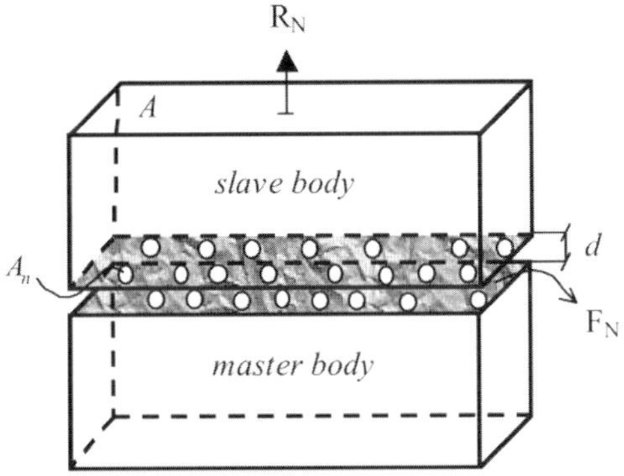

Fig. 2. Model to obtain an interface law.

where c_N is defined by the penalty parameters of the Augmented Lagrangian algorithmic. The following relationship correlates the current normal approach with the apparent mechanical pressure

$$p_N = c_N (g_n)^2. \tag{3}$$

The mechanical constants mentioned in (2) depend on the micromechanics of the surface. This constitutive equation was presented in [33].

3 A Simple Homogenization Method for Contact Interface

The basic aim of this paper is to derive constitutive contact laws as stated in Section 2 for a rough surface by using the finite element method. For this purpose one has to model and discretize the rough surface and then, by homogenization procedures develop an interface law for contact. This section summarizes the homogenization method leading to the contact interface laws. The interface law is obtained from numerical simulation using a model that consists of two deformable bodies in contact, see Figure 2. The contact surfaces of both bodies are rough. A representative surface at the contact interface is represented by Figure 3.

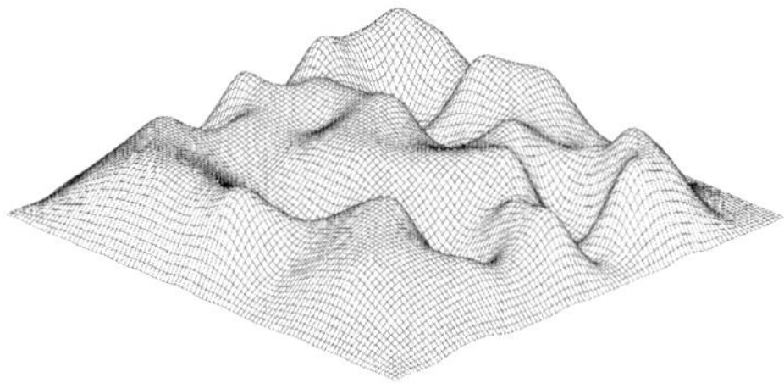

Fig. 3. Contact surface with macroscopic asperities.

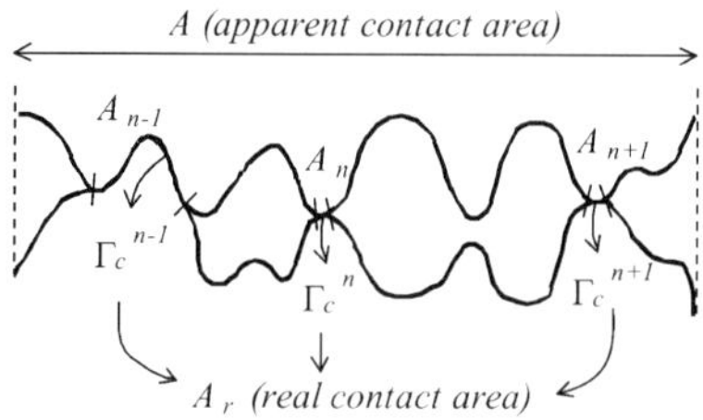

Fig. 4. Contact interface (cross section).

The procedure to find the associated interface law is applied in many steps. The bodies are discretized using standard hexahedral finite elements. One body is placed above another with initial distance to separate them. The inferior body is kept fixed in position and superior body is moved towards the inferior body by a displacement imposed on the top surface A. The fixed block is defined as master body and one that is in motion is defined as slave body. The prescribed displacement is applied in many small increments. In each increment, the resultant force R_N at the top of superior body is calculated by summing up the reaction forces R_{Nk} on the surface A related to each node k. Then, one has

$$R_N = \sum_k R_{Nk}. \tag{4}$$

The existence of a reaction force R_N indicates the first contact between the bodies. The maximum initial asperities height x between the two middle planes, which are in contact, is determined at this step; see Figure 1 for the geometrical relations.

The normal contact force F_N at the contact interface is obtained by taking into account the real contact area A_r. The actual contact occurs at n-discrete areas A_n on the discrete boundary G_c^n, as shown in Figure 4. This yields $A_r = \sum_{n=1}^{n_c} A_n$, when n_c asperities are in contact.

The real contact pressure t_{Nn} occurs on the discrete contact surface G_c^n. The actual normal contact force F_N is obtained by summing up the integrals of the real normal contact pressures t_{Nn} related to the discrete area A_n at each boundary G_n^c. Then, one may write

$$F_N = \sum_n \int_{\Gamma_c^n} t_{Nn} dA_n. \tag{5}$$

In general it is difficult to determine the actual discrete contact areas and to compute F_N from (5). Here a different procedure is followed. Considering the equilibrium of the bodies, it is clear that the normal contact force developed at the interface is equal to the normal reactions developed at the top of superior body. Therefore, within the finite element treatment it is sufficient, instead of computing (5), to calculate the normal reaction force R_N. Hence, one gets

$$F_N = R_N. \tag{6}$$

The total contact force F_N at the interface can be distributed on the apparent contact area A to yield a uniform apparent contact pressure p_N. Therefore, with Equation (6), one has

$$p_N = \frac{R_N}{A}. \tag{7}$$

Since R_N depends on the current mean plane distance d, the penetration law is displayed as

$$p_N = p_N(d). \tag{8}$$

At the end of each step the current mean plane distance d, the total reaction force R_N and the apparent contact pressure p_N are calculated. The analysis ends when the current mean plane distance d goes to zero. The numerical results after homogenization yield a microscopic contact law analogous to the theoretical law presented in Section 2. The homogenization is computed in terms of the average normal pressure p_N. The plotted curve describes the penetration law relating the apparent contact pressure p_N to the current mean plane distance d, see Equation (8).

The generation of the smoothen contact surfaces with Bézier interpolations [10], the procedures to obtain a statistical law and the techniques to obtain the maximum initial asperities height ξ for a generated surface were presented in detail in [3].

The same approach is adopted here for the computation of the tangential stress, when a block is displaced tangentially with respect to the other. The resultant tangential force is divided by the nominal area in order to get its value.

4 Constitutive Equation

The von Mises elastic-plastic constitutive law is based on the following multiplicative decomposition of the deformation gradient

$$\mathbf{F} = \mathbf{F}^e \mathbf{F}^p, \tag{9}$$

where the superscript e and p describe the elastic and the plastic part, respectively. The spatial elastic logarithmic strain tensor is given by

$$\mathbf{E}^e = \ln \mathbf{V}^e, \quad \text{where} \quad \mathbf{V}^e = (\mathbf{F}^e \mathbf{F}^{eT})^{1/2} \tag{10}$$

is the elastic left stretching tensor. The elasticity is described by the following strain energy function:

$$\psi(\mathbf{E}^e) = \frac{1}{2} k \vartheta^2 + \mu(\bar{\mathbf{E}}^e : \bar{\mathbf{E}}^e), \tag{11}$$

where $\vartheta = \operatorname{tr} \mathbf{E}^e$ and $\bar{\mathbf{E}}^e = \operatorname{Dev} \mathbf{E}^e$. Equation (11) leads to the following Kirchhoff–Treftz stress tensor:

$$\mathbf{\Sigma} = \mathbb{D}^e \mathbf{E}^e, \quad \text{where} \quad \mathbb{D}^e = \kappa \mathbf{I} \times \mathbf{I} + 2\mu \left(\mathbb{I} - \frac{1}{3} \mathbf{I} \times \mathbf{I} \right). \tag{12}$$

The logarithmic isotropic linear elastic material simplifies the volumetric-isochoric splitting. Note also that (12) is similar to the expression of the small strain Hooke's Law. The fourth-order tangent tensor is obtained from (12) as shown in detail in [22].

For computational purposes the classical radial return algorithm along with the von Mises plasticity with linear isotropic hardening is summarized below.

(1) Trial step:

1. $\mathbf{F}_t^e = \mathbf{F}_{i+1} \mathbf{U}_i^{p-1}$; $\mathbf{C}_t^e = \mathbf{F}_t^{eT} \mathbf{F}_t^e$;
2. $\mathbf{U}_t^e = (\mathbf{C}_t^e)^{1/2} = \lambda_i^e \mathbf{c}_i^e \times \mathbf{c}_i^e$; $\varepsilon_i = \ln \lambda_i^e$; $\mathbf{E}_t = \ln \mathbf{U}_t^e$;
3. $\mathbf{R}_t^e = \mathbf{F}_t^e \mathbf{U}_t^{e-1}$; $\mathbf{b}_i^e = \mathbf{R}_t^e \mathbf{c}_i^e$;
4. $\vartheta_t = \operatorname{tr}(\mathbf{E}_t)$; $\bar{\mathbf{E}}_t = \operatorname{Dev}(\mathbf{E}_t)$;
5. $\bar{\mathbf{\Sigma}}_t = 2\mu \bar{\mathbf{E}}_t$; $\bar{\sigma}_t = \sqrt{\frac{3}{2} \bar{\mathbf{\Sigma}}_t : \bar{\mathbf{\Sigma}}_t}$; $F_t = \bar{\sigma}_t - \sigma_y(\alpha_i)$.

(2) Radial return algorithm
if ($F_t < 0$) then
elastic step:

1. $\alpha_{i+1} = \alpha_i$;
2. $\mathbf{U}_{i+1}^p = \mathbf{U}_i^p$;
3. $\bar{\mathbf{S}}_{i+1} = \bar{\mathbf{S}}_t$;
4. $\bar{\mathbb{D}}_{i+1} = 2\mu \left(\mathbb{I} - \frac{1}{3} \mathbf{I} \times \mathbf{I} \right)$;

else if ($F_t > 0$) then
elastic-plastic step:

1. $\Delta\alpha = \dfrac{\bar{\sigma}_t - \sigma_y(\alpha_i)}{3\mu + h}$; $\alpha_{i+1} = \alpha_i + \Delta\alpha$; $\sigma_y(\alpha_{i+1}) = \sigma_y(\alpha_i) + h\Delta\alpha$;
2. $\Delta\mathbf{E}^p = \Delta\alpha\dfrac{3}{2\bar{\sigma}_t}\bar{\mathbf{S}}_t$; $\Delta\mathbf{U}^p = e^{\Delta\mathbf{E}^p}$; $\mathbf{F}^p_{i+1} = \Delta\mathbf{U}^p\mathbf{U}^p_i$; $\mathbf{U}^p_{i+1} = (\mathbf{F}^{pT}_{i+1}\mathbf{F}^p_{i+1})^{1/2}$;
3. $\bar{\mathbf{S}}_{i+1} = \dfrac{\sigma_{yi+1}}{\bar{\sigma}_t}\bar{\mathbf{S}}_t$;
4. $\bar{\mathbb{D}}_{i+1} = 2\mu\dfrac{\sigma_y(\alpha_{i+1})}{\bar{\sigma}_t}\left(\mathbb{I} - \dfrac{1}{3}\mathbf{I}\times\mathbf{I}\right) + \dfrac{3\mu}{\bar{\sigma}_t^2}\left(\dfrac{h}{3\mu + h} - \dfrac{\sigma_y(\alpha_{i+1})}{\bar{\sigma}_t}\right)\bar{\mathbf{\Sigma}}_t \times \bar{\mathbf{\Sigma}}_t$;

end if;

(3) add volumetric part:

1. $\mathbf{\Sigma}_{i+1} = k\vartheta_t\mathbf{I} + \bar{\mathbf{\Sigma}}_{i+1}$;
2. $\mathbb{D}_{i+1} = \kappa\mathbf{I}\times\mathbf{I} + \bar{\mathbb{D}}_{i+1}$.

The assumed linear isotropic hardening used in this paper is illustrated in Figure 5.

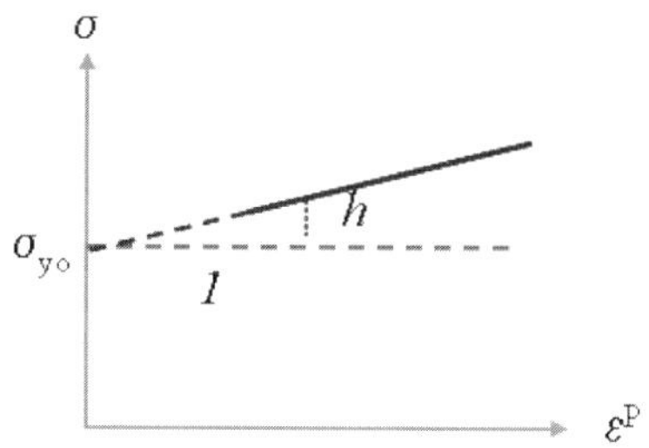

Fig. 5. Linear isotropic hardening.

Thus, one may write

$$\sigma_y(\alpha) = \sigma_{y0} + h\varepsilon^p, \tag{13}$$

where σ_{y0} is the initial yield stress, $\alpha = \varepsilon^p$ is the equivalent plastic strain and h is the linear hardening parameter.

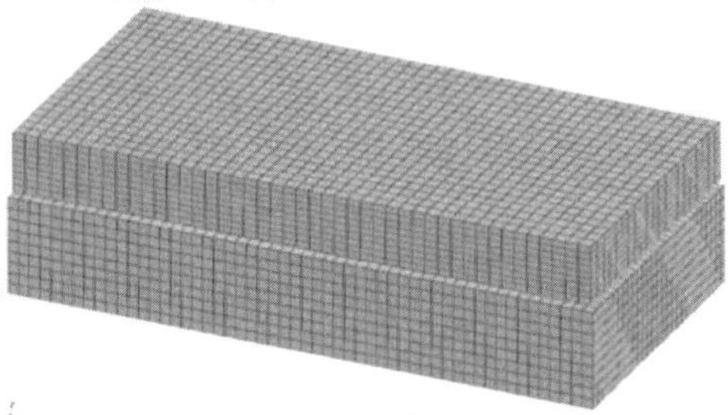

Fig. 6. Contact of two deformable bodies.

5 Numerical Simulation

In this section, three numerical examples are presented to obtain an interface law for rough surfaces numerically. In these examples two blocks are considered in contact as shown in Figure 6. This is done for three-dimensional bodies in contact. The homogenization method used was presented in Section 3. The Preconditioned Bi-Conjugate Gradient Method (PBCG) is used to solve the linear equations system [23]. The elastic-plastic material law presented in Section 4 is used in all examples. It is important to mention that each numerical laws are statistically computed curves resulted by 20 different random generated contact surfaces.

Example 1

In this first example, the number of master and slave surfaces is around 2,403 elements and the complete mesh is around 21,627 bricks elements. Each block has the same geometry of 90 mm $\times$ 45 mm $\times$ 15 mm and material properties defined by elasticity module of 70 GPa, Poisson coefficient of 0.3 and adopted initial yield stress of 200 MPa. The base of the master block is fixed and lateral displacements of both blocks are released. The used hardening parameter is defined by $h = E/100$. The contact surfaces are modified according to the theory presented by Bandeira et al. [3], such that the maximum initial asperities height ξ is 0.563444 mm. A uniform displacement of 4 mm is prescribed at the top of the slave block in several increments. Each analysis ends when the current mean plane distance d approaches zero. The mean plane distance goes to zero in the 93th increment of load.

After all generated surfaces were analyzed, the mean value curve of the normal pressures is depicted in Figure 7, which represents the constitutive interface law for different hardness.

The plastic zone developed at the master surface can be analyzed in each increment of loads by the equivalent plastic strain presented in the algorithmic for elastoplasticity stated in Section 4 (see Figure 8).

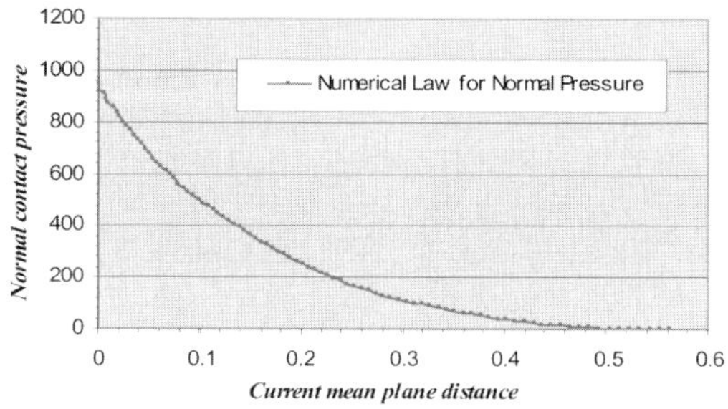

Fig. 7. Contact normal pressure.

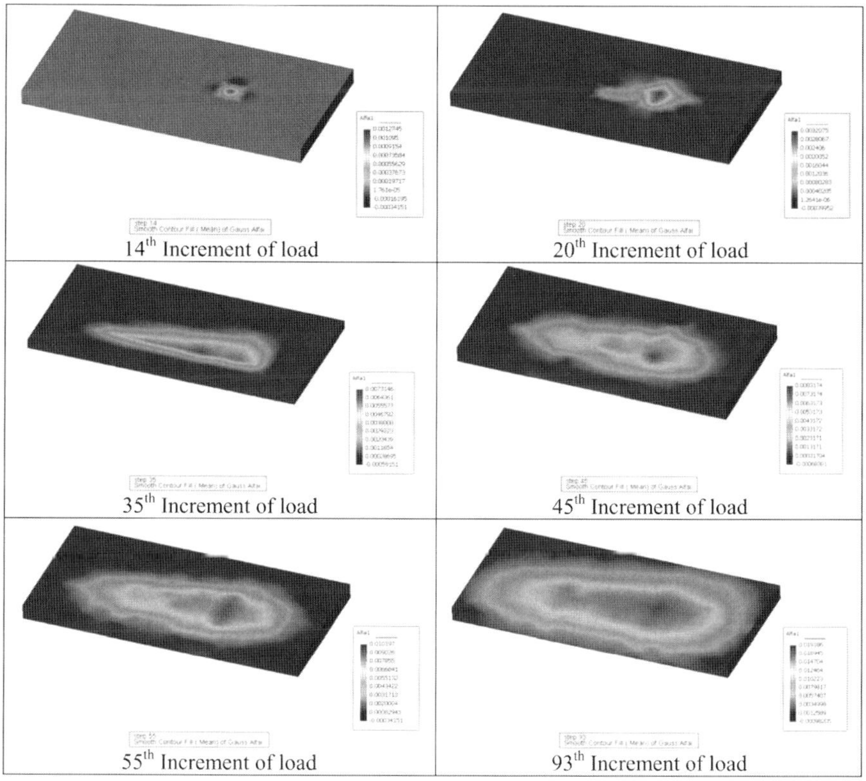

Fig. 8. Equivalent plastic strain (a) on the master surface.

Example 2

The second example is the same as the first one. The differences are that the maximum initial asperities height ξ is 0.180394 mm and the load consists of a uniform

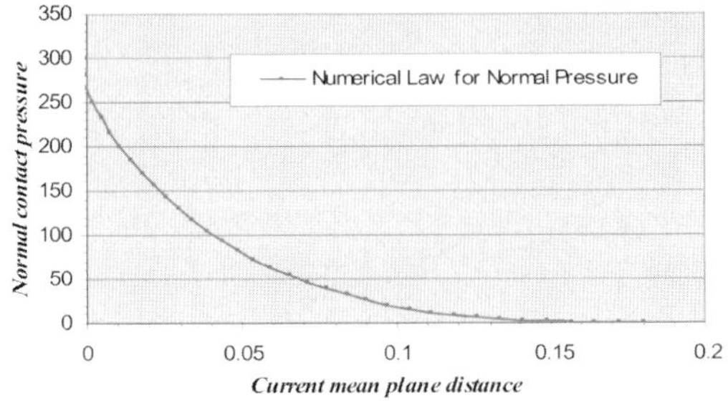

Fig. 9. Contact normal pressure.

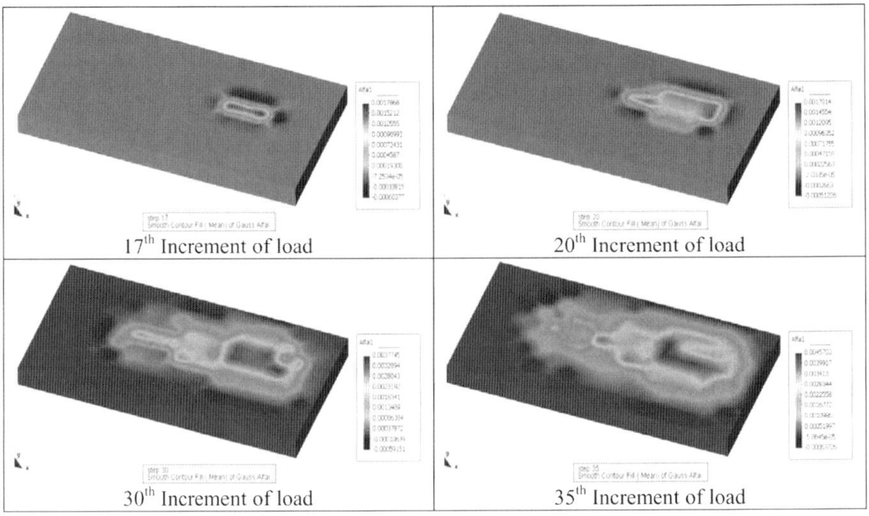

Fig. 10. Equivalent plastic strain on the master surface (a).

displacement of 8 mm prescribed at the top of the slave block in several increments. Each analysis ends when the current mean plane distance d approaches zero. The mean plane distance goes to zero in the 35th increment of load. After all generated surfaces were analyzed, the mean value curve of the normal pressures is depicted in Figure 9, which represents the constitutive interface law for different hardness.

The plastic zone developed at the master surface can be analyzed in each increment of loads by the equivalent plastic strain (see Figure 10).

Example 3

The third and final example has the same discretization of the first one. The difference is that the maximum initial asperity height ξ is 0.226265 mm and the applied load

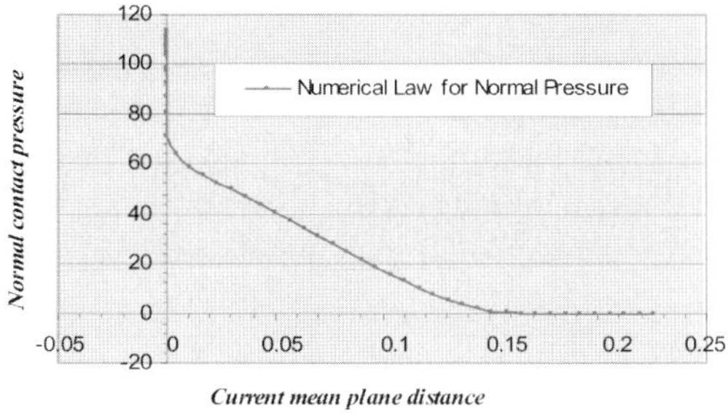

Fig. 11. Contact normal pressure for $h = E/10$.

is done in two steps. A uniform displacement of 0.24 mm is prescribed at the top of the slave block in several increments in the vertical direction. Each analysis ends when the current mean plane distance d approaches zero. The mean plane distance goes to zero in the 35th increment of load. After that, another increment of load of 0.10 mm is applied in the tangential direction, paralleled to the face of 90 mm, in 30 increments. The penalty parameters used for the normal direction is around 10^5 and for the tangential direction is around 5×10^4. The friction coefficient is 0.20. After this analysis, the mean value curve of the normal pressures is depicted in Figure 11, which represents the constitutive interface law for different hardness.

In this analysis the normal and tangential pressures are plotted over the displacement in each increment of load. The contact surfaces are smooth that makes the pressure very sensitive regarding the hardness. The behaviors of the normal pressure and the nominal tangential stress are presented in Figures 12a and 12b, respectively.

6 Conclusions

This work can be regarded as a complementary study to Bandeira et al. [5]. In that paper, the plasticity of the asperities is taken into account by assuming a constitutive equation based on associated von Mises yield function formulated in principal axes. The authors can concluded that is possible to modeling the micromechanical phenomena developed at the contact surfaces and obtained results for constitutive equations derived by numerical simulations with good agreements with the theoretical laws. In the numerical examples the normal and tangential contact pressures are plotted to show the contact surfaces behaviour. The plasticity evolution developed at the contact surfaces are also presented in the numerical results. It can be seen that the contact surfaces developed high plasticity effects during the contact mechanics.

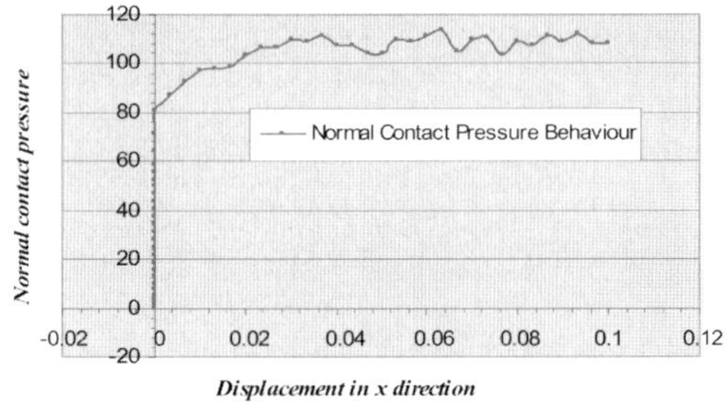

a) Normal pressure

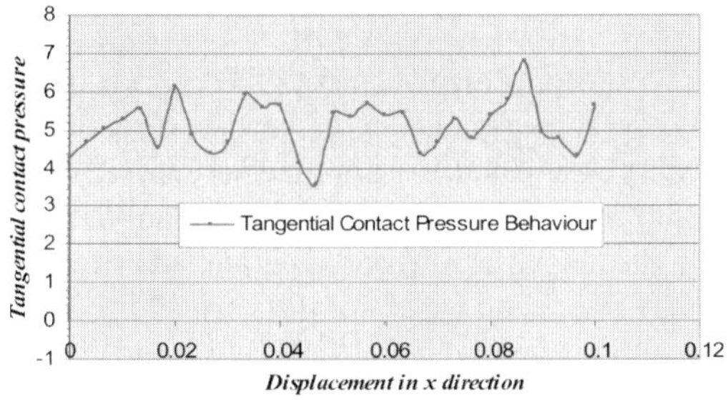

b) Tangential stress

Fig. 12. Constitutive law.

References

1. Alart, P. and Curnier, A. A mixed formulation for frictional contact problems prone to Newton like solution methods. *Computational Methods in Applied Mechanics and Engineering*, 92:353–375, 1991.
2. Bandeira, A. A., Wriggers, P. and Pimenta, P. M. Computational analysis of contact mechanics undergoing large 3D deformation. *European Conference on Computational Mechanics: ECCM*, Krakow, Poland, 2001.
3. Bandeira, A. A., Wriggers, P., Pimenta, P. M. Numerical derivation of contact mechanics interface laws using a finite element approach for large 3D deformation. *International Journal for Numerical Methods in Engineering: IJNME*, 2005 (presented at 3rd Contact Mechanics International Symposium: CMIS, Peniche, Portugal, 2001).
4. Bandeira, A. A., Wriggers, P., Pimenta, P. M. Numerical simulation of 3D contact problems under finite elastic-plastic deformation. *7th US National Congress on Computational Mechanics*, Albuquerque, New Mexico, USA, 2003.
5. Bandeira, A. A., Wriggers, P. and Pimenta, P. M. A 3D study of the contact interface behavior using elastic-plastic constitutive equations. In: Wriggers, P. and Nackenhorst, U. (Eds.), *Analysis and Simulation of Contact Problems*, Heidelberg, Vol. 27, pp. 313–324, 2006.

6. Bertsekas, D. P. *Constrained Optimization and Lagrange Multiplier Methods*. Academic Press, New York, 1984.
7. Bertsekas, D. P. *Nonlinear Programming*. Belmont, Athena Scientific, 1995.
8. Curnier, A. A theory of friction. *International Journal for Solids Structures*, 20:637–647, 1984.
9. Evseev, D. G., Medvedev, B. M. and Grigoriyan, G. G. Modification of the elastic-plastic model for the contact of rough surfaces. *Wear*, 150:79–88, 1991.
10. Farin, G. Curves and surfaces for computer aided geometric design: A practical guide. Department of Computer Science, Arizona State University, Arizona.
11. Fletcher, R. *Practical Methods of Optimization*, 2 vols. Chichester, John Wiley & Sons, 1980.
12. Greenwood, J. A. and Williamson, J. B. P. *Contact of Nominally Flat Surfaces*. Burndy Corporation Research Division, Norwalk, Connecticut, USA, 1966.
13. Heegaard, J. H. and Curnier, A. An augmented Lagrangian method for discrete large-slip contact problems. *International Journal for Numerical Methods in Engineering*, 36:569–593, 1993.
14. Kardestuncer, H. and Norrie, D. H. *Finite Element Handbook*. McGraw-Hill Book Company, 1987.
15. Kragelsky, I. V., Dobychin, M. N. and Kombalov, V. S. *Friction and Wear – Calculation Methods*. Pergamon Press, 1982 (Translated from Russian by N. Standen).
16. Laursen, T. A. and Maker, B. N. An augmented Lagrangian quasi-Newton solver for constrained nonlinear finite element applications. *International Journal for Numerical Methods in Engineering*, 38:3571–3590, 1995.
17. Laursen, T. A. and Simo, J. C. A continuum-based finite element formulation for the implicit solution of multibody, large deformation frictional contact problems. *International Journal for Numerical Methods in Engineering*, 36:3451–3485, 1993.
18. Laursen, T. A. and Simo, J. C. Algorithmic symmetrization of Coulomb frictional problems using augmented Lagrangians. *Computer Methods in Applied Mechanics and Engineering*, 108:133–146, 1993.
19. Luenberger, D. G. *Linear and Nonlinear Programming*, 2nd Edn. Addison-Wesley Publishing Company, Reading, MA, 1984.
20. Oden, J. T. and Pires, E. B. Algorithms and numerical results for finite element approximations of contact problems with non-classical friction laws. *Computer & Structures*, 19:137–147, 1983.
21. Parisch, H. A consistent tangent stiffness matrix for three-dimensional non-linear contact analysis. *International Journal for Numerical Methods in Engineering*, 28:1803–1812, 1989.
22. Pimenta, P. M. Finite deformation soil plasticity on principal axes. In: *Proceedings of the Third International Conference on Computational Plasticity, Fundamentals and applications*, D. R. J. Owen, E. Oñate and E. Hinton (Eds.), Pineridge Press Limited, Swansea, UK, pp. 859–870, 1992.
23. Press, W. H. *Numerical Recipes in C: The Art of Scientific Computing*, 2nd Edn. Cambridge University Press, 1995.
24. Simo, J. C. and Laursen, T. A. An augmented Lagrangian treatment of contact problems involving friction. *Computers & Structures*, 42:97–116, 1992.
25. Song, S. and Yovanovich, M. M. Explicit relative contact pressure expression: dependence upon surface roughness parameters and Vickers microhardness coefficients. AIAA Paper, pp. 87–152, 1987.

26. Tabor, D. Friction – The present state of our understanding. *Journal of Lubrication Technology*, 103:169–179, 1981.
27. Wriggers, P. Finite element algorithms for contact problems. *Archives of Computational Methods in Engineering*, 2(4):1–49, 1995.
28. Wriggers, P. and Simo, J. C. A note on tangent stiffness for fully nonlinear contact problems. *Communications in Applied Numerical Methods*, 1:199–203, 1985.
29. Wriggers, P., Simo, J. C. and Taylor, R. L. Penalty and augmented Lagrangian formulations for contact problems. In: *Proceedings of NUMETA 85 Conference*, J. Middleton and G. N. Pande (Eds.), Balkema, Rotterdam, 1985.
30. Wriggers, P., Vu Van, T. and Stein, E. Finite element formulation of large deformation impact-contact problems with friction. *Computers & Structures*, 37:319–331, 1990.
31. Wriggers, P. and Zavarise, G. On the application of augmented Lagrangian techniques for nonlinear constitutive laws in contact interfaces. *Comm. Num. Meth. Engng.*, 9:815–824, 1993.
32. Yovanovich, M. M. Thermal contact correlations. AIAA Paper, pp. 81–1164, 1981.
33. Zavarise, G., Schrefler, B. A. and Wriggers, P. Consistent formulation for thermomechanical contact based on microscopic interface law. In: *Proceedings of COMPLAS III*, D. R. J. Owen, E. Hilton and E. E. Oñate, Pineridge Press, 1992.

Frictional Contact for Pile Installation

Daichao Sheng

Centre for Geotechnical and Material Modelling, School of Engineering, The University of Newcastle, NSW, Australia
E-mail: daichao.sheng@newcastle.edu.au

Abstract. Penetration problems in geomechanics are extremely difficult to model, because they usually involve large deformation, frictional contact and large variation of material stiffness. This paper proposes a number of strategies to deal with the computational challenges involved in modelling pile installation in geomechanics. In particular, an Arbitrary Lagrangian–Eulerian method is proposed to handle the large deformation and mesh distortion problem. A smooth discretisation of the contact surfaces is proposed to handle the vertex problems of the penetrating body. An automatic time stepping scheme is proposed to solve the nonlinear stiffness equation. The effectiveness of these numerical enhancements will then be demonstrated in the simulation of installation of displacement piles.

Key words: Pile installation, Arbitrary Lagrangian–Eulerian method, frictional contact, Modified Cam Clay.

1 Introduction

Penetration problems in geomechanics involve the insertion of solid bodies into the ground. The solid body can be a displacement pile used to support a footing or a testing device used to measure soil properties. Numerical modelling of such problems is important, because it can improve our understanding of the physical processes involved, which in turn can lead to better interpretation of test results and more accurate estimation of pile capacity. However, such a simulation is extremely difficult to accomplish, mainly because the problem involves large deformation that challenges the design of an appropriate mesh. In principle, the soil elements beneath and around the penetrating body should be sufficiently small, to achieve good accuracy. However, using fine elements will inevitably lead to severe mesh distortion and likely negative Jacobian, due to the very large deformation in the soil around the penetrating body.

The frictional contact at the soil-structure interface also challenges a stably convergent solution, in particular when the penetrating body has corners or vertices. For

Peter Wriggers and Udo Nackenhorst (eds.), IUTAM Symposium on Computational Contact Mechanics, 239–255.

example, it is observed that the computed pile resistances are strongly oscillatory due to the cyclic release of the soil nodes from the pile end [5, 9]. The magnitude of the oscillation depends on the discretisation of the soil beneath and around the pile. In theory, the oscillation would vanish if the soil beneath the penetrating cone could be discretised into infinitely small elements.

Another computational challenge in numerical modelling of penetration problems arises from the large variation of the material stiffnesses. For example, a soil modelled by the Modified Cam Clay Model has basically a zero stiffness at zero mean stress, but a steel pile has a Young modulus in order of 100 GPa. Very often the penetrating body is therefore simply treated as a rigid body, to avoid the ill-conditioning of the stiffness matrix. However, sometimes we are also interested in the stresses in the penetrating body and thus have to consider its stiffness as well.

This paper proposes a number of strategies to deal with the computational challenges involved in numerical modelling of penetration problems. The effectiveness of the proposed numerical enhancements will be demonstrated through the simulation of installation of displacement piles.

2 Arbitrary Lagrangian–Eulerian method

The Arbitrary Lagrangian–Eulerian (ALE) method used here is based on the operator-split technique and is associated with the Updated Lagrangian (UL) formulation. We first briefly review the UL formulation of large deformations. We limit our discussion to time-independent problems that do not involve pore pressure dissipation. Generalisation to dynamic problems or consolidation problems requires some further research.

In order to accommodate the definitions of stress rates, we use index notations in this part of the paper. The index notations are then changed to vector and matrix notations in the next section, to facilitate the expressions on time stepping and iterations when solving the global equations.

The principle of virtual work states that if δu is a virtual displacement field satisfying the displacement boundary conditions, then equilibrium is satisfied provided:

$$\sum_{\alpha}\left(-\int_{V_\alpha}\sigma_{ij}\delta\varepsilon_{ij}\,\mathrm{d}V+\int_{V_\alpha}\delta u_i b_i\,\mathrm{d}V+\int_{S_\alpha}\delta u_i t_i\,\mathrm{d}S\right)$$
$$+\int_{S_c}(t_N\delta g_N+t_T\delta g_T)\,\mathrm{d}S=0, \tag{1}$$

where $\delta\varepsilon$ denotes the variation of the strain tensor derived from the virtual displacements, σ is the Cauchy stress tensor, b is the body force vector, t is the distributed force acting on the boundary S_α of the volume V_α, t_N and t_T are respectively the normal and tangential forces at the contact surface S_c, δg_N and δg_T are respectively the virtual normal and tangential gap, and the summation is over the number of bodies. For a nonlinear problem, Equation (1) is typically applied incrementally. We typically assume that the analysis starts at time 0 and all state variables that satisfy equilibrium are known up to time t. Further loading and deformation will require the equilibrium to be satisfied at time $t + \delta t$. The principle of virtual work becomes

$$\sum_{\alpha} \int_{V_{\alpha}^{t+\Delta t}} \sigma_{ij}^{t+\Delta t} \, \delta\varepsilon_{ij} \cdot \mathrm{d}V = R^{t+\Delta t} + C^{t+\Delta t}, \tag{2}$$

where R denotes the virtual work resulting from body forces and surface tractions, C denotes the virtual work resulting from the contact tractions, and the superscript denotes the time when the quantities are measured. For geometrically nonlinear problems, the configuration ($V^{t+\Delta t}$ and $S^{t+\Delta t}$) in Equation (2) is not known and has to be transferred to a known configuration, for example to the one at the start of the current time step t (Updated Lagrangian) or the one at time zero (Total Lagrangian). In either case, the second Piola–Kirchhoff stress and the Green–Lagrangian strain are usually introduced in places of σ and ε, respectively, to eliminate the effects of rigid body motion on the Cauchy stress tensor.

For problems involving large-slip contact, the contact constraints are always described based the current configuration and therefore the Updated Lagrangian formulation is preferable. Realising that constitutive laws for geomaterials are seldom written in terms of the second Piola–Kirchhoff stress and the Green–Lagrangian strain, we decompose the total stress rate into a frame-independent stress rate due to straining and a stress rate due to rigid body motion. Using the Jaumann stress rate, we have

$$\mathrm{d}\sigma_{ij}^{J} = \mathrm{d}\sigma_{ij} - \sigma_{ik} \, \mathrm{d}\Omega_{kj} - \sigma_{jk} \, \mathrm{d}\Omega_{ki} = C_{ijkl} \cdot \mathrm{d}\varepsilon_{kl}, \tag{3}$$

where C_{ijkl} is the stress-strain tensor derived from the constitutive relations in terms of the Cauchy stresses and the linear strains, and Ω is the spin tensor given by

$$\Omega_{ij} = \frac{1}{2}\left(\frac{\partial u_i}{\partial x_j} - \frac{\partial u_j}{\partial x_i}\right). \tag{4}$$

Introducing (3) into the virtual work equation, we can obtain the following equilibrium equation for the UL method based on the Jaumann stress rate

$$\begin{aligned} &\int_{V^t} C_{ijkl} \, \mathrm{d}\varepsilon_{kl} \, \delta(\mathrm{d}\varepsilon_{ij}) \mathrm{d}V^t + \int_{V^t} (\sigma_{ik}^t \, \mathrm{d}\Omega_{kj} + \sigma_{jk}^t \, \mathrm{d}\Omega_{ki}) \delta(\mathrm{d}\varepsilon_{ij}) \, \mathrm{d}V^t \\ &+ \int_{V^t} \sigma_{ij}^t \, \delta(\mathrm{d}\eta_{ij}) \, \mathrm{d}V^t = R^{t+\Delta t} + C^{t+\Delta t} - \int_{V^t} \sigma_{ij}^t \, \delta(\mathrm{d}\varepsilon_{ij}) \, \mathrm{d}V^t, \end{aligned} \tag{5}$$

where $\delta(\mathrm{d}\eta)$ is the variation of the nonlinear part of the incremental Green–Lagrange strain tensor. Linearisation of the terms on the left-hand side in the above equation will generally lead to a stiffness matrix due to elastoplatic materials (the first integration on the left-hand side) as well as a stiffness matrix due to the geometric nonlinearity (the second and third integration on the left-hand side). Linearisation of the virtual work due to contact tractions requires the discretisation of the contact surfaces, which will be discussed later.

At Gauss points, the stress increments are found by integrating $d\sigma$ in Equation (3) along given strain increments. Nazem et al. [3] have recently discussed alternative integration schemes for geomaterials that experience straining hardening.

2.1 ALE Solution

At the end of the UL solution outlined above, the mesh may be distorted since it moves along with the material. To avoid mesh distortion, the mesh and the material displacements can be separated from each other, thus allowing the mesh to move independently from the material. This assumption leads to the formulation of the ALE method, which adds an additional Euler step following each UL step. In the Euler step, a new mesh is generated for the deformed domain. All kinematic and static variables are then transferred from the old mesh to the new mesh using the relation between the material derivative and the mesh derivative

$$\underline{\dot{f}} = \dot{f} + (\nu_i - \nu_i^r)\frac{\partial f}{\partial x_i}, \tag{6}$$

where f is an arbitrary function, ν_i is the material velocity, ν_i^r is the mesh velocity, $\dot{f}$ denotes the time derivative of f with respect to mesh (grid points) coordinates, and $\underline{\dot{f}}$ represents the time derivative of f with respect to material coordinates. The term $\nu_i - \nu_i^r$ is called the convective velocity.

Realising that the mesh displacements are decoupled from the material displacements, we understand that the new mesh to be established in the Euler step can be arbitrary, except two basic conditions are satisfied:

1. The new mesh should conform to the deformed boundaries of the domain and the materials, and
2. Its topology and connectivity should remain the same as the old mesh.

These requirements are to ensure that the mapping of variables between the new and old meshes can be carried out material by material and element by element. The first requirement also implies that the nodes on a material or domain boundary should remain on it. We also notice that, for an initially optimal mesh of a homogeneous domain, if the displacements at all the boundary nodes are prescribed and such displacements ensure an optimal division of the boundaries, an elastic analysis based on the prescribed displacements should result in an optimal distribution of internal nodes. Because of the displacements at all boundary nodes are prescribed, the elastic properties used in such an analysis would be arbitrary as long as the material is assumed to be homogeneous and isotropic. With these in mind, our first goal in the Euler step is then to obtain the optimal division of the material and domain boundaries obtained in the UL step, or in other words, to relocate the nodes on the boundaries

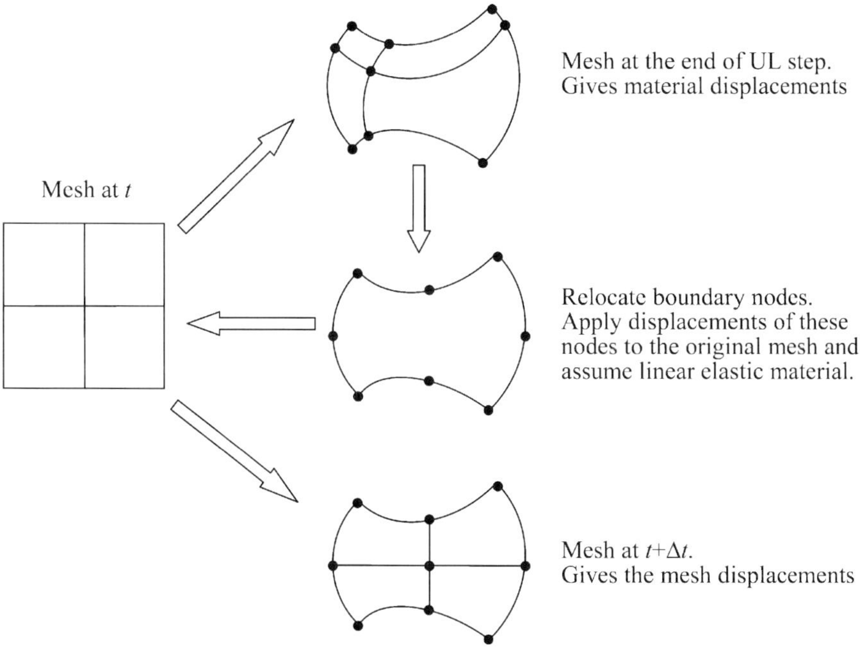

Fig. 1. General procedure of the ALE method.

so that they are optimally located. We also assume that the initial mesh before the UL step is optimal. Comparing the relocated nodes with the nodes in the initial mesh gives us the mesh displacements of the nodes on all the boundaries. With the boundary displacements now known, we carry out an elastic analysis on the initial mesh, to obtain the optimal mesh displacements of internal nodes. Because we prescribe the displacements along all boundaries and material interfaces, the actual values of the elastic parameters used in this analysis are not important and one set of elastic parameters can be used for the entire domain, regardless of the presence of real material interfaces. The new mesh so obtained also shares the same connectivity and topology as the old mesh. The procedure outlined above is illustrated in Figure 1.

Once the mesh displacements are obtained, the state variables such as stresses and hardening parameters are then transferred to the new mesh. The general procedure for the ALE method consists of the following steps.

1. Perform an UL to find the material displacements.
2. Relocate the nodes on all boundaries of the domain and between materials.
3. Optimise the mesh by performing an elastic analysis.
4. Remap historical variables from the old mesh onto the new mesh.

The total nodal displacements are then the summation of the material and mesh displacements. The historical variables remapped to the new mesh can be state variables such as stresses and hardening parameters or nodal variables such as displacements. If the stresses and hardening parameters are remapped, the consistency condition may not be satisfied and the objectivity is not guaranteed. If the displacements are remapped, the stresses and hardening parameters have to be re-integrated over the accumulated strains and it is hence very difficult to deal with initial stresses and stress-path dependence. Neither method will guarantee the equilibrium at the end of Step 4. In addition, when friction contact is involved, historical variables such as frictional slip have to be remapped to the new mesh, giving another source of loss of equilibrium. If equilibrium is indeed not satisfied, additional iterations may be carried out at the end of Step 4, to reduce the unbalanced forces. These iterations can be done with the fixed mesh, i.e. under the assumption of small deformation. The resulting displacements from these iterations are then added to the total nodal displacements.

3 Smooth Discretisation of Contact Surfaces

Linearisation of the virtual work equation (5) requires the linearisation of the virtual work due to contact forces, which in turn requires the discretisation of the contact surfaces. In this paper, the so-called node-to-segment contact element is used, where contact constraint is enforced for each slave node while the master surface is discretised into curved segments. The smooth discretisation of the master surface uses Bézier polynomials [6, 10]. The discretised virtual work due to normal and tangential contact forces respectively takes the form

$$C_N = \int_{S_c} t_N \, \delta g_N \, \mathrm{d}S \approx \sum_{s=1}^{n_c} \delta \mathbf{u}_s^T [\varepsilon_N \, A_s \, g_{N_s} \mathbf{B}_N(\xi)] = \sum_{s=1}^{n_c} \delta \mathbf{u}_s^T \, \mathbf{F}_{Ns}^c, \tag{7}$$

$$C_T = \int_{S_c} t_T \cdot \delta g_T \, \mathrm{d}S \approx \sum_{s=1}^{n_c} \delta \mathbf{u}_s^T \, \mathbf{B}_\xi(\xi) \left\| \frac{\partial \bar{\mathbf{x}}_s}{\partial \xi} \right\| \gamma A_s = \sum_{s=1}^{n_c} \delta \mathbf{u}_s^T \, \mathbf{F}_{Ts}^c, \tag{8}$$

where n_c is the total number of slave nodes, $\delta \mathbf{u}_s$ is the nodal virtual displacements of the contact element (consisting of one slave node and two master nodes), A_s is the contact area associated with the slave node, ε_N is the penalty parameter for normal contact, g_{Ns} is the normal gap, γ is the tangential stress which will depend on the constitutive law for slip contact and on the tangential penalty parameter for stick contact, $\bar{\mathbf{x}}_s$ are the coordinates of the projection of the slave node on the master segment, ξ is the distance between $\bar{\mathbf{x}}_s$ and the master node that is closest to the slave node, $\mathbf{B}_N$ and $\mathbf{B}_\xi$ are Bézier polynomials, and $\mathbf{F}_{Ns}^c$ and $\mathbf{F}_{Ts}^c$ are the nodal forces of the contact element.

4 Automatic Load Stepping Scheme

Combining Equations (1), (7), and (8) leads to a set of global equations in matrix form

$$\sum_{\alpha}\left(-\int_{V^\alpha} \delta\boldsymbol{\varepsilon}^T \boldsymbol{\sigma}\, \mathrm{d}V + \int_{V^\alpha} \delta\mathbf{u}^T\, \mathbf{b}\, \mathrm{d}V + \int_{S_\sigma^\alpha} \delta\mathbf{u}^T\, \mathbf{t}\mathrm{d}S\right)$$
$$+\int_{S_c} (t_N\, \delta g_N + t_T\, \delta g_T)\, \mathrm{d}S = \delta\mathbf{U}^T (\mathbf{G}(\mathbf{U}) + \mathbf{F}_N^c(\mathbf{U}) + \mathbf{F}_T^c(\mathbf{U}))$$
$$= \delta\mathbf{U}^T (\mathbf{G}(\mathbf{U}) + \mathbf{G}^c(\mathbf{U})) = \delta\mathbf{U}^T\, \mathbf{R}(\mathbf{U}) = \mathbf{0}, \tag{9}$$

where $\mathbf{U}$ is used in place of $\mathbf{u}$ to indicate the discretised global displacement field, $\mathbf{G}(\mathbf{U})$ denotes the domain contributions to the residual vector, $\mathbf{G}^c(\mathbf{U})$ denotes the contact contributions, and $\mathbf{R}(\mathbf{U})$ is the global residual vector. The global tangent matrix is obtained by linearising (9) at a given $\mathbf{U}$ according to

$$\mathbf{K}(\mathbf{U}) = \frac{\partial\mathbf{R}(\mathbf{U})}{\partial\mathbf{U}} = (\mathbf{K}_{ep}(\mathbf{U}) + \mathbf{K}_{nl}(\mathbf{U}) + \mathbf{K}_{Ns}(\mathbf{U}) + \mathbf{K}_{Ts}(\mathbf{U})), \tag{10}$$

where $\mathbf{K}_{ep}$ is the stiffness matrix due to the material stiffness, $\mathbf{K}_{nl}$ is the stiffness matrix due to geometric nonlinearity, and $\mathbf{K}_{Ns}$ and $\mathbf{K}_{Ts}$ are the tangent matrices due to the normal and tangential contact.

Instead of using the standard Newton–Raphson method, an alternative method to solve the system of nonlinear equations is to use the incremental Euler methods with different orders of accuracy, e.g. the first-order accurate Euler method and the second-order accurate modified Euler method. The difference between such a pair of solutions naturally constructs a measure of the load path error, and can be used to control the load step sizes. Abbo and Sloan [1] and Sheng and Sloan [8] have successfully applied such methods to solve quasi-static problems for elastoplastic soils. One of such methods is here generalised to solve the large deformation frictional contact problem.

To apply the automatic load stepping method, we first note that the residual vector consists of the external force vector $\mathbf{F}^{\mathrm{ext}}$ and the internal force vector $\mathbf{F}^{\mathrm{int}}$,

$$\mathbf{R}(\mathbf{U}) = \underbrace{\sum_{\alpha}\int_{V^\alpha} \mathbf{N}^T\mathbf{b}\,\mathrm{d}V + \sum_{\alpha}\int_{S_\alpha^\alpha} \mathbf{N}^T\mathbf{t}\,\mathrm{d}S}_{\mathbf{F}^{\mathrm{ext}}} - \underbrace{\left(\sum_{\alpha}\int_{V^\alpha} \mathbf{B}^T\boldsymbol{\sigma}\,\mathrm{d}V + \sum_{s=1}^{n_c}(\mathbf{F}_{Ns}^c + \mathbf{F}_{Ts}^c)\right)}_{\mathbf{F}^{\mathrm{int}}},$$
$$= \mathbf{F}^{\mathrm{ext}} - \mathbf{F}^{\mathrm{int}}, \tag{11}$$

where the matrix $\mathbf{N}$ contains the shape functions, and the matrix $\mathbf{B}$ contains the derivatives of the shape functions. Note that the contact forces are considered as a part of the internal force in Equation (11). The internal forces are always a function of the displacement field. The external forces also depend on the displacement field in large deformations (or in small deformations where the reactions at non-zero prescribed displacements are treated as external forces).

With the separation of the external and internal forces, we can compute the two displacement increments based on the known displacement $\mathbf{U}_{n-1}$

$$\Delta\mathbf{U}_1 = [\mathbf{K}(\mathbf{U}_{n-1})]^{-1}(\mathbf{F}_n^{\text{ext}} - \mathbf{F}_{n-1}^{\text{ext}}), \tag{12}$$

$$\Delta\mathbf{U}_2 = [\mathbf{K}(\mathbf{U}_{n-1} + \Delta\mathbf{U}_1)]^{-1}(\mathbf{F}_n^{\text{ext}} - \mathbf{F}_{n-1}^{\text{ext}}). \tag{13}$$

With these displacement increments, the first-order accurate Euler and the second-order accurate modified Euler updates are then given by

$$\mathbf{U}_n = \mathbf{U}_{n-1} + \Delta\mathbf{U}_1, \tag{14}$$

$$\hat{\mathbf{U}}_n = \mathbf{U}_{n-1} + \frac{1}{2}(\Delta\mathbf{U}_1 + \Delta\mathbf{U}_2), \tag{15}$$

respectively. The difference of the two updates normalised by the current displacement norm constitutes a relative error measure

$$R = \frac{\|\hat{\mathbf{U}}_n - \mathbf{U}\|}{\|\mathbf{U}_n\|} = \frac{1}{2}\frac{\|\Delta\mathbf{U}_2 - \Delta\mathbf{U}_1\|}{\|\mathbf{U}_n\|}. \tag{16}$$

This relative error is then used to control the size of the load subincrement. The current load subincrement is accepted if R is less than a user-defined tolerance, *DTOL*, and rejected otherwise. In either case, the size for the next subincrement is calculated according to

$$q = 0.7\sqrt{DTOL/R}, \tag{17}$$

$$\Delta T_{n+1} = q\Delta T_n, \tag{18}$$

where ΔT is a pseudo-time step and is used to control the increment of external forces. The coefficient of 0.7 is included to prevent the step control mechanism from choosing load subincrements which just fail to meet the local error tolerance and may be regarded as a safety factor.

When the relative error does not exceed the tolerance *DTOL* and the substep is accepted, the first-order accurate Euler solution is further iterated to reduce the unbalanced forces, using the Newton iteration:

$$\mathbf{U}_n^i = \mathbf{U}_{n-1} + \Delta\mathbf{U}_1^i, \tag{19}$$

$$\Delta\mathbf{U}_1^i = \mathbf{U}_1^{i-1} + \delta\mathbf{U}^i, \tag{20}$$

$$\delta\mathbf{U}^i = \left(\mathbf{K}(\mathbf{U}_n^{i-1})\right)^{-1}\mathbf{R}(\mathbf{U}_n^{i-1}), \tag{21}$$

where $\delta \mathbf{U}^i$ represents the iterative displacement correction. To start the procedure, the displacements $\mathbf{U}_n^0$ are set to $\mathbf{U}_{n+1} + \Delta \mathbf{U}_1$. The iteration is stopped once the condition $\|\mathbf{R}(\mathbf{U})\| \leq ITOL \cdot \|\mathbf{F}_n^{\text{ext}}\|$ is satisfied; if this condition is not satisfied within the maximum number of iterations (typically set to 10), the substep is regarded as a failure. The step size is then reduced according to Equation (18) and the step is restarted. The iteration tolerance *ITOL* is typically set to $ITOL = DTOL/10$, to minimise the load path error for a given setting of *DTOL* and avoid needless iteration [8].

Compared to the Newton–Raphson scheme (supposing it requires only one iteration in each load increment), the automatic scheme requires one extra tangent matrix formation/factorisation and one extra Equation back-substitution. In reality, the number of iterations in the Newton–Raphson scheme may be well above one, and hence the true saving of the automatic scheme lies in the reduction of the number of iterations.

5 Numerical Examples

In the first example, we study the mesh effects on pile resistance. A relatively short pile is chosen for convenience to reduce the computational work. The finite element meshes and material properties adopted in the analyses are shown in Figure 2. Linear triangular elements are used both for the soil and the pile. Four types of mesh are analysed, with the width of the soil elements beneath the pile varying from one to one-eighth of the pile radius.

An elastic pile of radius 0.4 m is pushed into the soil to a depth of 2.0 m. The penetration is achieved by imposing a total vertical displacement at the two top nodes of the pile shaft. The soil is modelled as a non-associated Mohr–Coulomb material, with the properties given in Figure 2. The pile is modelled as an elastic material with a Young's modulus $E = 2 \times 10^8$ kPa, which is 2×10^4 times the modulus of the soil. Other material properties defined in Figure 2 are: $\nu =$ Poisson's ratio, $c' =$ cohesion, $\phi' =$ frictional angle, $\psi' =$ dilation angle, and $\gamma =$ unit weight of the soil. The pile-soil interfacial friction coefficient is set to 0.01. The penalty parameters in the normal and tangential directions are set to 10^6 kN/m^3. It was found that this value can be increased or decreased by an order of one, without causing significant change in the numerical results.

The UL method was employed to analyse mesh A and mesh B shown in Figures 2a and 2b, respectively. The deformed meshes from the UL analyses are illustrated in Figures 2e and 2f, respectively. In addition, the ALE method was also used to analyse the same meshes. Figure 2g shows the deformed mesh from the ALE analysis for mesh B. This is not much different from the mesh in Figure 2f, except that the nodes along the boundaries (as well as the internal nodes) are relocated. The meshes shown in Figures 2h and 2i were also obtained from ALE analyses.

Figure 3 shows the predicted pile resistances as a function of the depth of penetration for the four meshes. As expected, the UL and ALE analyses give almost the

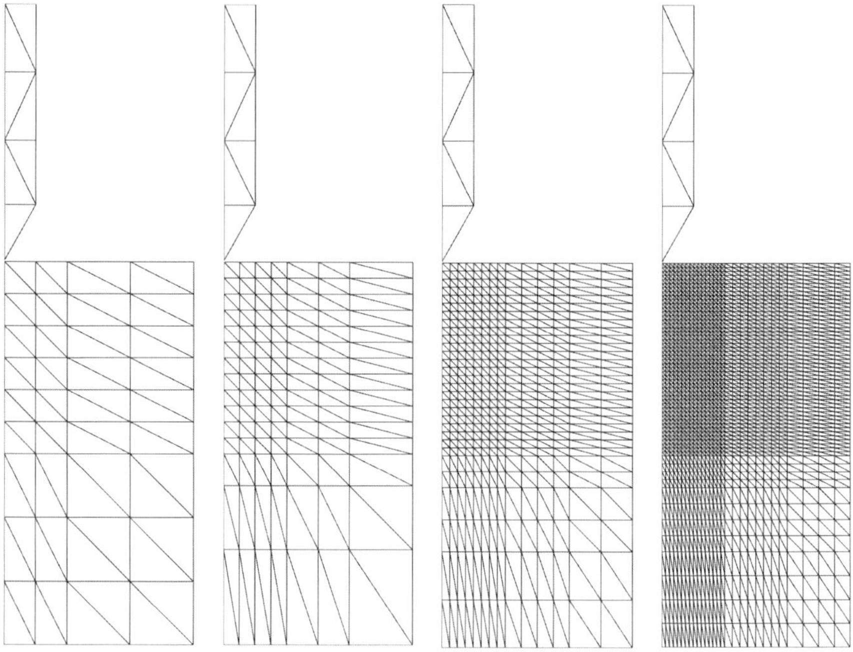

(a) Mesh A (b) Mesh B (c) Mesh C (d) Mesh D

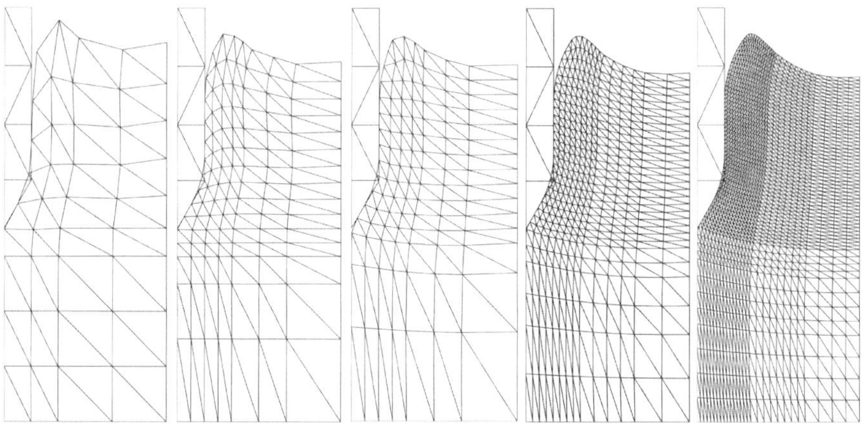

(e) UL: Mesh A (f) UL: Mesh B (g) ALE: Mesh B (h) ALE: Mesh C (i) ALE: Mesh D

Fig. 2. Original and deformed meshes (Mohr–Coulomb soil: $E = 10^4$ kPa, $\nu = 0.3$, $c' = 1$ kPa, $\phi' = 30°$, $\psi' = 20°$, $\gamma = 20$ kN/m^3. Elastic pile: $E = 1.0 \times 10^8$ kPa, $\nu = 0.3$. Dimension of initial soil domain: 2.4 m $\times$ 4.8 m. Dimension of pile shaft: 0.4 m $\times$ 3 m. Cone angle: 60°. Pile-soil interfacial friction coefficient: 0.01. Penetration: 2.5 m).

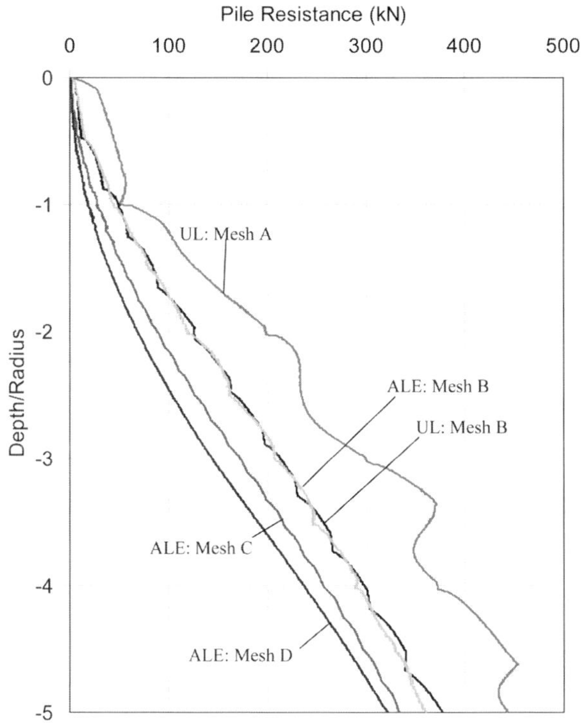

Fig. 3. Effects of element size on the pile resistances.

same curve for Mesh B. At any penetration the predicted pile resistance decreases as the mesh becomes finer. The predicted resistances for mesh A differ from those for mesh D by up to a factor of 2. However, the differences between Mesh C and Mesh D are not as pronounced as those between mesh A and mesh B, indicating that the numerical solution is converging as the mesh is refined. It is also observed that the numerical oscillations in the predicted pile resistances become less pronounced for the finer meshes such as C and D.

In the second example, a steel pile of radius 0.2 m is pushed into a Mohr–Coulomb soil to a depth of 2.5 m. The finite element mesh shown in Figure 4, where the width of the finest soil elements is roughly one quarter of the pile radius, was used for the ALE analysis. In this case, a UL analysis can also be carried out, provided that the width of the smallest soil elements increases to about one half of the pile radius (a coarse mesh that uses a quarter of the soil elements is shown in Figure 4). Figure 4 indicates that the ALE analysis can simulate the penetration of the pile to a depth of 12.5 pile radii without generating any significant or unwanted mesh distortion. This figure also shows that, due to the dilatant behaviour predicted by the

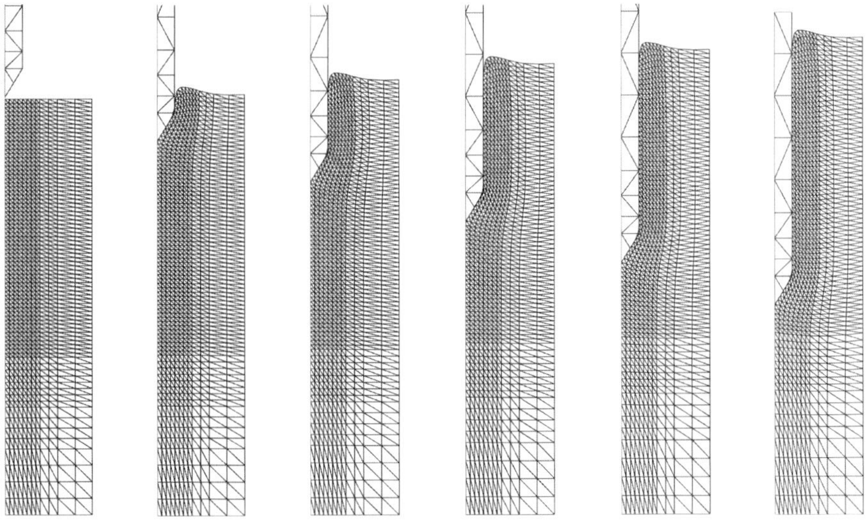

Fig. 4. Deformed meshes (Mohr–Coulomb soil: $E = 10^4$ kPa, $\nu = 0.3$, $c' = 10$ kPa, $\phi' = 35°$, $\psi' = 20°$, $\gamma = 20$ kN/m^3. Elastic pile: $E = 1.0 \times 10^8$ kPa, $\nu = 0.3$. Dimension of initial soil domain: 1 m × 5 m. Dimension of pile shaft: 0.2 m × 3 m. Cone angle: 60°. Pile-soil friction coeffcient: 0. Penetration: 2.5 m).

Mohr–Coulomb model under shear, the volume of the soil increases significantly as the pile is inserted.

In Figure 5, the predicted total resistances (the reactions at the two top nodes where displacements are prescribed) are plotted against the normalised penetration depth. For the UL analysis that uses a coarse mesh and the straight contact discretisation, the predicted total resistance shows strong oscillations. Comparing the curves denoted by UL ($\mu = 0.1$, straight contact segment) and UL ($\mu = 0.1$), we see that the smooth discretisation of the contact surfaces reduces the oscillations significantly. Comparing the curves denoted by UL ($\mu = 0.1$) and UL ($\mu = 0.0$), we see that the degree of oscillation depends on the amount of pile-soil interfacial friction.

The ALE solutions illustrated in Figure 5 were obtained using the finer mesh and the smooth discretisation of the contact surfaces. These solutions, even though still somewhat oscillatory, are much smoother than the UL solutions. In theory, to completely remove the oscillations, we need to use very fine (infinitesimally small) soil elements. A key advantage of the proposed ALE method is that it can effectively solve the oscillation problem by using a fine mesh.

Figure 6 shows stress contours in the soil and the pile. The stress bulbs beneath the pile are clearly shown in the figures for radial, vertical and circumferential stresses. The shear stress shows two stress concentration areas (bulbs). In the pile, the maximum radial, vertical and circumferential stresses occur at its tip, while the smal-

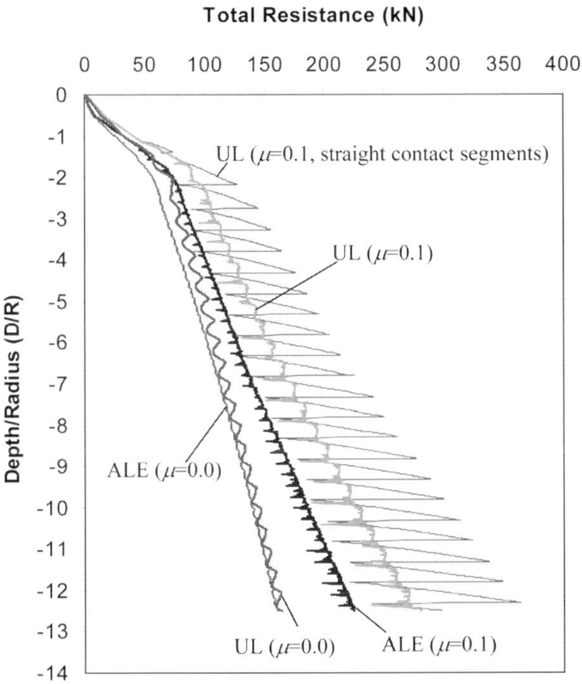

Fig. 5. Predicted load-displacement curves (ALE solutions obtained with a fine mesh; UL solutions obtained with a coarse mesh; smooth contact discretisation used in all solutions except the one indicated; μ: interfacial friction coefficient).

lest vertical stress occurs at its shoulder. The key advantage here is that the pile is treated as a deformable body, compared to the rigid piles assumed in the analyses of Simo and Meschke [9] or Sheng et al. [5]. Some minor element overlapping, which is caused by the penalty method and the smooth discretisation of the pile surface, can be observed near the tip.

All the solutions were obtained with the automatic scheme described previously using 1000 coarse load steps. The ALE analyses for the fine mesh require approximately 6–8 hours CPU time on an IBM T41 laptop with a 1.6 GHz Pentium processor and 1.5 GB RAM. The UL analyses for a coarse mesh, with one quarter of the elements in the fine mesh, require approximately 2–3 hours of CPU time on the same machine. Allowing for the difference in the degrees of freedom in these meshes, we see the ALE method is almost as efficient as the UL method. A standard Newton–Raphson scheme (see [6]) was also tried with 10,000 load steps, but failed to provide a reasonable solution. The failures of the Newton–Raphson scheme

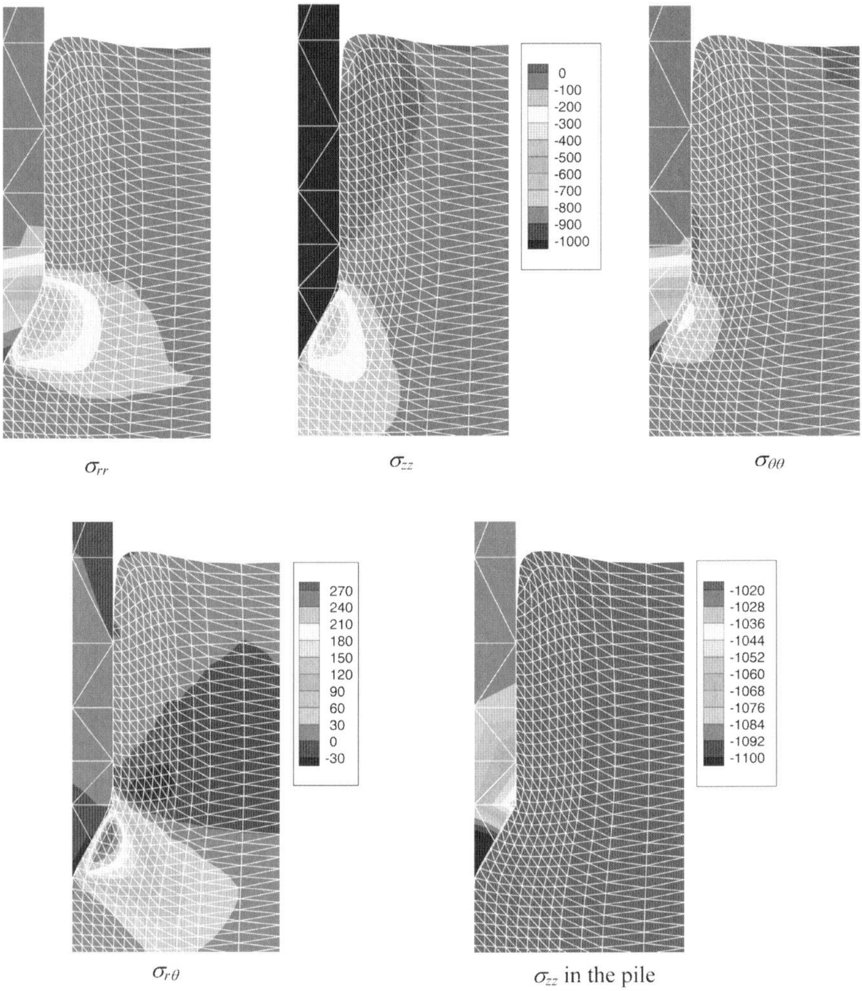

Fig. 6. Stress distributions obtained from ALE analysis (σ_{rr}: radial stress; σ_{zz}: axial stress; $\sigma_{\theta\theta}$: circumferential stress; $\sigma_{r\theta}$: shear stress; legends in kPa and tension positive; pile-soil interfacial friction coefficient: 0.1).

were caused either by lack of convergence, or by numerical breakdown if the non-converged steps were allowed to continue.

Figure 7 shows the ALE mesh for a pile penetrating slowly into a Modified Cam Clay (MCC) soil. The soil is fully drained and its properties are given in the figure: λ is the slope of the normal compression line (NCL), κ is the slope of the unloading-reloading line, e_N is the void ratio on the NCL when the mean stress is one (kPa),

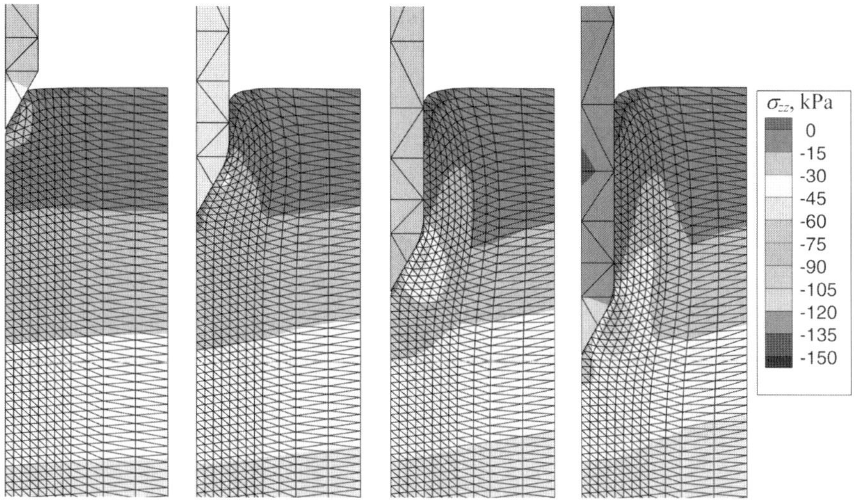

(a) Axial stress distribution (pile-soil interfacial friction: 0.0)

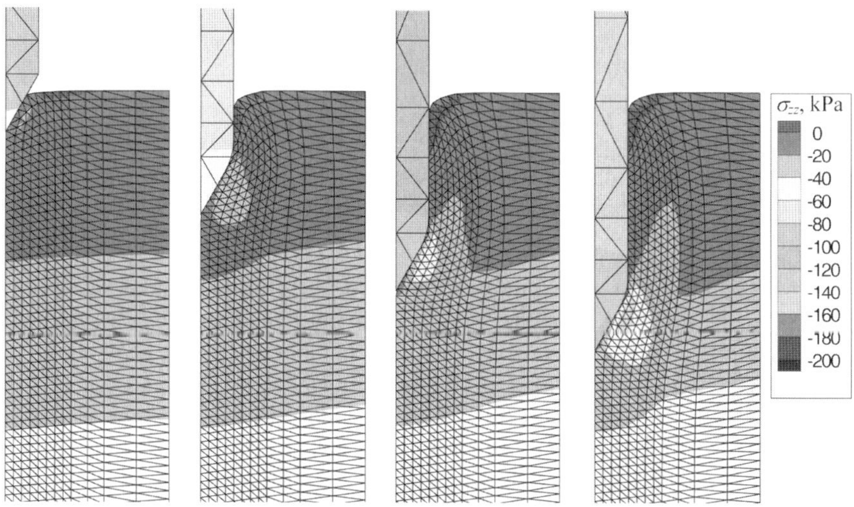

(b) Vertical stress distribution (pile-soil interfacial friction: 0.02)

Fig. 7. Pile pushed into Modified Cam Clay (elastic pile: $E = 10^8$ kPa, $\nu = 0.3$, shaft radius = 0.2 m, length = 3.5 m. Mohr–Coulomb soil: 0.05 m thick, $E = 10^4$ kPa, $\nu = 0.3$, $c' = 10$ kPa, $\phi' = 30°$, $\psi' = 20°$, $\gamma = 20$ kN/m^3. Modified Cam Clay soil: 4.95 m thick, $\nu = 0.3$, $\phi' = 30°$, $\gamma = 20$ kN/m^3, $\lambda = 0.2$, $\kappa = 0.05$, $e_N = 2.0$, OCR = 3).

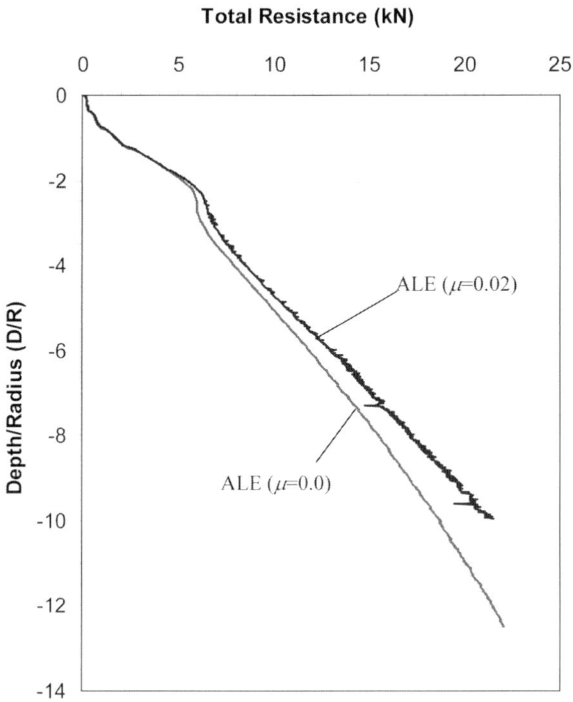

Fig. 8. Predicted load-displacement curves in MCC soil (μ: interfacial friction coefficient).

and OCR is the initial over-consolidation ratio. The description and implementation of the MCC constitutive model in the finite element procedure can be found in [7] or in [4]. Because the MCC soil has a zero elastic bulk modulus at zero mean stress, the top layer of soil elements (0.05 m thick) was modelled using the Mohr-Coulomb constitutive model. The deformed meshes in Figure 7 show that a part of the initial ground surface is actually in contact with the pile, and that the ALE meshes sustain the optimal form of the initial undeformed mesh. We can also notice that the ground surface remains more or less at the same level as the pile penetrates, indicating that overall the soil is compressive rather dilatant. The load-displacement curves shown in Figure 8 further demonstrate the effectiveness of the new algorithms in reducing oscillations.

6 Conclusions

The ALE method presented in this paper is effective for handling the mesh distortion in penetration problems and does not cause much increase in the CPU time.

The smooth discretisation of contact surfaces can reduce the oscillation in the load-displacement curves significantly, while the automatic load stepping scheme is more robust in solving the systems of nonlinear equations. These algorithms advance the simulation of penetration problems in geomechanics considerably. Additional enhancements that are possible include the incorporation of a mortar-based frictional contact formulation for higher order elements in the soil domain [2], as well as the development of coupled displacement and pore pressure methods with dynamic effects.

References

1. Abbo, A.J. and Sloan, S.W. (1996). An automatic load stepping algorithm with error control. *International Journal for Numerical Methods in Engineering*, 39:1737–1759.
2. Laursen, T.A. (2002). *Computational Contact and Impact Mechanics*. Springer, Berlin.
3. Nazem, M., Sheng, D. and Carter, J.P. (2006). Stress integration and meshing refinement for large deformation in geomechanics. *International Journal for Numerical Methods in Engineering*, 65:1002–1027.
4. Potts, D.M. and Zdravkovic, L. (2001). *Finite Element Analysis in Geotechnical Engineering, Vol. 1: Theory*. Thomas Telford, London.
5. Sheng, D., Axelsson, K. and Magnusson, O. (1997). Stress and strain fields around a penetrating cone. In *Numerical Models in Geomechanics*, Pietruszczak, S. and Pande, G.N. (Eds), Balkema, Rotterdam, pp. 653–660.,
6. Sheng, D., Sun, D.A. and Matsuoka, H. (2006). Cantilever sheet-pile wall modelled by frictional contact. *Soils and Foundations*, 46:29–37.
7. Sheng, D., Sloan, S.W. and Yu, H.S. (2000). Aspects of finite element implementation of critical state models. *Computational Mechanics*, 26:185–196.
8. Sheng, D. and Sloan, S.W. (2001). Load stepping methods for critical state models. *International Journal for Numerical Methods in Engineering*, 50:67–93.
9. Simo, J.C. and Meschke, G. (1993). A new class of algorithms for classical plasticity extended to finite strains. Application to geomaterials. *Computational Mechanics*, 11:253–278.
10. Wriggers, P. (2006). *Computational Contact Mechanics*. Springer, Berlin.

The Influence of Contact Pressure on the Dynamic Friction Coefficient in Cylindrical Rubber-Metal Contact Geometries

M.A. Jiménez, J.M. Bielsa, R. Rodríguez and S. Dobón

Instituto Tecnológico de Aragón (Materials Department), C/ María de Luna 7–8, E-50018 Zaragoza, Spain
E-mail: majimenez@ita.es

Abstract. As it is commonly know, classic Coulomb's and Amonton's friction laws, which mainly establish that the friction coefficient is independent of the area of contact, are proven to be not valid in the case of rubber-like materials. In this particular case, and due to their specific mechanical properties, the friction coefficient should be expressed as a function of contact pressure, sliding speed, temperature and lubrication regime, if the latter were the case.

The dependence with the contact pressure is associated to the varying ratio of real (microscopic level) to apparent (macroscopic level) area of contact when the vertical load (contact pressure) is rising. The problem increases in complexity when neither the contact pressure distribution nor the ratio of real to apparent area of contact are uniform along the apparent area of contact, being the cylindrical contact geometry a typical example of this situation. In the present paper, the dependence of the dynamic friction coefficient value with the contact pressure in cylindrical rubber-aluminium contact geometries is analysed in detail.

As commented before, contact pressure distribution is not constant along the whole cylindrical area of contact and, thus, friction coefficient value has to be calculated indirectly from friction force results obtained in tribotesting by means of a mathematical method. In addition, the apparent area of contact in the cylindrical case is also not constant when the vertical load increases. Thus, the robustness of the method has to be improved by comparison of experimental measurements of the apparent area of contact with FEM results of the tribotesting, combining them with the adjustment of the rubber material model. As it will be explained along the paper, the method consists on combining FEM simulations of the tribotesting to obtain contact pressure distributions along the cylindrical area of contact for different vertical loadings, and then on developing a mathematical procedure for obtaining a final analytical expression for the dynamic friction coefficient vs. the contact pressure. Finally, it is checked that this method provides good correlations with already existing friction models which can be found in the literature [2, 3].

Key words: Tribology, polymers, dynamic friction coefficient, contact pressure, finite element method, cylindrical contact geometry.

Peter Wriggers and Udo Nackenhorst (eds.), IUTAM Symposium on Computational Contact Mechanics, 257–275.

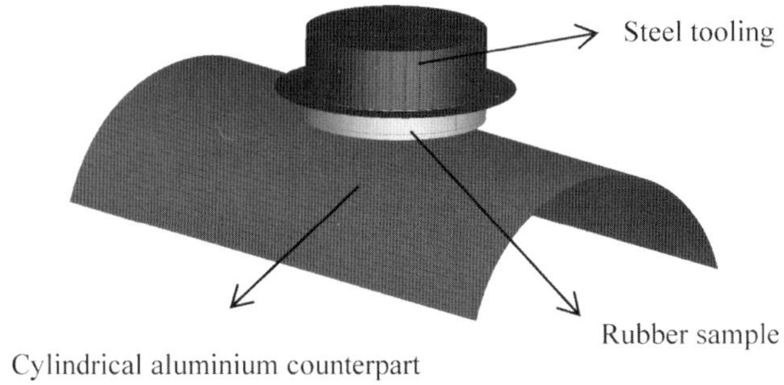

Fig. 1. "Flat-on-cylinder" contact geometry (tribometer test configuration).

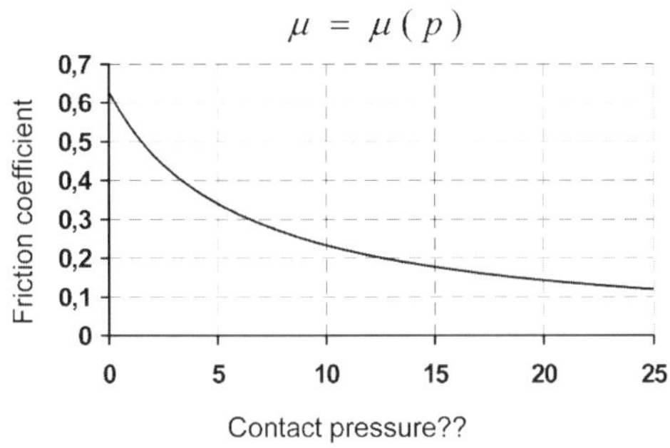

Fig. 2. Dependency of the friction coefficient on rubbers with contact pressure (sample graph).

1 Antecedents and Scope

In the present paper, the dependence of the dynamic friction coefficient value with the contact pressure in rubber-aluminium cylindrical contact geometries is analysed in detail. In this "flat-on-cylinder" configuration, which is also selected for tribometer testing with reciprocating motion, contact pressure distribution is namely not uniform along the whole area of contact if it is compared to a classical "flat-on-flat" tribometer test configuration. Since contact pressure plays a substantial role in rubber friction as dependency of the friction coefficient, to obtain values of the friction coefficient calculated directly as the ratio between the tangential and vertical forces for a given average value of the contact pressure along the area of contact, without detailed postprocessing of the experimental results, may lead to severe errors. As it will be shown later, contact pressure distribution changes significantly along the area of contact, showing high scatter with respect to the mentioned average value.

Due to the previous, the friction coefficient value vs. contact pressure has to be calculated indirectly from friction force results obtained in tribotesting by means of a mathematical method.

In addition, the apparent area of contact is also not constant when the vertical load increases. Thus, the robustness of the method has to be improved by comparison of experimental measurements of the apparent area of contact with FEM results of the tribotesting, combining them with the adjustment of the rubber material model.

As it will be explained in this paper, the method consists on combining FEM simulations of the tribotesting to obtain contact pressure distributions along the cylindrical area of contact for different vertical loadings, and then on developing a mathematical procedure for obtaining a final analytical expression for the dynamic friction coefficient as dependent with the contact pressure.

The main objectives of the present paper are the following:

- to analyse the dependency of the friction coefficient with contact pressure on rubber-metal sliding contacts;
- to compare the contact pressure distribution in "flat-on-flat" as well as in "flat-on-cylinder" configurations.

To study the friction coefficient dependency with contact pressure in cylindrical metallic counterparts. The main expected result is

- to develop $\mu = \mu(p)$ experimental laws from tribometer tests results on 'flat-on-cylinder" configurations.

According to the previous objectives, the next approach has been followed:

- planning and execution of tribometer testing on "flat-on-cylinder" configurations using rubber as sliding sample;
- FE analysis of area of contact and contact pressure distribution vs. applied vertical load;
- mathematical procedure for obtaining real $\mu = \mu(p)$ experimental laws.

2 Tribilogical Behaviour of Rubbers

As it is commonly known, classic Coulomb's and Amonton's friction laws, which mainly establish that the friction coefficient is independent of the area of contact, are proven to be unvalid in the case of rubber-like materials. In this particular case, and due to their specific mechanical properties, the friction coefficient should be expressed as a function of contact pressure, sliding speed, temperature and lubrication regime, if the latter were the case.

$$\mu = \mu(L, T, \dot{\gamma}, \upsilon, R_a, \ldots). \tag{1}$$

The dependence with the contact pressure is associated to the varying ratio of real (microscopic level) to apparent (macroscopic level) area of contact when the vertical load (contact pressure) is rising. The problem increases in complexity when neither the contact pressure distribution nor the ratio of real to apparent area of contact are uniform along the apparent area of contact, being the cylindrical contact geometry a typical example of this situation.

In the above proposed expression (1), several dependencies on external parameters such as vertical load (contact pressure), temperature, sliding speed, etc, are introduced as key variables. Rubber-like materials, and rubbers in particular, have high friction characteristics, a consequence of their low elastic modulus and their viscoelasticity. Thus, under contact pressure their deform in a large extent, resulting in high values of the real area of contact. Hence, classical model for metals are no longer valid for the case of rubber friction.

The high friction coefficient has been exploited in many applications, for example: tyres, shoe soles, bicycle brake blocks, etc. However, there many other applications in which the frictional behavior of the rubber is expected to be on the opposite, as for example in the case windscreen wipers and seals. In such cases the rubber must e treated to show low frictional properties, in the case of dry friction, or the working conditions must be ensured to be in the hydrodynamic lubrication regime.

It is commonly known that friction coefficient values are difficult to find in literature. This is because the friction coefficient can rarely assumed to be constant and, as stated in expression (1), depends on several factors such as contact pressure (vertical load), sliding velocity, temperature, surface roughness and lubrication regime, in such cases where applicable.

2.1 Dependence on Vertical Load

As it was described in previous section, Amonton and Coulomb established that friction force is proportional to the vertical load and independent of the geometry of the contact. Coulomb defined the friction coefficient μ as the ratio between friction to vertical load. For materials obeying this law, μ is independent of the vertical load and thus of the normal stress. Rubber does not obey Amonton's and Coulomb's laws since the friction coefficient falls markedly when increasing of normal stress. For this particular behaviour, an analytical law, which later would be of wide use, is due to Thirion [2]:

$$\frac{1}{\mu} = a + b\left(\frac{P}{E}\right), \tag{2}$$

where μ is the friction coefficient, P is the normal stress, E is the elastic modulus of the rubber and a and b are empirical constants. Schallamach [3] showed later how behaviour described in Equation (2) may be explained on the assumption that the friction force is proportional to the true area of contact, resulting in:

$$\mu = \text{const.}\left(\frac{P}{E}\right)^{-1/n}, \tag{3}$$

where the value of n is derived from a model which considers the deformation of the rubber on the asperities of the metallic counterpart, and depends on the geometry and distribution considered for peaks and valleys. In general, n depends on the nominal normal stress, but for restricted ranges is considered to be constant. At sufficiently high normal stresses, the real area of contact becomes equal to the apparent area of contact, so that the frictional force becomes constant and μ is inversely proportional to P, as described in (3). This particular condition is referred to as "saturation".

3 Tribometer Testing on "Flat-on-Cylinder" Configurations

As commented in Section 1, first step on the proposed approach for the presented work consists on the planning and execution of tribometer testing on "flat-on-cylinder" configurations with reciprocating motion, using rubber as sliding sample and aluminium as cylindrical countermaterial. Tests will be run under lubricated conditions.

The rubber sample consists on a cylinder block with 6 mm height and 12 mm diameter, which is fixed to the tribometer reciprocating actuator by means of an encapsulated tooling, which avoids the use of adhesive for fixing. The aluminium countermaterial consists on a 22 mm diameter cylinder. Figures 3 and 4 show, respectively, a sketch of the test configuration on tribometer and a picture of the tribometer itself.

The tribometer used for testing is a Phoenix Tribology TE77 tribometer, with reciprocating motion, which has been adapted for the measurement of friction on rubbers. In addition, an specific tooling has been machined to fix the aluminium cylinder to the lower fixed part of the tribometer. The test plan which has been arranged for the present work consists on applying a sequence of six different vertical loads on ascending order (100 N to 500 N), for an oscillating frequency of the reciprocating movement of 2 Hz (being 15 mm the forwards and backwards stroke of the rubber sample over the aluminium cylinder).

Figure 5 shows the friction force curves obtained for two tests carried out as commented above. Prior to run the testing, a 30 minutes run-in has been applied to the rubber at 120% of both the nominal oscillating frequency and maximum vertical load.

As commented before, these experimental results will be used as input data for the postprocessing methodology for obtaining the friction coefficient as a function of the contact pressure which is presented in this paper.

4 FE Analysis of the Rubber-Metal Cylindrical Configuration

Next step on the proposed approach for the presented work consists on the development of a FE model for the analysis of the tribometer test configuration.

For the present step, the following main objectives have been established:

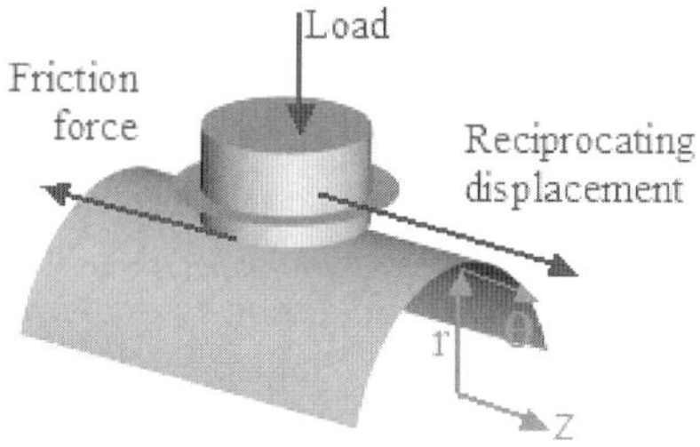

Fig. 3. Sketch of the "flat-on-cylinder" tribometer test configuration.

Fig. 4. Phoenix tribology TE77 tribometer.

- to calculate the macroscopic areas of contact vs. applied vertical load;
- to calculate the contact pressure distribution over the cylindrical area of contact (θ and Z coordinates) in further steps.

As commented before, the output of the FE simulations to be carried out within this step (area of contact vs. applied vertical load and contact pressure vs. θ and Z coordinates), will be used as input data for the postprocessing methodology which will be explained in next sections.

Two sets of FE simulations have been performed using the commercial code ABAQUS. In these simulations, a vertical load is applied to the rubber capsule in static condition (rubber sample not moving) up to each load level which has been previously tested. Next, the areas of contact and contact pressure distributions are analysed in this static condition for each load level. Finally, a calculation restart is performed again for each load level, in which displacement is imposed to the rubber sample so as to consider the redistribution of the contact pressure distributions due to the sample stroke.

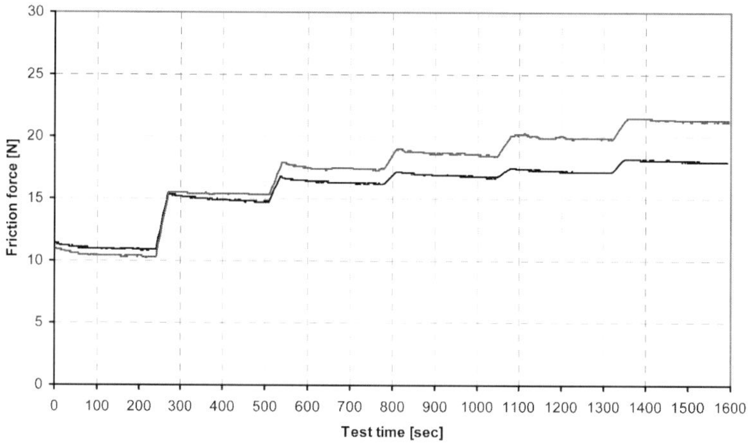

Fig. 5. Results obtained in terms of friction force.

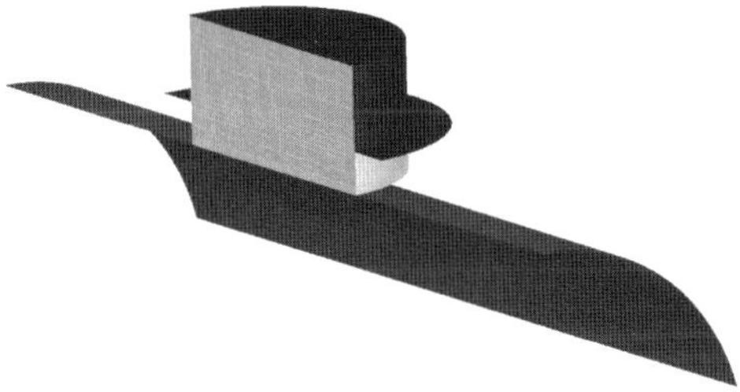

Fig. 6. FE model.

For this, a half symmetry FE model has been used in order to save computational time (Figure 6). The test sample has been simulated with solid continuum hybrid elements (C3D8H in ABAQUS library), while the cylinder and the capsule fixing tool have been modelled as rigid surfaces. In order to obtain more accurate results of the contact pressure distribution, but maintaining the computational cost, the test sample has been divided into two zones with different element sizes fixed together through a tied contact. The refined zone consists of two rows of elements in the contact zone (Figure 7).

In the "static" simulations (no sample stroke), a constant friction coefficient value of 0.1 is assigned to the cylinder-rubber contact. In the "dynamic" simulations (sample stroke) this value is modified to the average dynamic friction coefficient obtained in the tribometer tests.

Figure 8 shows the application of the vertical load, the simulation is divided into 6 steps of each 10 seconds. From each step, a restart simulation, in which the displacement is applied, is performed. These simulations consist of two steps, in the

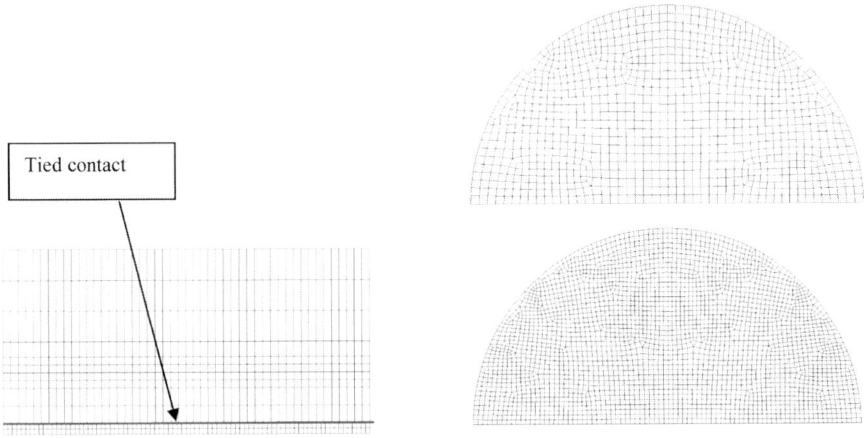

Fig. 7. FE model of the test sample.

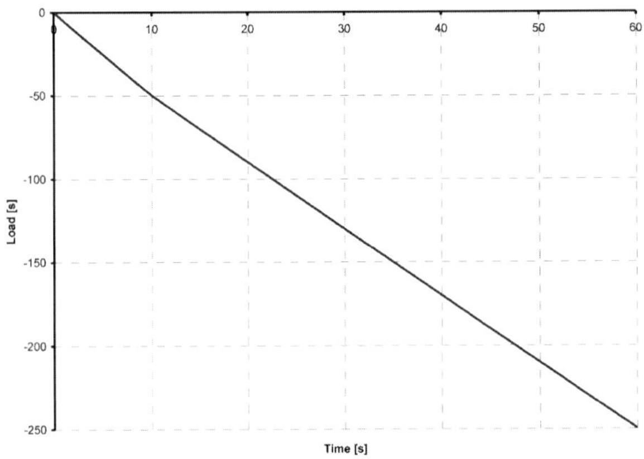

Fig. 8. Vertical load.

first one, the static friction coefficient used in the loading simulation is transformed into the dynamic friction coefficient obtained experimentally. In the second step a 15 mm displacement is applied to the capsule in 10 seconds. As can be observed in Figure 9 the displacement and the time in which it is applied are long enough so that the tangential reaction force (frictional force) can stabilize.

In the following sections, the contact pressure distributions as well as the area of contacts are obtained for both the "static" and the "dynamic" simulations.

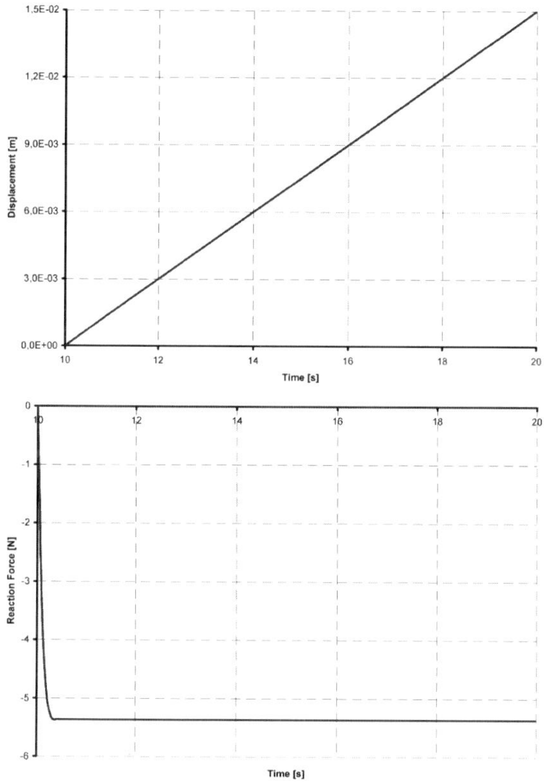

Fig. 9. Axial displacement and tangential reaction force.

4.1 Contact Pressure Distributions

Figure 10 shows the contact pressure distributions obtained from both FE simulations. It can be observed that in the "dynamic" simulations the load is partially transferred to the zone contrary to the movement. This is due to the accumulation of rubber under the capsule.

4.2 Validation of the Simulations Through the Contact Area Measurement

The next step consists on obtaining correlations between the area of contact obtained experimentally and that one obtained in simulation. The experimental area of contact has been measured by means of analysing the trace which the rubber leaves on a printing paper.

Figure 12 shows the comparison of the contact area obtained experimentally and through Abaqus. In can be observed that the bigger the load the lower the error

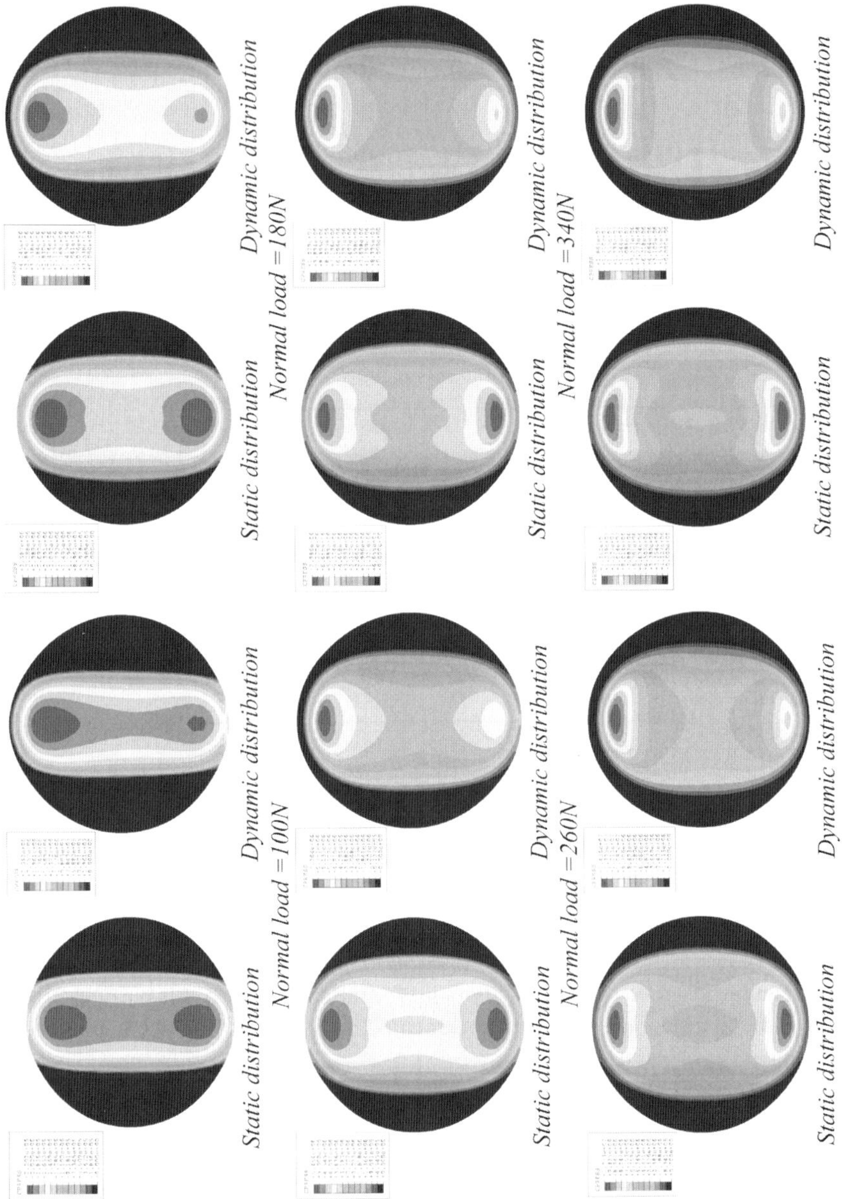

Fig. 10. Static and dynamic contact pressure distributions.

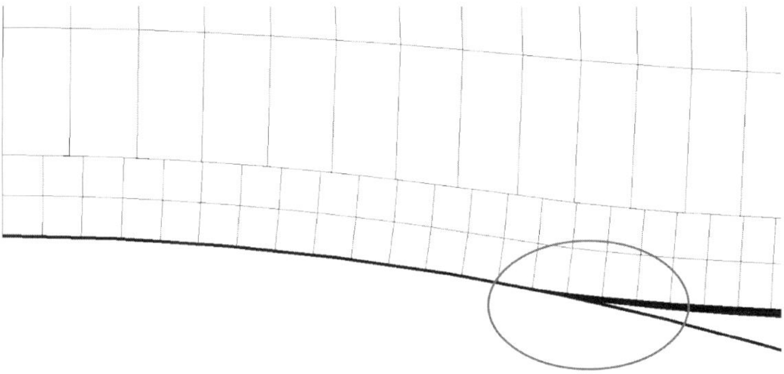

Fig. 11. Proximity between rubber and aluminium cylinder at low loads.

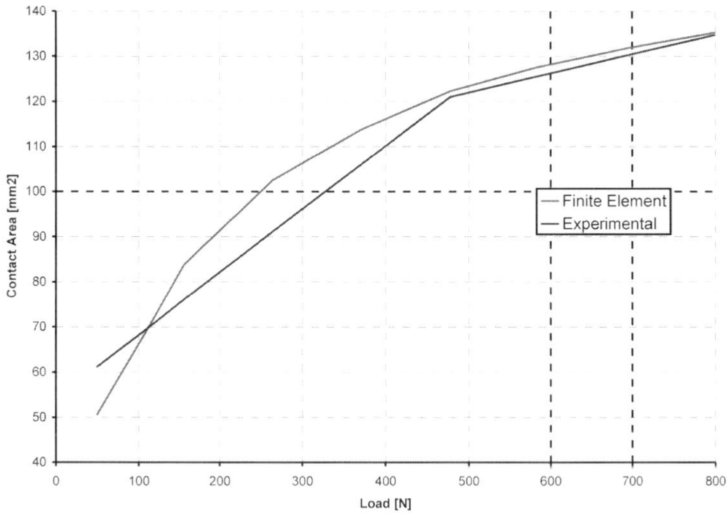

Fig. 12. Comparison of experimental and numerical contact areas.

(Table 1). In can be explained by a transfer of ink from the test probe to the paper in zones that are not actually in contact but that are quite close due to the curvature of the aluminium cylinder (Figure 11).

Since relatively good correlations are obtained, the FE model is validated for providing contact pressure distributions and area of contact as input data for the mathematical postprocessing methodology. This methodology will be explained in the next section.

Table 1. Experimental and numerical contact areas.

Load [N]	FE Area [mm^2]	Experimental Area [mm^2]	% Error
50	50.68	61.20	–17.19
478	122.29	121.00	1.06
800	135.26	134.70	0.41

5 Methodology for Friction Coefficient Calculation

At this moment, the frictional force is known due to the tribological tests results, and the contact pressure distribution is known due to the FE simulations, as well as the area of contact for each load level. It is also known the relation between the frictional force and the contact pressure by the following expression (4):

$$F_f = \int_A \mu(P) \cdot P \cdot \mathrm{d}A \approx \sum_i \mu(P_i) \cdot P_i \cdot A_i, \tag{4}$$

where

- $F_f \rightarrow$ frictional force measured by the tribometer;
- $A \rightarrow$ contact area;
- $P \rightarrow$ contact pressure;
- $\mu(P) \rightarrow$ friction coefficient dependent of the contact pressure.

The integral can be transformed into a summatory due to the finite element discretization. Next step is to approximate the friction coefficient by an analytical function. Choosing a polynomial allows the friction coefficient curve to take the form it desires and gives some advantages when manipulating the sum above (4). This way, approximating the friction coefficient by an order n polynomial (5) and substituting in (4) a system of $n + 1$ unknowns is obtained:

$$\mu(P) = \sum_{j=0}^{n} k_j \cdot P^j, \tag{5}$$

$$F_f = \sum_i \left(\sum_{j=0}^{n} k_j \cdot P_i^j \right) \cdot P_i \cdot A_i. \tag{6}$$

Manipulating adequately Equation (6):

$$F_j = \sum_{j=0}^{n} \left(\sum_i A_i \cdot P_i^{j+1} \right) \cdot k_j. \tag{7}$$

Table 2. Number of equation systems relying on the polynomial order.

Polynomial order	Number of possible systems
2	20
3	15
4	6
5	1

Table 3. Equations system combinations.

Equations system	Code	Load [N]
1256	1	100
2345	2	180
1234	3	260
3456	4	34
1246	5	420
1356	6	500

Taking into account the number of different load levels used in the tribological test, and the order of the polynomial chosen, a different number of equations systems can be obtained, see Table 2.

In general, N being the order of the chosen polynomial, to solve the system it is necessary to consider $N + 1$ tests (all of them at different values of normal load).

5.1 Polynomial Solutions

The polynomial order chosen was 3, it is, a third order function (4 constants). From the 15 different equations systems only 6 will be used. They are shown in Table 3. Each equation system combination is represented by a four digit code. Each digit represents one of the six load levels which will be used to shape the system (in ascending order).

Figures 13 and 14 show the results obtained using first the "static" contact pressure distribution (from "static" FE simulation results), and afterwards the "dynamic" contact pressure distribution (from "dynamic" FE simulation results), respectively. As commented in the previous section, the "static" simulations are performed using a constant value of the friction coefficient (0.1), whereas the "dynamic" simulations are fed with average friction coefficient values obtained in tribotesting. The third order polynomials are compared with the fifth order polynomial obtained using all load levels and the experimental results (normal load/measured area vs. friction coefficient).

The vertical lines delimit the validity range of the solutions, and they correspond to the minimum and the maximum value of the averaged contact pressure calculated

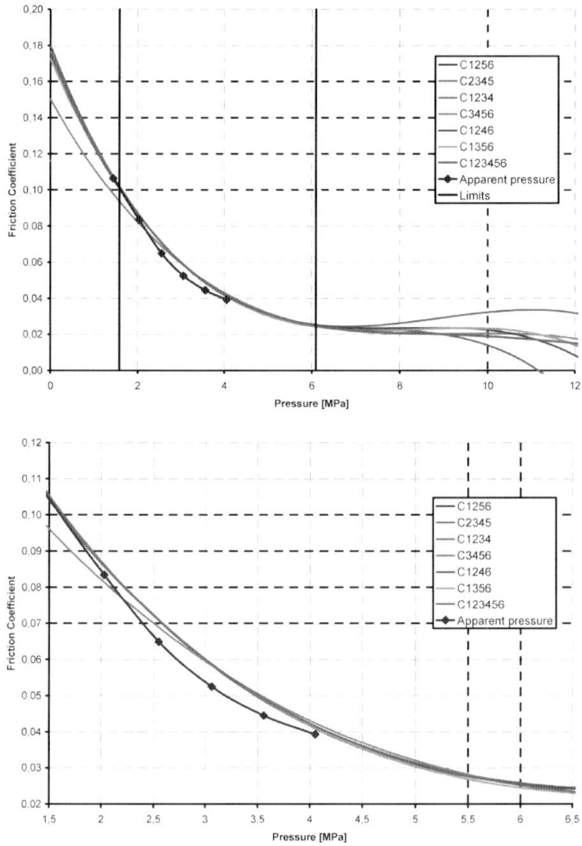

Fig. 13. Polynomial solutions obtained from static contact pressure distributions.

from the contact pressure distribution obtained in FE simulation, corresponding to the 100 N and 500 N calculations.

Each curve plotted in the figures above corresponds to the solution from solving a given four-equations linear system (see Table 3). In addition, these curves are compared to the one (plotted with square markers) that is obtained using experimental results only (calculated as the ratio of tangential to vertical load measured in tribometer).

Once the solution is obtained in the "static" condition, the methodology is applied again in the "dynamic" condition, which proves closer to reality since the contact pressure distribution is expected to change when movement is applied to the rubber sample.

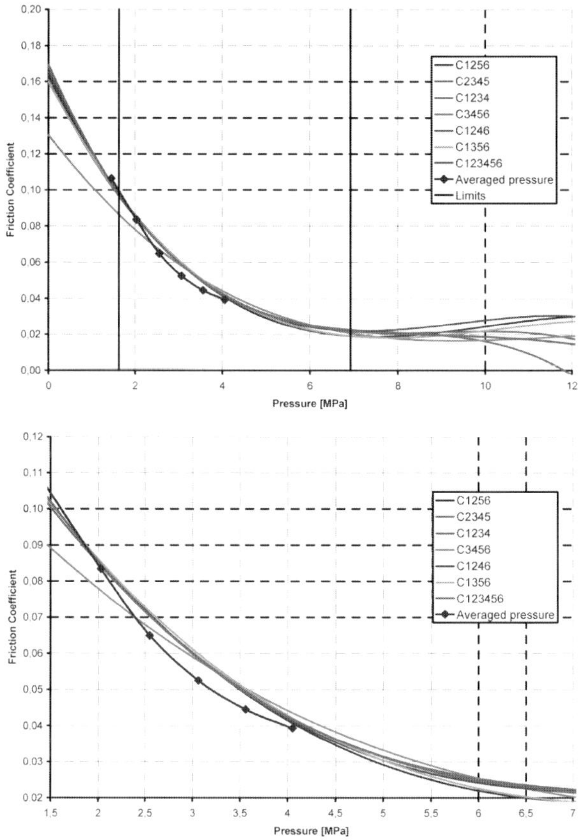

Fig. 14. Polynomial solutions obtained from dynamic contact pressure distributions.

Curve C3456 deviates from the others in the low pressure zone, because the data used for the obtaining of that curve are corresponding to the high pressure zone.

It can be observed that using this methodology the range of the obtained friction coefficient is wider (almost double) than using experimental results only (calculated as the ratio of tangential to vertical load measured in tribometer). Also some deviations (up to 10%) can be found between both methods. Finally it can be also observed that the experimental curve (plotted with square markers) seems to tend to a value of around 0.03 while the polynomial curves (red) seems to tend to 0.02 what would suppose a 50% of deviation.

In addition, one important remark is that although each solution of the different linear system results into a third order polynomial form (which could lead to very different shapes from one to nother), the six linear systems solutions are showing

Table 4. Number of equation systems relying on the polynomial order.

Custom function	Equation	a	b	c	$\mathbf{R}^2$
1	$\mu = \frac{1}{a+b\cdot P^2}$	7.78976	0.969635	–	0.99718
2	$\mu = \frac{1}{a+b\cdot P^c}$	7.1941	1.2245	1.8633	0.99768

Table 5. Friction models parameters.

	a	b	c	$\mathbf{R}^2$
Thirion	–0.40289	6.07683	–	0.97563
Schallamach	0.178286	1.05243	–	0.97762
Custom function 1	7.78976	0.969635	–	0.99718
Custom function 2	7.1941	1.2245	1.8633	0.99768

the same tendency an shape within the given range of validity, even the fifth order polyomial.

5.2 Obtaining a Simplified Function

The polynomial function are easily derivable and are accurate in the valid range, but out of that range there is no certainty of their behaviour, they can even take negative values. In order to avoid that effect, a different function, which approaches accurately the values given by the polynomial, is sought. Figure 15 shows the best functions found. They have few parameters, are derivable, tends to zero and have a controlled value at low pressure values.

5.3 Comparison with Classical Models

Carrying out the same approximation used in Section 5.2 but with the classical Thirion (8) and Schallamach (9) models the following results are obtained:

Thirion model:

$$\mu = \frac{1}{a + b \cdot P}, \tag{8}$$

Schallamach model:

$$\mu = a \cdot P^{-b}. \tag{9}$$

It must be noted that due to the swarm of points in the high pressure zone, the $\mathbf{R}^2$ parameter cannot take values very close to one. In Figure 16 the expression of the custom functions is shown.

It can be observed that with both classical functions, the adjustment obtained is worse, besides, both functions take big values at low pressures.

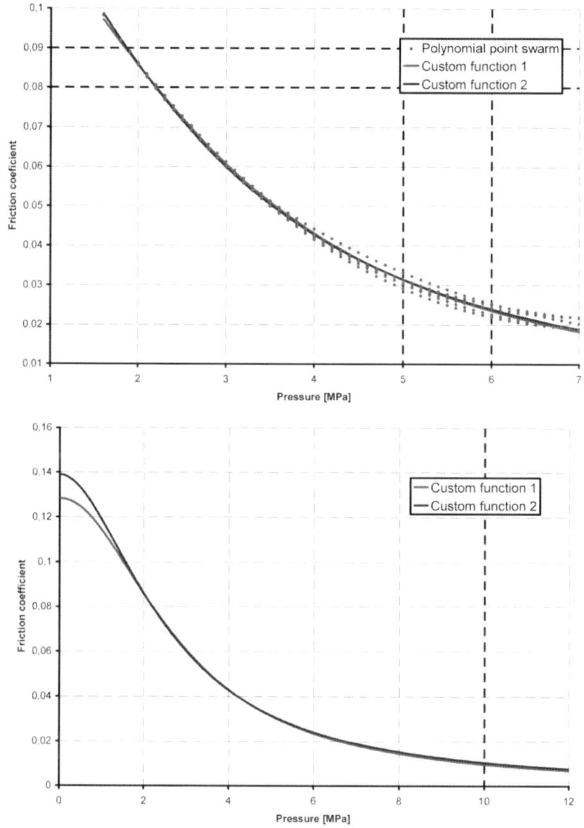

Fig. 15. Simpler function approximation.

6 Conclusions

Since contact pressure distribution is not uniform along the cylindrical contact area, as well as the dependency of area of contact with vertical load, friction coefficient should not be expressed by its dependency with the average contact pressure, and thus as the ratio of tangential to vertical forces measured in tribometer, as it is done in "flat-on-flat" configurations.

Due to the previous conclusion, a mathematical procedure has been developed to obtain $\mu = \mu(p)$ experimental laws from friction force results obtained by tribometer testing, and from area of contact and contact pressure distributions obtained by numerical simulation.

The measurement of the contact area validates the FE simulations. The errors at lows loads may be caused by errors in the measurements, not in the FE model.

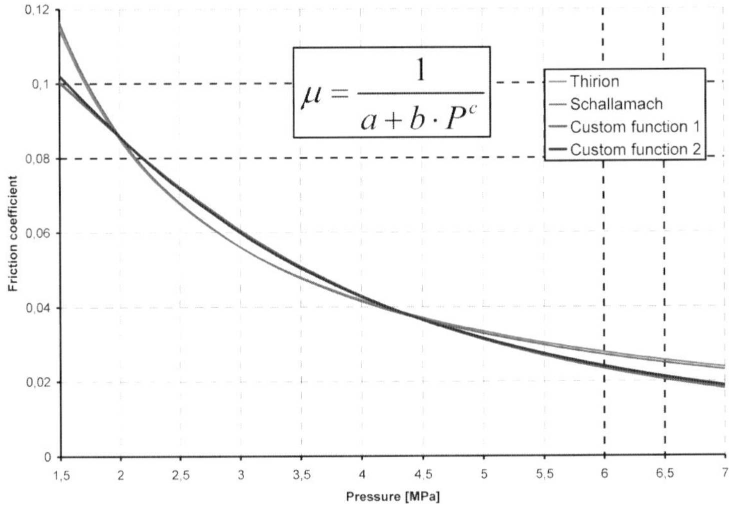

Fig. 16. Comparison of friction models.

Experimental $\mu = \mu(p)$ laws obtained following the herewith presented method can be fully integrated into commercial FEA by fitting of common friction laws existing in literature.

Finally, this methodology eliminates any geometrical dependency of the rubber sample with the calculated results in terms of friction coefficient vs. contact pressure.

Acknowledgements

This work is co-funded by the Integrated Project "KRISTAL" (VI F.P.) and the "Torres Quevedo Program" from the Spanish Ministry for Education and Science.

References

1. E. Rabinowicz, *Friction and Wear of Materials*, 2nd edition. Wiley-Interscience, 1995.
2. P. Thirion, Les coefficients d'adhérence du caoutchouc. *Rubber Chem. & Tech.*, 21:505–515, 1948.
3. A. Schallamach, Friction and abrasion of rubber. *Wear*, 1:384–417, 1958.
4. S. Bouissou, J.P. Petit, and M. Barquins. Normal load, slip rate and roughness influence on the polymethylmethacrylate dynamics of sliding: 1. Stable sliding to stick-slip transition. *Wear*, 214:156–164, 1998.
5. S. Bouissou, J.P. Petit, and M. Barquins. Normal load, slip rate and roughness influence on the PMMA dynamics of sliding: 2. Characterisation of the stick-slip phenomenon. *Wear*, 215:137–145, 1998.

6. Hanbum Cho and J.R. Barber. Stability of the three-dimensional Coulomb friction law. *Proc. Roy. Soc. (London) A*, 455:839–861, 1999.
7. T. Baumberger, P. Berthoud, and C. Caroli. Physical analysis of the state- and rate-dependent friction law. II. Dynamic friction. *Phys. Rev. B*, 60(6), 1999.
8. H. Olsson, K.J. Astrom, C. Canudas de Wit, M. Gafvert, and P. Lischinsky. Friction models and friction compensation. *J. Eur. Control*, 4(3), 2003.
9. O. Ronsin and K. Labastie Coeyrehourcq. State, rate and temperature-dependent sliding friction of elastomers. *Proc. Roy. Soc. (London) A*, 457:1277–1294, 2001.
10. J.J. Thomsen and A. Fidlin. Analytical approximations for stick-slip vibration amplitudes. *Internat. J. Non-Linear Mech.*, 38:389–403, 2003.
11. F. Xia. Modelling of a two-dimensional Coulomb friction oscillator. *J. Sound Vibration*, 265:1063–1074, 2003.
12. J. Neter, M.H. Knutner, C.J. Nachtscheim and W. Wasserman. *Applied Linear Statistical Models*, 4th edition. McGraw-Hill, 1996.

Modular Modelling of Dynamic Systems with Elastomer Contacts

M. Kröger, P. Moldenhauer and G. Gäbel

Institute of Dynamics and Vibrations, Leibniz Universität Hannover, Appelstr. 11, 30167 Hannover, Germany
E-mail: kroeger@ids.uni-hannover.de

Abstract. The physical effect of friction appears in many technical applications. One of the most interesting fields is rubber friction which depends on a variety of parameters, e.g. normal pressure, temperature, relative velocity, surface roughness, lubrication and wear. These properties often lead to operating conditions where the coefficient of friction decreases with respect to the relative velocity. As a consequence unwanted friction induced self-excited vibrations may occur. A model is set up to describe tread block vibrations under consideration of the complex contact properties and their interactions. Therefore, modules are developed describing each single effect. The Hurty/Craig/Bampton transformation is applied to the linear finite element model of the tread block. By considering only some vibration modes the number of degrees of freedom is significantly reduced. The simulation results show the typical tread block vibrations and the attending wear effect which are observed in experiments.

Key words: Rubber friction, tyre tread block, non-linear contact stiffness, wear.

1 Introduction

In a lot of applications frictional contact effects technical problems. This is especially the case for contacts with at least one component made of elastomers and if relative motion occurs between the contact partners. Typical unwanted effects are juddering windscreen wipers or side windows of cars, stick-slip vibrations of seals and the noise from squealing tyres. Elastomer contacts are difficult to describe, because a large number of influencing parameters must be considered when modelling these contacts.

The aim of this research is to separate the complex modelling problem in smaller sub-problems and afterwards link the single results in a reasonable way with the respective interactions. A tyre tread block is used to exemplify the modelling approach. Tyre tread blocks are always subjected to changing operating conditions such as acceleration or antiblock braking as well as cornering. Even for a constant

Peter Wriggers and Udo Nackenhorst (eds.), IUTAM Symposium on Computational Contact Mechanics, 277–290.

vehicle velocity the tread block experiences an excitation due to the rotation of the tyre and the roughness of the road surface. The contact time of a tread block for a vehicle velocity of 100 km/h takes at most 5 ms. During this contact phase the block passes a sticking phase and snaps out with a high acceleration also under free rolling conditions. This leads to an unsteady behaviour of the tread blocks which has to be covered in models to achieve a realistic simulation.

Present tread block models usually show a high grade of complexity. For example, in [1] a three-dimensional finite element approach with non-linear material description and thermo-mechanical coupling is applied. Such models are only applicable within limits for dynamic calculations.

On the other hand there are models that treat the tread block as a simple spring or elastic beam [2]. Here, the block geometry and important structural effects are neglected. The aim of this approach is to compromise between these approaches. This means to model the dynamic properties of the tread block and its interaction including the frictional contact without a loss of numerical efficiency.

2 Dynamic Model

In this work a model with point contacts is built based on the concept of Sextro [3]. It is characterised by a fast and efficient calculation procedure. The basic idea is to treat the non-linear contact mechanics and the linear assumed structural mechanics separately. The object of this investigation is to divide the complex modelling problem in smaller sub-problems (modules) and afterwards combine the respective results in a reasonable way, see Figure 1. According to this concept the system is parted into four modules:

- Module 1: Dynamic description of block layer,
- Module 2: Friction characteristic,
- Module 3: Non-linear contact stiffness,
- Module 4: Wear process.

The aim of this model is the calculation of the resulting deformation shape, the vibration behaviour and local wear of the tyre tread block. If the global load configuration of the tyre is known, which can be the tyre load and the slip with respect to time the model provides the deformation shape, the vibration behaviour and the local wear of the tread block. The results of the simulations can in turn be used as input for a global tyre model.

For the modelling of the tread block the elastomer contact is of special interest. The surface roughness leads to a non-linear normal force-displacement relation and the tangential contact forces depend on the local friction coefficient which is a function of the contact parameters. The contact forces excite the structure to vibrations which can be described numerically efficient by finite elements and an adequate modal reduction method. Furthermore, the sliding process of the tread block changes the block geometry due to high local contact stresses which in turn influence the contact forces. To model elastomer contacts the consideration of the interaction of each module is of great importance. In Figure 2 the dynamic tyre tread block model with the arrangement of the modules is shown. The friction characteristic, contact stiffness and tread block geometry describe the contact and influence the dynamic behaviour of the tread block. The properties of the single modules are discussed in detail in the following subsections.

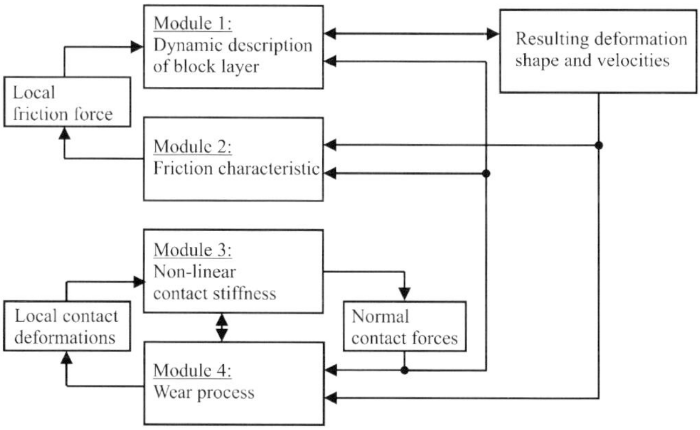

Fig. 1. Modular modelling of dynamic systems with elastomer contacts.

2.1 Dynamic Description of Block Layer – Module 1

The block layer considers the dynamics resulting from structural effects like inertia, elasticity and damping. To achieve a high numerical efficiency it is necessary to reduce the model with respect to its degrees of freedom (DOFs) by an adequate method. For the presented model the Hurty/Craig/Bampton transformation was chosen, cf. [4, 5]. Based on a linear finite element model the system is subjected to this reduction method. It has originally been developed for applications in the field of structural dynamics and can be interpreted as the combination of a static and a modal condensation. The static condensation by itself, index sc in Equation (1), neglects the velocity or acceleration dependent terms of the secondary DOFs. However, this error can be arranged arbitrarily small by the number of additional modal ansatz functions, index mod in Equation (1). Coupling matrices, index c in Equation (1), complement the system matrices:

$$\begin{bmatrix} \mathbf{M}_{\mathrm{mod}} & \mathbf{M}_{\mathrm{c}} \\ \mathbf{M}_{\mathrm{c}}^T & \mathbf{M}_{\mathrm{sc}} \end{bmatrix} \begin{bmatrix} \ddot{\mathbf{q}} \\ \ddot{\mathbf{x}}_1 \end{bmatrix} + \begin{bmatrix} \mathbf{D}_{\mathrm{mod}} & \mathbf{D}_{\mathrm{c}} \\ \mathbf{D}_{\mathrm{c}}^T & \mathbf{D}_{\mathrm{sc}} \end{bmatrix} \begin{bmatrix} \dot{\mathbf{q}} \\ \dot{\mathbf{x}}_1 \end{bmatrix} + \begin{bmatrix} \mathbf{K}_{\mathrm{mod}} & \mathbf{0} \\ \mathbf{0} & \mathbf{K}_{\mathrm{sc}} \end{bmatrix} \begin{bmatrix} \mathbf{q} \\ \mathbf{x}_1 \end{bmatrix} = \begin{bmatrix} \mathbf{0} \\ \mathbf{F}_1 \end{bmatrix}. \quad (1)$$

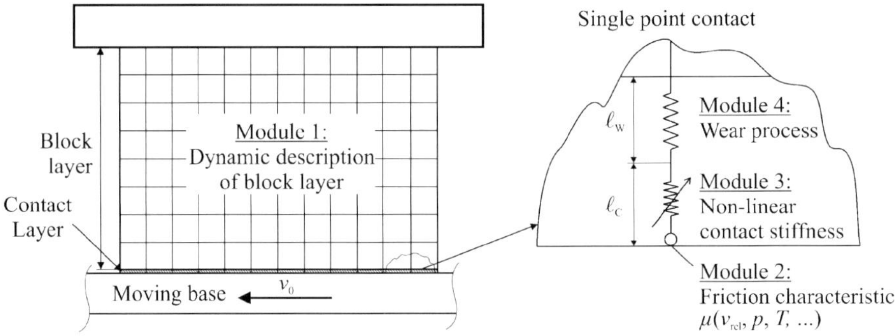

Fig. 2. Dynamic tyre tread block model.

Here, the actual advantage of this condensation is revealed: Physically important nodes where the normal and tangential contact forces $\mathbf{F}_1$ are applied are treated as primary DOFs $\mathbf{x}_1$ and thus are explicitly retained within the simulation without inverse transformation. All other degrees of freedom are interpreted as secondary degrees of freedom.

In the present case of the tread block these are the DOFs of all nodes above the contact layer. They are modally condensed and described by the modal coordinates $\mathbf{q}$. The actual DOF reduction is achieved by considering in addition to the primary DOFs $\mathbf{x}_1$ only a small number of important mode shapes. An interaction between primary and modal DOFs is guaranteed by respective coupling matrices $\mathbf{M}_c$ and $\mathbf{D}_c$. In case of the present example with 16 contacting nodes and ten additional mode shapes a reduction rate of 87% is achieved. Young's modulus E of the rubber material used here has been determined by a quasi-static tensile test to $E = 12.0$ N/mm^2 and the Poisson ratio ν has been set to the typical value of $\nu = 0.49$. In the following Rayleigh damping is assumed which leads to a damping matrix $\mathbf{D} = \alpha\mathbf{M} + \beta\mathbf{K}$. For the simulations shown here the coefficients were set to $\alpha = 0$ and $\beta = 10^{-4}$ s.

2.2 Friction Characteristic – Module 2

The local friction characteristic is important for the entire system dynamics. For example, a decreasing friction coefficient with respect to the relative velocity can lead to stick-slip vibrations [6]. In the case of rubber friction there are in general four physical effects that contribute to the friction process according to Kummer [7]:

- Adhesion is caused by inter-molecular binding forces between the surfaces of the contact partners and dominates on dry smooth surfaces.
- Hysteresis originates from the internal material damping of rubber. The energy put into the rubber by deformation due to the roughness asperities and the sliding velocity can only partially be regained. This effect is dominant on rough surfaces.

- Viscous friction occurs if there is a sheared liquid layer between the contact partners such as grease or water.
- Cohesion arises on abrasive surfaces that lead to cracks and wear in the rubber material. For the generation of these new surfaces energy is dissipated.

In technical applications these effects always occur in combination resulting in a complex friction behaviour. It depends on many parameters like relative velocity, normal pressure, temperature, surface roughness, lubrication and wear. As a consequence many types of friction characteristics can be observed for rubber contacts. To determine a local friction characteristic it is possible to use mathematical models [8, 9], simulations [10], a multi-scale approach [11] or experiments [12]. The adequacy of the single methods for a given application depends on the respective problem. The theoretical approaches are not subject of this publication. The advantage of an experimental investigation is that all physical effects are measured with their interactions. In case of sliding friction the direction of the friction force is always directed opposed to the sliding direction.

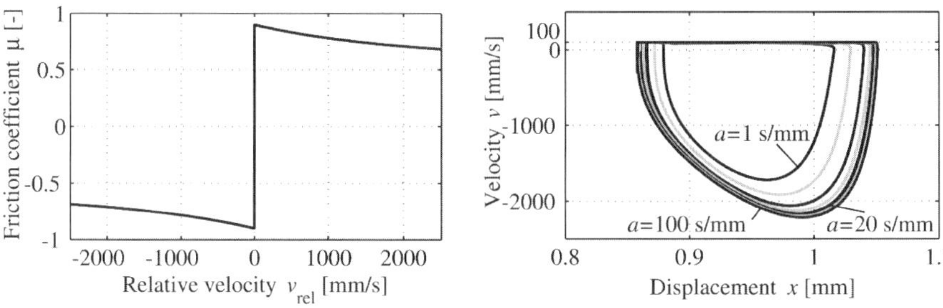

Fig. 3. Friction characteristic and simulated limit cycles for a varying parameter a.

In the following, simulations are performed with a velocity dependent decreasing local friction characteristic based on an exponential function. For numerical reasons the non-smooth friction characteristic is approximated by an arctan function to avoid the distinction between sticking and sliding, cf. Equation (2).

$$\mu(v_{\text{rel}}) = \frac{2}{\pi} \arctan(a\, v_{\text{rel}})(\mu_\infty + (\mu_0 - \mu_\infty)\, e^{(\gamma\, |v_{\text{rel}}|)}). \tag{2}$$

The coefficients were set to $\mu_\infty = 0.6$, $\mu_0 = 0.9$ and $\gamma = -0.0005$ s/mm to realise a decreasing friction characteristic with respect to the relative velocity. The factor a in the arctan argument is a numerical parameter whose value has to be reasonably chosen. Therefore, a parameter study has been carried out. Figure 3 shows the limit cycles of the horizontal DOF at the leading edge of the tread block for the parameter $a = 1, 2, 5, 10, 20, 50$ and 100 s/mm for $v_0 = 100$ mm/s. With an increasing value of a the sticking phase is extended and higher rebound velocities are observed. For a sufficiently high value of the parameter a there are only slight differences in the limit cycles. In the following investigations the parameter a has been set to $a = 20$ s/mm.

2.3 Non-Linear Contact Stiffness – Module 3

The road surface is composed of stones, sand, filler and binder and comprises a large range of length scales of surface roughness. To describe the contact between tread block and road surface, the information about the road texture is necessary. Because of the surface roughness contact only exists at discrete points. That leads to an area of real contact which is only a small fractional amount of the nominal contact area. The area of real contact depends on the texture of the road surface, on the normal and tangential load, on the static and dynamic material properties and on the contact time. Because of the visco-elastic material behaviour of rubber the area of real contact increases with increasing contact time. In this study the area of real contact has been measured using a pressure sensitive film for a tread block interacting with two different patches on a concrete surface shown in Figure 4.

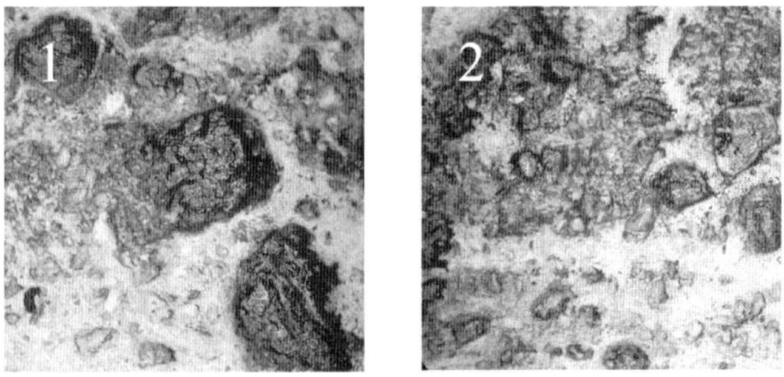

Fig. 4. Different contact patches on concrete surface.

If a local pressure more than 0.2 N/mm^2 is exceeded, the microcapsules break and by a chemical reaction, a red color impression is generated. In this study a quadratic tread block with a nominal contact area of 225 mm^2 is used. Figure 5 shows the measured area of real contact between the tread block and the concrete patch 2 under a static normal force of 49.5 N. With increasing normal load the tread block is pressed more and more into the road surface and the area of real contact increases. Figure 5 also shows the influence of the normal load F_N on the measured percentage contact area A/A_0 for both contact patches.

The contact stiffness c_N is defined as the ratio of the applied load to the relative displacement of the contacting bodies. In this case the displacement comprises the global deformation of the rubber block and the local penetration into the rough road

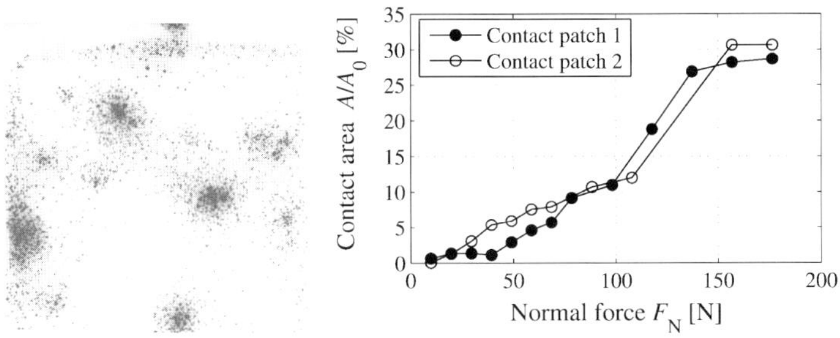

Fig. 5. Measured area of real contact (Contact patch 2, $F_N = 49.5$ N) and its load dependency between tread block and concrete surfaces.

surface. The apparent contact stiffness at the interface is a function of the area of real contact, cf. [13]. If only a few number of asperities interact with the soft rubber block and an additional normal load is applied, the system reacts with a large displacement. With an increasing number of active junctions a higher additional normal load has to be applied to achieve the same change of displacement. As a consequence the contact stiffness increases with a larger area of real contact. In this study the test rig shown in Figure 6 is used to measure the static contact stiffness between a tread block and different surfaces in normal direction. Defined weights are used to apply different normal forces. The setup is mounted on air bearings which avoid dry friction to measure very small contact forces accurately.

The measurement procedure of the normal force-displacement relation starts with the smallest and ends with the highest normal contact forces. The rubber block is not separated from the rough surface during one complete series of measurements. The normal force-displacement relation $F_N(s_N)$ has been analysed for both contact patches on the concrete surface and also for a corundum paper with grit 400 and for a glass surface. The measured force-displacement relations depicted in Figure 7 show a strong non-linearity for the two concrete areas over the entire measurement range.

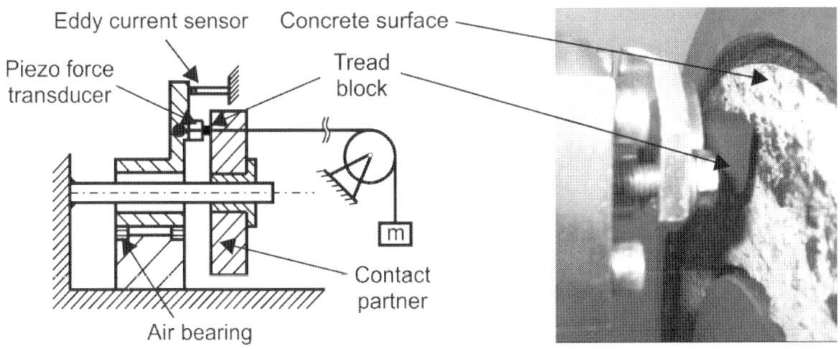

Fig. 6. Test rig for measuring contact stiffness.

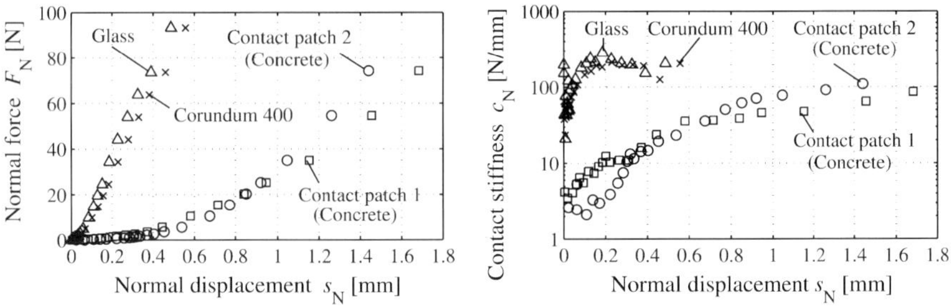

Fig. 7. Measured force-displacement relation and resulting contact stiffness.

The characteristics are quite similar. For the smoother contact partners corundum and glass, a normal force more than 5 N leads to more or less linear characteristic curves. It is assumed, that the higher percentage contact area is responsible for the higher contact stiffness c_N on corundum and on glass [14].

The measured non-linear force-displacement relations can now be used as an input for the dynamic tread block model. The contact layer considers the local contact behaviour such as the local friction coefficient with its dependence on the contact pressure and the relative velocity, see Figure 2. The height of the contact layer is small compared to the height of the block layer. Furthermore the contact layer considers the non-linear contact stiffness between the rubber and the rough surface by non-linear springs.

These non-linear springs are coupled to the block layer. Since the friction characteristic and the non-linear stiffness due to the surface roughness are already described by the contact layer it is possible to model the rough surface in the simulation as perfectly smooth. This leads to a simple and fast contact algorithm. In addition, by means of the compression u of the non-linear springs with the stiffness $c_{K,i}(u)$ the local normal contact forces are determined. Via the local friction coefficient that can be gained by models, simulations or experiments (see Section 2.2) the tangential forces within the contact zone are calculated and applied to the system.

As an example quantitative investigations on a corundum surface with grit 400 are shown in Figure 8. The respective measurement of the normal force-displacement characteristic shows in the range of small displacements a progressive behaviour, for larger displacements it is nearly linear. For the simulation of this static normal force-displacement relation a local contact stiffness $c_{K,i}(u)$ of the following form is used which depends on the local deformation of the contact layer at each point contact i with the parameters c_∞ and k:

$$c_{K,i}(u) = c_\infty(1 - e^{-ku}). \tag{3}$$

The variable u represents the compression of one single contact spring and results from the interactions between block layer and contact layer in the simulation.

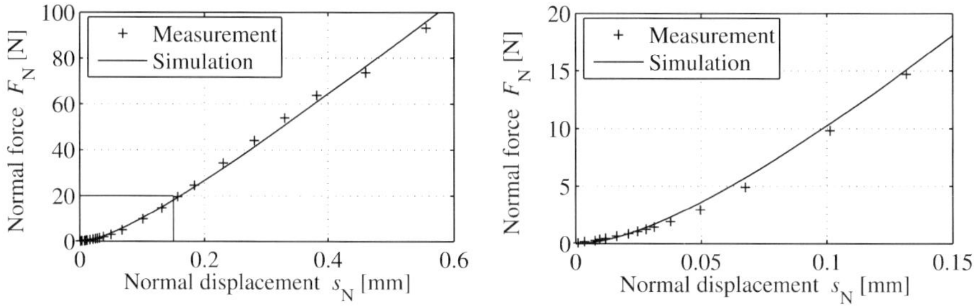

Fig. 8. Measured and simulated static force-displacement relation (left) and magnification (right).

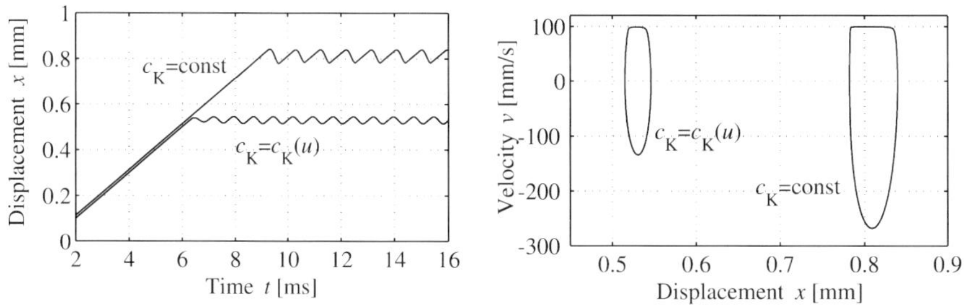

Fig. 9. Comparison of limit cycles with constant and non-linear contact stiffness.

The entire model can thus be interpreted as a series connexion of block elasticity and contact stiffness in the normal direction. The contact stiffness in turn is arranged by a parallel connexion of the non-linear springs $c_{\mathrm{K},i}(u)$ for every point contact. Figure 8 shows the measured and simulated force-displacement characteristic in normal direction for the entire block. The coefficients c_∞ and k are found by an approximation to achieve a good agreement with the respective measurement ($c_\infty = 43$ N/mm^2, $k = 4$ mm^{-1}).

To show the influence of the non-linear contact stiffness simulations were carried out with a constant ($c_{\mathrm{K},i} = c_\infty$) and the non-linear contact stiffness of Equation (3). The length of the contact layer has been set to $\ell_\mathrm{c} = 0.3$ mm, see Figure 2, and the block is compressed 0.2 mm in normal direction. The velocity between the tread block and the moving base is $v_0 = 100$ mm/s. The friction characteristic described in Section 2.2 is used. For the constant and the non-linear contact stiffness stick-slip vibrations occur, see Figure 9. However, in case of a non-linear contact stiffness a higher stick-slip frequency and smaller stick-slip amplitudes are observed. Beyond, the mean displacement is smaller. These effects can be explained by a smaller resulting normal force because the non-linear force-displacement relation rises gradually with increasing normal displacement.

2.4 Wear Process – Module 4

During the braking or acceleration process the tread blocks slide over the rough road surface. To analyse the sliding contact of a tread block on a concrete surface a tribometer test rig is used, see Figure 10. The tread block is pressed on the rotating friction disc whereas piezo force transducers placed near to the contact measure normal and tangential contact forces. The abrasion of rubber results in a change of the block geometry which effects a different contact situation. In the beginning of the sliding process a lip is observed at the leading edge of the tread block resulting in higher local contact forces. Therefore, the maximum mass loss takes place at this lip. Figure 11 shows edges of a tyre tread block after handling maneuvers with the characteristic abraded shape.

The wear rate is the mass loss per sliding distance. For a quadratic tread block sliding on concrete surface the wear rate is analysed for two different nominal contact pressures and two sliding velocities and printed in Figure 11. In the beginning of the friction process a certain maximum of the wear rate is measured which can be explained with the fast abrasion of the leading edge. With increasing sliding distance also the trailing area gets abraded and a typical s-shape of the tread block surface

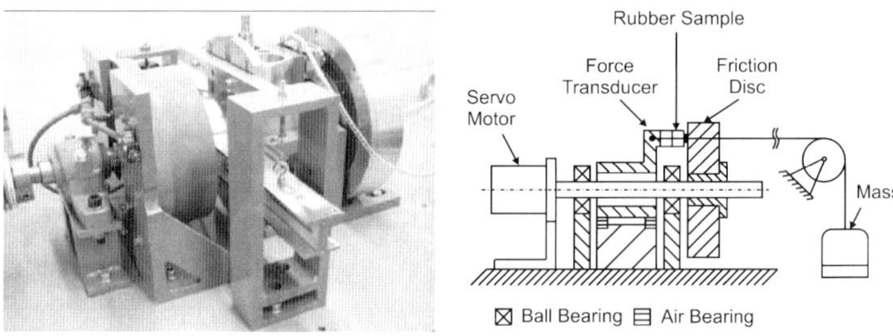

Fig. 10. Tribometer test rig.

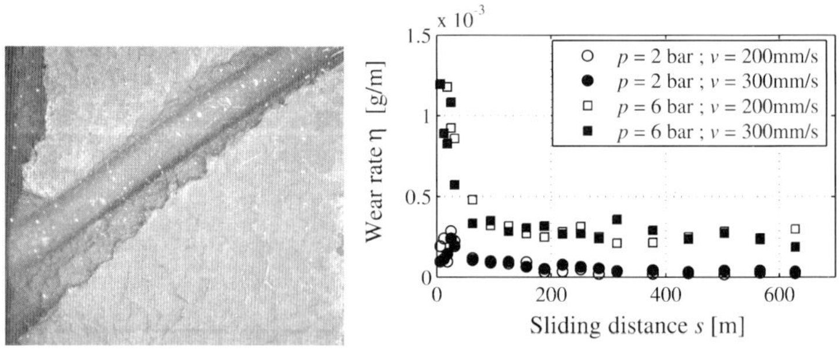

Fig. 11. Typical wear at tread block edges after handling maneuvers and measured wear rate.

is generated under high loading. After a sliding distance of about 200 m a constant wear rate is reached for all investigated configurations.

The wear rate depends on the contact parameters, e.g. normal contact pressure, sliding distance, temperature and surface roughness. With increasing normal contact pressure the wear rate increases. For the measurements conducted on the concrete surface the wear rate is nearly independent of the sliding velocity. The friction coefficient was simultaneously measured during the experiments for these contact conditions. However, in case of the applied high normal pressures and sliding velocities the coefficient of friction is not strongly influenced by the wear process.

To consider wear in the tread block model it is necessary to find an adequate law which covers the wear behaviour of elastomers in a realistic way. In the literature there are a number of equations for the wear calculation that mostly describe the mass loss per friction time $\dot{m} = \mathrm{d}m/\mathrm{d}t$ or sliding distance $\eta = \mathrm{d}m/\mathrm{d}x$, see [1, 15–18]. If the velocity v is known both notations $\dot{m}$ and η can be transferred:

$$\dot{m} = \frac{\mathrm{d}m}{\mathrm{d}t} = \frac{\mathrm{d}m}{\mathrm{d}x}\frac{\mathrm{d}x}{\mathrm{d}t} = \frac{\mathrm{d}m}{\mathrm{d}x}\, v = \eta\, v. \tag{4}$$

According to these laws the wear rate is related in general to the local pressure p, the velocity v and the friction coefficient μ. Depending on the respective wear laws in the literature there are coefficients b_j to consider non-linear wear behaviour. A generalised wear law can be written as

$$\dot{m} = b_0 \mu^{b_1} {p_\mathrm{N}}^{b_2} {v_\mathrm{rel}}^{b_3}. \tag{5}$$

To implement wear in the tread block model an additional wear zone with the height ℓ_w is added to the contact layer, see Figure 2. This is realised by linear springs in normal direction at each point contact, cf. [19]. The stiffness of one spring $c_{\mathrm{w},i}$ depends on the height of the wear zone $\ell_{\mathrm{w},i}$ and increases with time and sliding distance due to the wear process:

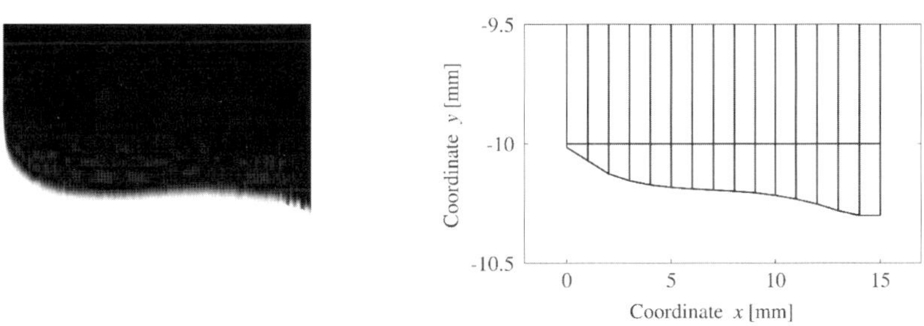

Fig. 12. Measured and simulated abraded tread block shape (scaled).

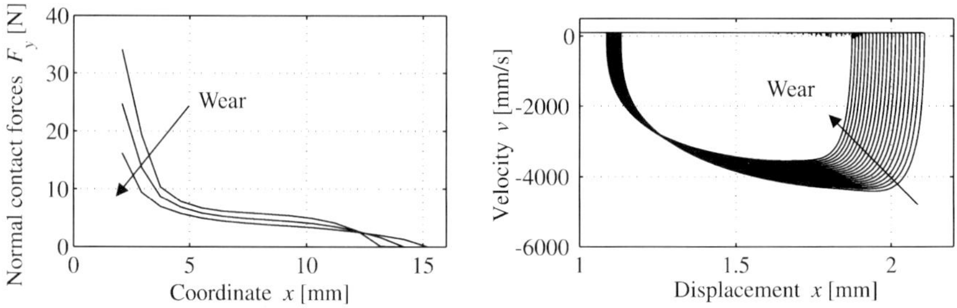

Fig. 13. Measured force-displacement relation and resulting contact stiffness.

$$c_{\mathrm{w},i}(t) = \frac{E A_i}{\ell_{\mathrm{w},i}(t)}. \tag{6}$$

Here, A_i represents the corresponding area of one point contact. Assuming a homogeneous material the mass loss per time can be transferred to a decreasing height of the wear zone by the density ρ and the point contact area A_i:

$$\dot{\ell}_{\mathrm{w},i}(t) = \frac{1}{\rho A_i} \dot{m}_i(t). \tag{7}$$

The remaining height of the wear zone can be calculated during the simulation by time integration of Equation (7) at each point contact. As a first approach the law of Fleischer ($b_1 = b_2 = b_3 = 1$ in Equation (5)) is used for the wear calculation in the module. Figure 12 shows the resulting block shapes with the typical s-shape for the experiment (left) and the simulation (right). The experiment was conducted on a corundum surface with grit 400 and a normal pressure of $p_{\mathrm{N}} = 0.25$ N/mm^2 and a sliding velocity of $v_0 = 100$ mm/s. The simulation results which also show the s-shape are calculated with a wear coefficient of $b_0 = 6 \cdot 10^{-6}$ s^2 and a velocity of the moving base of $v_0 = 100$ mm/s. The local friction characteristic depicted in Figure 3 is used.

As a consequence of the exceeding mass loss at the run-in edge the pressure peak at the run-in edge is lowered with time due to the decreasing lengths $\ell_{\mathrm{w},i}$, cf. Figure 13. Although the simple wear law of Fleischer has been employed for the simulations both shapes agree qualitatively. The change of the block geometry also leads to a different dynamic behaviour which can be observed as well at the limit cycles which are shown in Figure 13. The displacement amplitudes decrease during the sticking phase due to smaller contact forces.

3 Conclusion

The function of many technical applications is interfered by contact problems with elastomers. The contact behaviour of elastomers depends on many parameters which requires an extensive modelling. At these components sliding friction frequently occurs due to the relative motion of the contacting bodies. The modelling approach presented in this investigation is to section the complex problem of rubber behaviour into smaller sub-problems which are treated as single modules. In case of the shown tyre tread block application the dynamic description of the tread block layer, the friction characteristic, the non-linear contact stiffness and the change of the tread block geometry due to wear are considered. The Hurty/Craig/Bampton reduction method is applied to the linear finite element model of the tread block which leads to numerical efficiency. By considering only some vibration modes the number of degrees of freedom is significantly reduced without neglecting the important system dynamics. The condensed model is then complemented by a discrete number of point contact elements that cover the local contact behaviour by the friction characteristic module, the non-linear contact stiffness module and the wear module. The simulation results can be used as an input parameter for global tyre models. Moreover, special interest is focused on the experimental measurement of the non-linear normal contact stiffness and the wear behaviour of the sliding tread block on rough surfaces.

Acknowledgements

The investigations receive financial support from the European Commission within the integrated project SILENCE (www.silence-ip.org) and from the Deutsche Forschungsgemeinschaft as part of the DFG-Forschergruppe Dynamische Kontaktprobleme mit Reibung bei Elastomeren (www.ibnm.uni-hannover.de/FOR492).

References

1. Hofstetter K (2004) *Thermo-mechanical Simulation of Rubber Tread Blocks during Frictional Sliding*. PhD Thesis, Vienna University of Technology, Austria.
2. Bschorr O, Wolf A, Mittmann J (1981) *Theoretische und experimentelle Untersuchungen zur Abstrahlung von Reifenlärm*. MBB-Bericht Nr. BB-48381-Ö.
3. Sextro W (2000) *Dynamical Contact Problems with Friction*, Lecture Notes in Applied Mechanics. Springer, Berlin.
4. Hurty W C (1965) Dynamic analysis of structural systems using component modes. *AIAA Journal*, 3(4):678–685.
5. Craig R, Bampton M (1968) Coupling of substructures for dynamic analyses. *AIAA Journal*, 6(4):1313–1319.
6. Popp K, Rudolph M, Kröger M, Lindner M (2002) Mechanisms to generate and to avoid friction induced vibrations. *VDI-Berichte*, 1736:1–15.
7. Kummer H W (1966) *Unified Theory of Rubber and Tire Friction*. Engineering Research Bulletin B-94, Pennsylvania State University, USA.
8. Klüppel M, Heinrich G (2000) Rubber friction on self-affine road tracks. *Rubber Chem. Technol.*, 73:578–606.

9. Persson B N J (2006) Rubber friction: Role of the flash temperature. *J. Phys.: Condens. Matter*, 18:7789–7823.
10. Moldenhauer P, Lindner M, Kröger M, Popp K (2005) Modelling of hysteresis and adhesion friction of rubber in time domain. In: *Constitutive Models for Rubber IV*, P-E Austrell and L Kari (Eds), Balkema, Rotterdam, pp. 515–520.
11. Wriggers P, Nettingsmeier J (2004) Frictional contact of elastomer materials on rough rigid surfaces. *Proc. Appl. Math. Mech.*, 4:360–361.
12. Lindner M (2005) *Experimentelle und theoretische Untersuchungen zur Gummireibung an Profilklötzen und Dichtungen*. Fortschritt-Berichte VDI, Reihe 11, Nr. 311, PhD Thesis, University of Hannover, Germany.
13. Andersson P (2005) *Modelling Interfacial Details in Tyre/Road Contact – Adhesion Forces and Non-linear Contact Stiffness*. PhD Thesis, Chalmers University of Technology, Gothenburg, Sweden.
14. Gäbel G, Kröger M (2006) Non-linear contact stiffness in tyre-road interaction. In: *The 6th European Conference on Noise Control (EURONOISE)*, Paper 118, pp. 1–6.
15. Fleischer G (1973) Energetische Methode der Bestimmung des Verschleißes. *Schmierungstechnik*, 4(9):269–274.
16. Archard J F (1953) Contact and rubbing of flat surfaces. *J. Appl. Phys.*, 24:981–988.
17. Zhang S (2004) *Tribology of Elastomers*, Tribology and Interface Engineering Series, Vol. 47. Elsevier, Amsterdam.
18. Viswanath N (1995) Development of an equation for the wear of polymers. *Wear*, 181–183:42–49.
19. Moldenhauer P, Kröger M (2006) Vibrations of a tyre tread block under consideration of local wear. In: *Proceedings of the European Solid Mechanics Conference*, Budapest, Hungary.

Two FEM Approaches for the Prediction and Quantification of "Stick-Slip" Phenomena on Rubber-Metal Sliding Contacts

J.M. Bielsa, C. Bernad, R. Rodríguez and M.A. Jiménez

Instituto Tecnológico de Aragón (Materials Department), C/ María de Luna 7–8, E-50018 Zaragoza, Spain
E-mail: jmbielsa@ita.es

Abstract. The stick-slip phenomenon constitutes a challenge when referring to tribological design of rubber sealing components in a wide range of pneumatic and hydraulic applications. Several systems such as brake and clutch servo actuators, hydraulic gearshifts and other actuation systems are influenced by this phenomenon which is commonly associated to system control problems (system vibrations & noise) and to the reduction of the service life of rubber sealing components. Under a macroscopic point of view, the stick-slip consists on the sudden and successive change of the state of relative movement between two sliding surfaces in contact from "static" to "sliding".

In the present paper, two FEM-based approaches are presented with the final objective of developing a numerical predictive tool for the analysis and quantification of this undesired phenomenon. First approach is based on a common existing in literature mass-spring system over a moving surface and implemented into a parametrized FE modelling. The methodology based on the parametrization of the FE simulations allows to study the influence of several system variables such as the frictional force evolution, the mass of the system, the system stiffness and damping and also sliding speed on the frequency and amplitude of the stick-slip instability. All the system variables corresponding to the simplified mass-spring model are connected to the real system in the way that an analysis or proposed modification on any of them can be directly translated into effective modifications of the real system to prevent "stick-slip" events.

As an alternative approach to the direct FE simulation, a more affordable technique like the complex eigenvalue analysis of the problem is also applied to this example in order to obtain system instabilities. Even though this technique is more suitable for the analysis of mode-coupling friction instabilities, it is shown that the results in terms of eigenvalues can compared and correlated to the time domain simulation results. Finally, a comparison of both approaches is carried out with the objective of developing a low-time-consuming tool with reasonable accuracy for predicting and quantifying stick-slip phenomena on real rubber sealing components.

Peter Wriggers and Udo Nackenhorst (eds.), IUTAM Symposium on Computational Contact Mechanics, 291–309.

Key words: Tribology, polymers, "stick-slip", friction, finite element method, user subroutine, parametrization, complex eigenvalue analysis.

1 Antecedents and Scope

Frictional instabilities, commonly named as "stick-slip", constitute a challenge when referring to tribological design of polymer components in a wide range of industrial applications. Several systems such as hydraulic gearshifts, elevator sliders and many other hydraulic and pneumatic actuation systems are influenced by this phenomenon commonly associated to system control problems (system vibrations & noise) and to the reduction of the service life of polymeric components. From a macroscopic point of view, the "stick-slip" phenomenon consists on the sudden and successive change of the state of movement between two sliding surfaces in contact, lubricated or not, from "static" to "sliding". With the objective of developing a numerical-based tool to analyse and to predict this phenomenon, a rubber-metal sliding system will be modelled by a simplified mass-spring system and simulated with the commercial FE software ABAQUS.

Thus, the present paper has been divided into two main parts: first part comprises the development of a simplified mathematical "mass-spring" model (based on [1] and [2]) to be applied in the analysis of "stick-slip" phenomena on rubber-metal sliding contacts through a dynamic Finite Element simulation. The second part carries out a similar study of the "mass-spring" case through a Complex Eigenvalue Analysis, which proves more time consuming than the former FE analysis. Finally, a correlation between the results obtained following both methods is sought, being the final objecive the use of both methods as a predictive tool.

2 Stick-Slip: Description of the Phenomenon and Analytical Modelling

Stick-slip is a phenomenon of unsteady sliding resulting from a negative slope in the friction force to sliding speed relationship, in combination with the elasticity of the mechanical system of which the friction contact is part. This phenomenon consists on the sudden and successive change from "stick" state to "sliding" state, provoking the apparition of vibration and noise.

The stick-slip system can be mathematically simplified to a mass-spring group on a moving surface (Figure 1) where the spring represents the stiffness of the real system. The explanation of the stick-slip through this simplification is that, when the force of the spring equals the friction force the relative movement starts, and due to the decrease of the friction coefficient from static to dynamic, the mass suffers an acceleration in the direction contrary to the surface velocity.

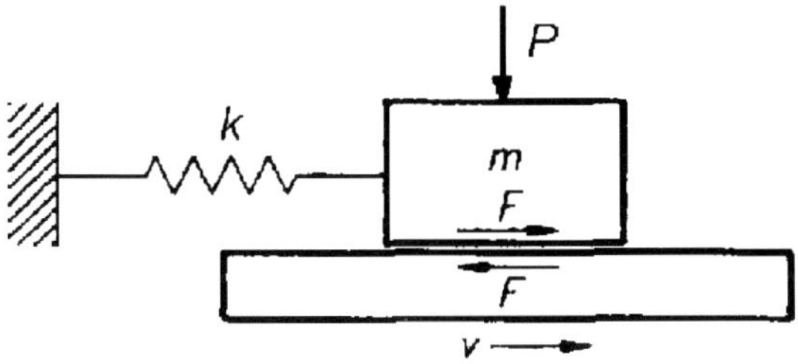

Fig. 1. Stick-slip numerical model.

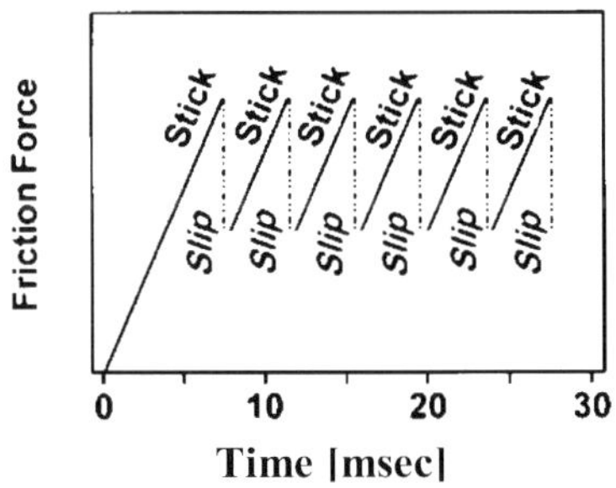

Fig. 2. Frictional force during stick-slip phenomenon.

With this model the frictional force that is expected to obtain is shown in Figure 2. While the mass is sticking the frictional force equals the spring force, it is, increases lineally with time (the velocity of the surface is constant). When the spring force reaches the maximum value of frictional force that the contact is able to transmit, the state of the mass changes from sticking to sliding, then suddenly the maximum frictional force changes from the static to the dynamic value, and the mass accelerates. Finally when the spring force equals again the dynamic maximum frictional force the contact is able to transmit, and the inertia force is compensated, the mass goes back to sticking and the cycle starts again. If the stick-slip phenomenon is severe enough, the mass can achieve the compression of the spring, at this point the relative velocity is cero and the mass can continue to accelerate to the right (Figure 1) and a reversal of the relative sliding velocity and friction force can take place. Figure 3 shows the evolution of the mass relative displacement and velocity during one stick-slip cycle with velocity reversal.

However the model represented in Figure 1 can present some instabilities due to the lack of damping in the system. To solve that problem a dashpot has been included in parallel with the spring. This represents the damping inherent to the system.

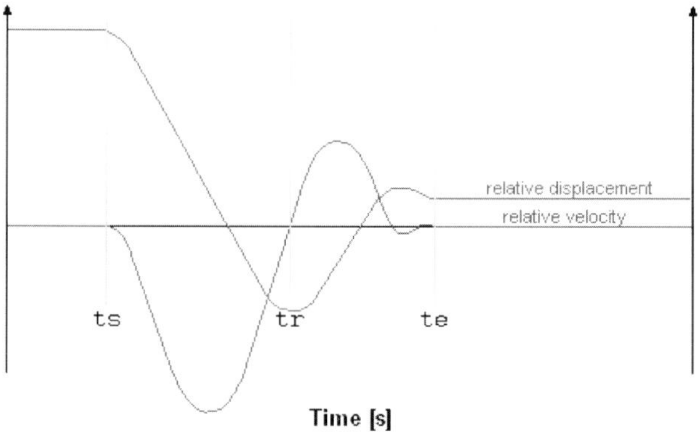

Fig. 3. Velocity reversal during stick-slip.

3 Application of FEM to the Study of Stick-Slip Phenomenon

Using the ABAQUS parametric capabilities a whole set of simulations, in which the parameters of the numerical stick-slip model as well as the parameters of the friction model are varied, have been carried out. The results obtained have been post-processed using a statistical tool (MiniTab) [15–17] and in order to analyse the influences of these parameters on the stick-slip phenomenon. This procedure will allow to get a deeper understanding of the stick-slip phenomenon and to develop a methodology to predict and minimize the stick-slip phenomenon in real cases (model parameters can be translated into real variables of the real system).

3.1 Parametric Studies of "Stick-Slip"

A FE model will be used to simulate the phenomenon of stick-slip through a simplified 1D "mass-spring" system. After performing a set of parametric simulations varying the input parameters of the model by means of Python scripting (spring stiffness, friction coefficient, vertical load, sliding speed, etc.), the influence of the input parameters on the output parameters corresponding to amplitude and frequency of the mass movement due to "stick-slip" will be analysed using a statistical tool (Minitab software).

In order to save computational resources, the problem has been split into two parametric studies in which three different parameters have been varied.

The FE model used in the parametric studies can be observed in Figure 4. It consists of a 11 mm × 4 mm pill meshed with CPS4R elements (4-node bilinear, reduced integration with hourglass control plane stress element) using a total amount of 387 elements. The material model used on the pill elements is the Marlow hyperelastic model, fed using uniaxial tensile test data of rubber material. The moving

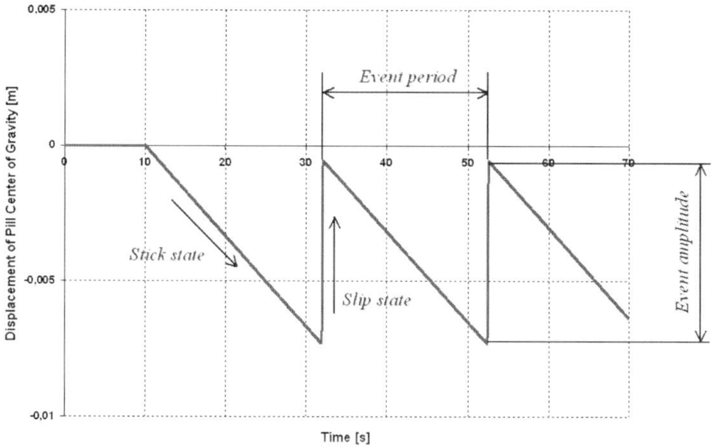

Fig. 4. Output variables of the simplified "stick-slip" FE model.

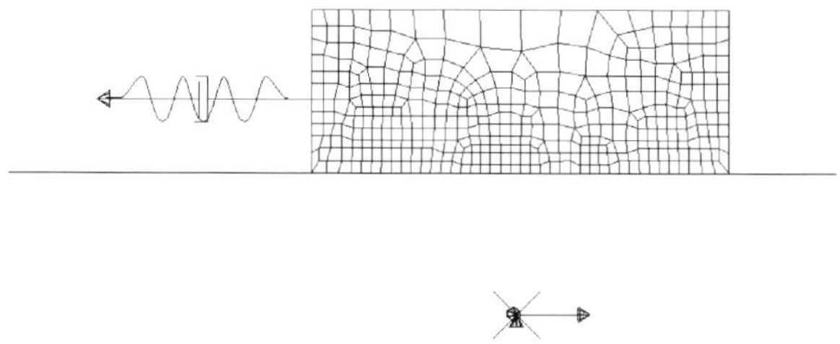

Fig. 5. Stick-slip FE model.

surface (horizontal flat line in Figure 5) has been modeled as an analytical rigid surface. The friction modelling is performed using the user subroutine FRIC with elastic stick formulation (in order to improve convergence) and with friction coefficient dependency on the contact pressure and on the dynamic state (static-kinetic transition and relative sliding speed).

The simulation consists of two steps. During the first step the load is applied on the pill. This is a non-linear geometric, unsymmetrical static step in which the rigid surface is kept still. During the second step the rigid surface starts moving at constant velocity. This step is no longer static but dynamic.

The parameters of the stick-slip simplified model that have been used in each simulation are listed in Table 1. Only the damping value of the dashpot has been kept constant along all simulations. In further steps, a more complex and complete study should be performed using all significant parameters of the friction and stick-slip model.

Table 1. Parameter values in the parametric studies.

	STUDY 1	STUDY 2
μ_s	*0.115*	*(0.115, 0.183, 0.25)*
μ_d / μ_s	*(0.2, 0.4, 0.6)*	*0.33*
A_0	*(50, 60, 70)*	*60*
k (N/m)	*4000*	*(2000, 4000, 6000)*
P (N)	*(80, 100, 120)*	*100*
c (Ns/m)	*0.007*	*0.007*
D (mm)	*20*	*(20, 50, 80)*

In Table 1, μs is the static friction coefficient, $\mu d/\mu s$ is the ratio between dynamic and static friction coefficient, A_0 is the coefficient of the exponential function describing the static-kinetic transition and k, P and c are, respectively, the spring constant, the vertical load and the damping constant of the stick-slip simplified model (Figure 1). D is the total displacement of the rigid surface at the end of the second step, as the simulation time is kept constant along all simulations, the displacement D is directly related to the velocity of the rigid surface. This way, a displacement of 20 mm is equivalent to a velocity of 0.33 mm/s, 50 mm is equivalent to 0.83 mm/s and 80 mm is equivalent to 1.33 mm/s.

3.2 FEA Results

Next the results of the parametric studies are shown. The statistical post-processing of the results gives two sets of results, the Pareto chart of the effects, that shows how important is the influence of independent variables and their cross product on the dependent variables, and the main effects plot, that gives an idea of the tendency of the dependency of the variables and the response surface, that allows to predict the value of the dependent variables for non-simulated (or non-tested) cases.

Table 2 shows the combination of parameters simulated and the results obtained in terms of the dependent variables previously mentioned, it is the period in seconds and the amplitude in meters of the displacement curve of the centre of gravity of the pill.

In the next sub-sections, the influence of the input parameters detailed in the table above is analysed for each output variable of the "stick-slip" phenomenon separately (from one side, the influence on the period, and from the other one, the influence on the amplitude). This separate analysis on each output variable is performed statistically by means of the software Minitab. However, if one desires to extract general conclusions of the dependencies of the phenomenon, it should be taken into account

Table 2. Parameter values in the parametric studies.

STUDY 1					STUDY 2				
A0	μ_d / μ_s	P [N]	Period [s]	Amplitude [m]	μ_s	K [N/m]	D [mm]	Period [s]	Amplitude [m]
50	0.2	80	*5.9282*	*0.0037*	0.115	2000	20	*20.0030*	*0.0051*
60	0.2	80	*5.7497*	*0.0037*	0.1825	2000	20	*30.0095*	*0.0061*
70	0.2	80	*5.8888*	*0.0037*	0.25	2000	20	*37.5841*	*0.0168*
50	0.4	80	*8.0181*	*0.0028*	0.115	4000	20	*10.0011*	*0.0032*
60	0.4	80	*8.0340*	*0.0028*	0.1825	4000	20	*16.8368*	*0.0061*
70	0.4	80	*8.0649*	*0.0028*	0.25	4000	20	*21.9861*	*0.0084*
50	0.6	80	*5.5208*	*0.0018*	0.115	6000	20	*7.4859*	*0.0026*
60	0.6	80	*5.5919*	*0.0018*	0.1825	6000	20	*11.4807*	*0.0041*
70	0.6	80	*5.6405*	*0.0018*	0.25	6000	20	*15.3819*	*0.0056*
50	0.2	100	*7.1954*	*0.0046*	0.115	2000	50	*8.5722*	*0.0066*
60	0.2	100	*7.3338*	*0.0046*	0.1825	2000	50	*13.4657*	*0.0123*
70	0.2	100	*7.5508*	*0.0046*	0.25	2000	50	*15.0330*	*0.0168*
50	0.4	100	*9.9902*	*0.0034*	0.115	4000	50	*4.5450*	*0.0039*
60	0.4	100	*10.0074*	*0.0034*	0.1825	4000	50	*6.8806*	*0.0061*
70	0.4	100	*10.0291*	*0.0035*	0.25	4000	50	*8.7911*	*0.0084*
50	0.6	100	*7.0244*	*0.0023*	0.115	6000	50	*3.0332*	*0.0026*
60	0.6	100	*7.1344*	*0.0023*	0.1825	6000	50	*4.6603*	*0.0041*
70	0.6	100	*7.0925*	*0.0023*	0.25	6000	50	*6.1522*	*0.0056*
50	0.2	120	*8.8879*	*0.0055*	0.115	2000	80	*5.6494*	*0.0077*
60	0.2	120	*9.0623*	*0.0055*	0.1825	2000	80	*8.4175*	*0.0122*
70	0.2	120	*9.3983*	*0.0055*	0.25	2000	80	*9.4082*	*0.0120*
50	0.4	120	*11.9829*	*0.0041*	0.115	4000	80	*2.8295*	*0.0035*
60	0.4	120	*11.9125*	*0.0041*	0.1825	4000	80	*4.3598*	*0.0061*
70	0.4	120	*11.9953*	*0.0041*	0.25	4000	80	*5.4979*	*0.0084*
50	0.6	120	*8.5130*	*0.0027*	0.115	6000	80	*2.4844*	*0.0021*
60	0.6	120	*8.5453*	*0.0028*	0.1825	6000	80	*2.9133*	*0.0041*
70	0.6	120	*8.5679*	*0.0028*	0.25	6000	80	*3.8467*	*0.0056*

that both the amplitude and the period of "stick-slick" are, in fact, coupled in the presence of friction and, thus, valid conclusions about dependencies of the phenomenon are those which consider how a given parameter (for instance, the stiffness of the system) is affecting both amplitude and frequency of "stick-slip" at the same time.

Study 1. Output variable: Period

Figures 6 and 7 show the results obtained in study 1 for the period as output variable. In other words, these figures show how important is the influence of the single parameters which are varying in the FE simulations, and their cross products, on this output variable.

Pareto chart of the effects shows, in qualitative terms, which single and cross-product parameters have greater influence in the given output variable. Main effects

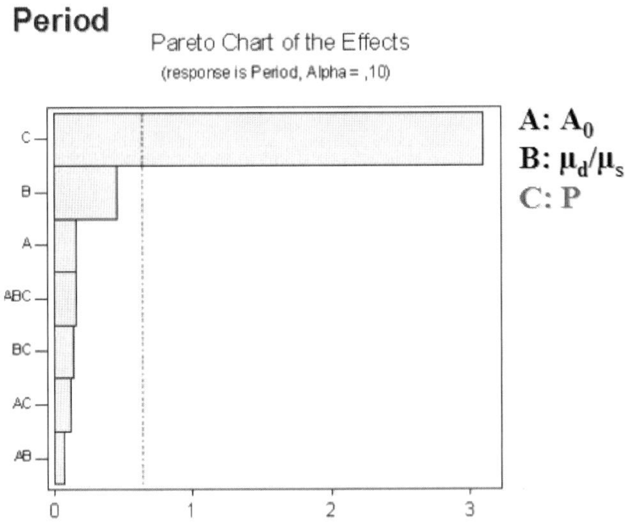

Fig. 6. Pareto chart for study 1, with period as output variable.

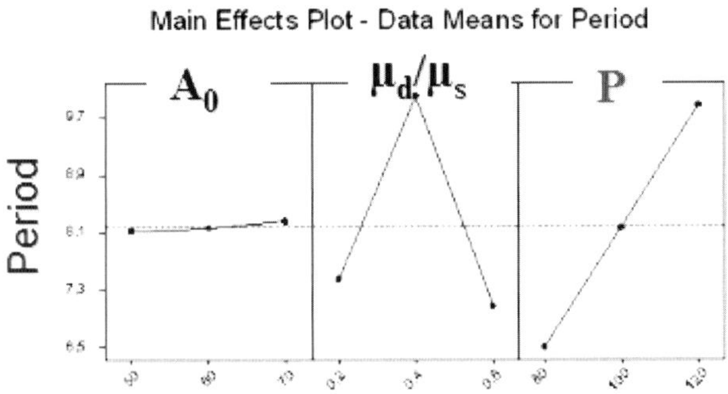

Fig. 7. Main effects plot for study 1, with period as output variable.

plot show the tendency of the direct relation (whether it is direct or inverse) between each single parameter and the given output variable.

It can be observed in Figure 6 that only the normal load (P), and in less extent the static-dynamic friction coefficient ratio ($\mu d/\mu s$), have significant influence on the period of the "stick-slip". This information is extracted from a 23 factorial design of experiments (DOE). Next Figure 7 was obtained from an 23 factorial with central points design of experiments. It can be observed in Figure 7 that including the central point the parameter $\mu d/\mu s$ gets greater importance while A_0 has no influence.

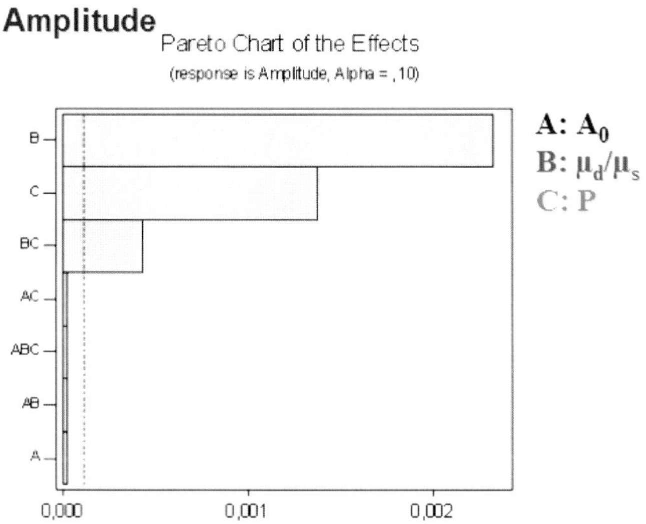

Fig. 8. Pareto chart for study 1, with amplitude as output variable.

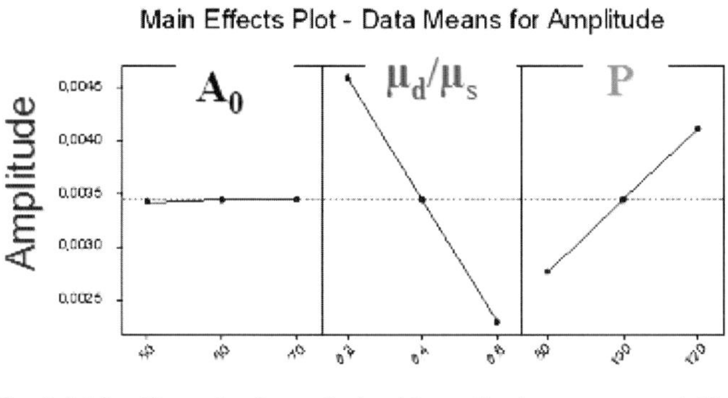

Fig. 9. Main effects plot for study 1, with amplitude as output variable.

Study 1. Output variable: Amplitude

The next figures show the results obtained in study 1 for the amplitude of the "stick-slip" as output variable.

It can be observed in Figure 8 that only the static-dynamic friction coefficient ratio ($\mu d/\mu s$), the normal load (P), and in less extent their product, have significant influence on the period of the stick-slip. Figure 9 confirms the conclusions obtained from Figure 8.

It can be also observed that in this case all dependencies are linear.

From study 1, it can be concluded that, for the given ranges of study:

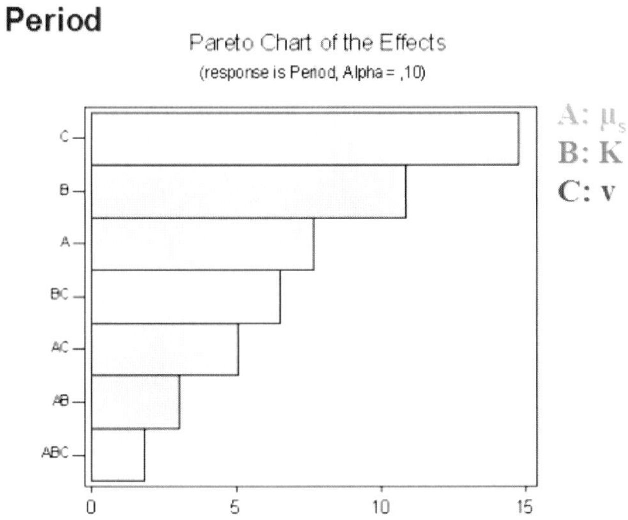

Fig. 10. Pareto chart for study 2, with period as output variable.

- The frequency of "*stick-slip*" is most inversely affected by the vertical load (P).
- The amplitude of "*stick-slip*" is directly affected by the vertical load (P) and inversely affected by the dynamic to static ratio ($\mu d/\mu s$).

The final conclusion is that only the influence of the vertical load (P) in the whole "stick-slip" phenomenon is clear, according to the fact that both the amplitude and frequency of "stick-slip" are coupled in the presence of friction. It seems, however, that the $\mu d/\mu s$ ratio has also importance, but with the present analysis is not possible to extract a more precise conclusion, thus, a more detailed analysis of the influence of this parameter should be carried out.

Study 2. Output variable: Period

Figures 10 and 11 show the results obtained in study 2 for the amplitude of the "stick-slip" as output variable. As in study 1, the figures shown are Pareto chart and main effects plot.

From the study it can be concluded that all parameters, μs (static friction coefficient), K (spring stiffness of the equivalent system) and the countermaterial velocity, their two-by-two product and even the product of all of them have some importance on the output variable, the period of the "stick-slip".

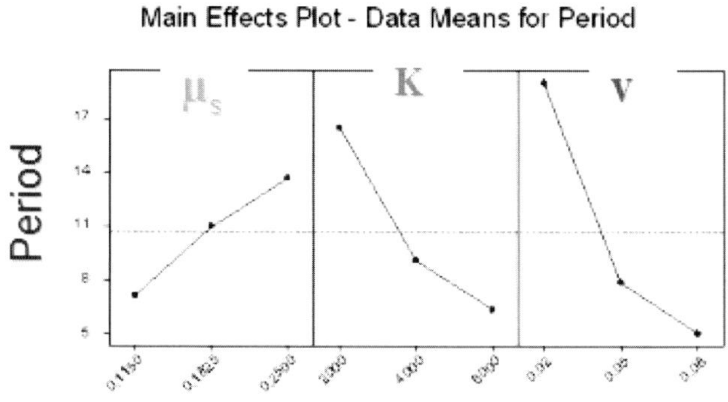

Fig. 11. Main effects plot for study 2, with period as output variable.

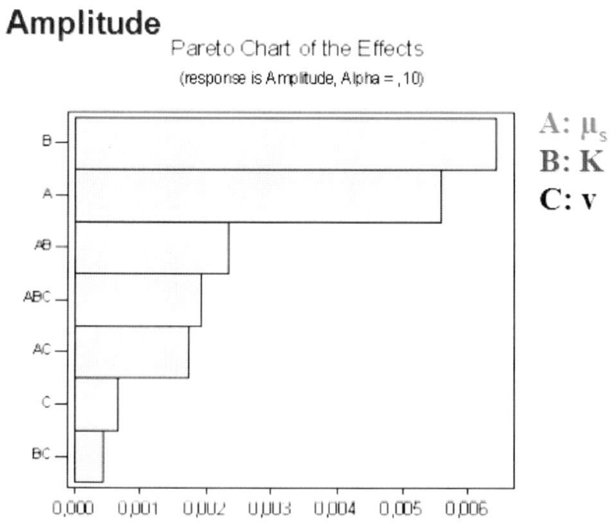

Fig. 12. Pareto chart for study 2, with amplitude as output variable.

Study 2. Output variable: Amplitude

Figures 12 and 13 show the results obtained in study 2 for the amplitude of the "stick-slip" as output variable. As in study 1, the figures shown are Pareto chart and main effects plot.

In this case, results show that the influence of the stiffness of the spring (K) and the static friction coefficient (μs) are the main parameters affecting the amplitude of the "stick-slip".

Finally, from study 2 it can be concluded that within the range of study:

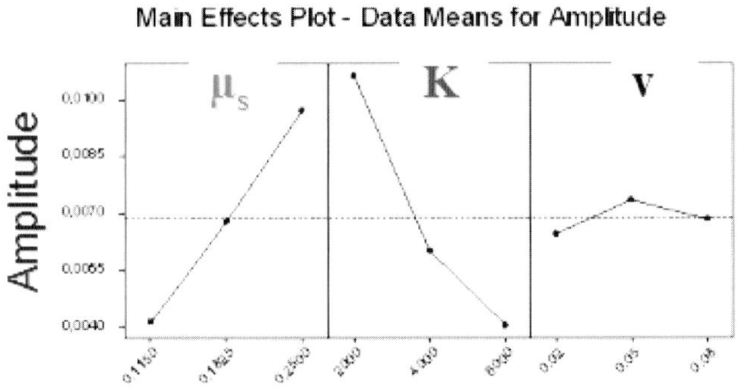

Fig. 13. Main effects plot for study 2, with amplitude as output variable.

- The frequency of "*stick-slip*" is directly affected by both the relative sliding speed (v) and the stiffness of the system (K).
- The amplitude of "*stick-slip*" is most inversely affected by the stiffness of the system (K).

Again, the final conclusion is that the influence on the whole "stick-slip" phenomenon of the stiffness of the system (K), and also of the static friction coefficient (μs), are clear.

4 Application of CEA to the Study of Stick-Slip Phenomenon

An alternative and simplified approach for numerical studies of the "stick-slip" phenomenon is the complex eigenvalue analysis (CEA) of the system (1D mass-spring system in the present case). As it will be shown later, it will be attempted to define a CEA output variable or index that, by means of this methodology, allows the prediction or at least to obtain the probability of appearance of "stick-slip". Computational cost for CEA reduces significantly that of the FEA approach.

As an initial basis for the aplication of CEA methodology, the "stick-slip" phenomena can be regarded as a system dynamic instability provoked by friction effects. In this approach, the non-linear equations of motion that describe the problem are linearized around a prescribed configuration and solved for the steady state sliding equilibrium position.

FE codes (ABAQUS) implement CEA solution based on the discretized system allowing the simulation of complex systems taking into account non-linear effects in a static analysis previous to the complex eigenvalue extraction. Developments in commercial software are mainly devoted to mode coupling effects of friction rather than in stick-slip, where it plays a secondary role.

In this basis, CEA solves the following differential equation, by means of projecting the corresponding matrices into the subspace of eigenmodes and extracting

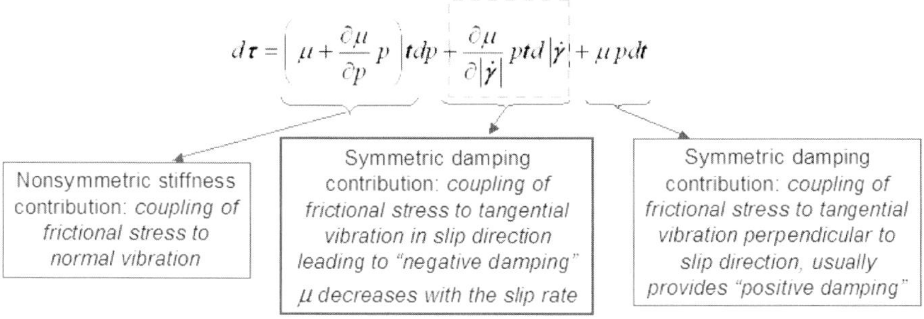

Fig. 14. Contributions to pertubatiuons in τ.

their eigenvalues. In this equation, "stick-slip" can be considered as a perturbation in τ, in which it can be derived the contributions from stiffness and damping (see Figure 14).

As solution for the previous, CEA gives as output variable a complex number, in which the real part is related to the damping of the associated movement (perturbation), and the imaginary part is related to the frequency of that movement (see Figure 15).

Study 2, performed with a dynamic FE calculation, was also studied using the complex eigenvalue analysis. Additionally to this study 2, several stable cases (no "stick-slip" appeared accordingly to dynamic FE calculations) were generated (study 3) to analyse CE results for stable and unstable cases, and thus, to better definition of the stability index.

The cases in which "stick-slip" appeared (study 2), resulted after CEA in complex eigenvalues with no imaginary part. In order to have an unique output variable for all cases, it has been decided to study the sum of all real parts with no imaginary part ($\Sigma\sigma$ with $\omega = 0$) as the namely stability index.

Next, Table 3 shows the cases simulated (study 2 and study 3) and the CEA results obtained:

From these results, and their equivalent dynamic FE simulations, it can be concluded that:

- A complex eigenvalue with positive real part and no imaginary part indicates "*stick-slip*".
- A complex eigenvalue with positive real part and non-zero imaginary part indicates an oscillatory or sub-damped response.
- A complex eigenvalue with negative real part and no imaginary part indicates an over-damped response.

The previous can be checked in Figure 16, in which the displacement-time response (FE analysis) of the center of gravity of the 1D mass-spring system is plotted for the previous sub-cases.

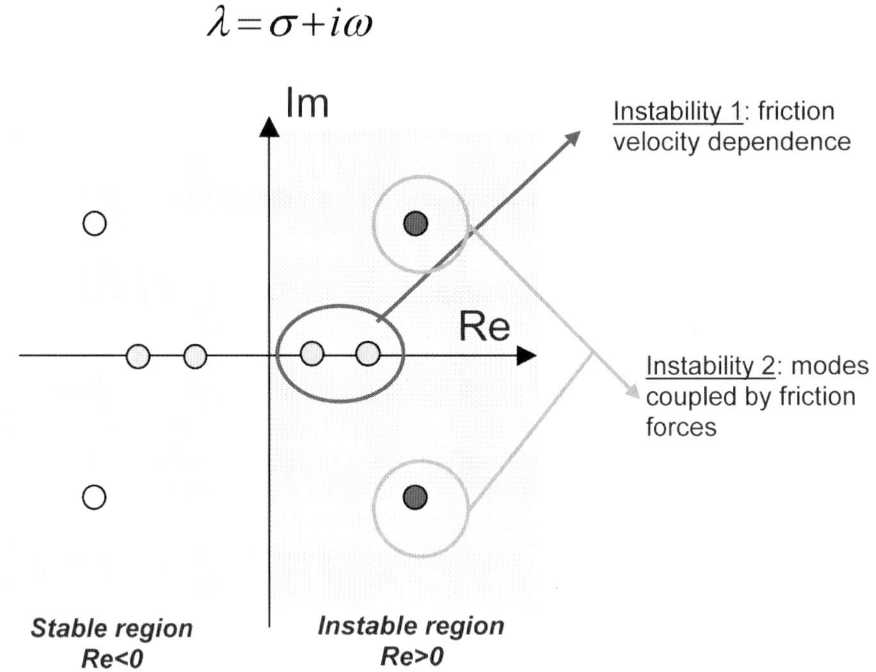

Fig. 15. Possible solutions of CEA.

As detailed above, dark curve corresponds to a over-damped response (negative real part and no imaginary part), yellow curve corresponds to a sub-damped response (positive real part and non-zero imaginary part) and finally red curve corresponds to a typical "stick-slip" response (positive real part and no imaginary part). First conclusion to this study is that the lower the sumatory of the real parts is, the more damped and thus the more stable vs. "stick-slip" the system is. In the previous figure, curves from cases 1 to 3 correspond to increasing values of $\Sigma\sigma$.

Applying the same statistical post-processing used to the dynamic FE simulations for the analysis of the index proposed as "stability" variable (sumatory of real parts), the results as presented in Figure 17 are obtained.

In this case, results analysed are those corresponding to study 2, with the objective of attempting to find a correlation between FEA results and CEA ($\Sigma\sigma$ as stability index) results.

From this statistical post-processing to the study 2, the following conclusions can be extracted from the Pareto charts of both FEA (Figures 10 and 12) and CEA (Figure 17) approaches:

- The $\Sigma\sigma$ is most influenced by the static friction coefficient (μ_s).

Table 3. Numerical results of the CEA parametric studies.

STUDY 2				STUDY 3			
μ_s	K [N/m]	v [m/s]	$\Sigma\sigma$	μ_d / μ_s	K [N/m]	c [Ns/m]	$\Sigma\sigma$
0.115	2000	0.00033	14809	0.5	5000	5000	-7382.58
0.115	4000	0.00033	14794	0.5	5000	2000	-134.22
0.115	6000	0.00033	14790	0.5	5000	1500	-72.52
0.115	2000	0.00083	13782	0.5	5000	1000	-7.85
0.115	4000	0.00083	13772	0.5	5000	500	-20.07*
0.115	6000	0.00083	13770	0.95	5000	200	-6.12
0.115	2000	0.00133	12890	0.5	10000	1000	-94.27
0.115	4000	0.00133	12883				
0.115	6000	0.00133	12881				
0.1825	2000	0.00033	114172				
0.1825	4000	0.00033	114655				
0.1825	6000	0.00033	114846				
0.1825	2000	0.00083	108369				
0.1825	4000	0.00083	108819				
0.1825	6000	0.00083	108998				
0.1825	2000	0.00133	48336				
0.1825	4000	0.00133	47854				
0.1825	6000	0.00133	47666				
0.25	2000	0.00033	186123				
0.25	4000	0.00033	187183				
0.25	6000	0.00033	187655				
0.25	2000	0.00083	178546				
0.25	4000	0.00083	179541				
0.25	6000	0.00083	179981				
0.25	2000	0.00133	171171				
0.25	4000	0.00133	172104				
0.25	6000	0.00133	172514				

* v =0.033m/s instead of 0.00033 m/s

- Stiffness of the system (K) is critical vs. frequency and amplitude of "stick-slip", but not described within $\Sigma\sigma$ with $\omega = 0$

And finally, from the main effects plot of both FEA (Figures 11 and 13) and CEA (Figure 18) approaches:

- The $\Sigma\sigma$ is directly affected by the static friction coefficient (μ_s)
- May be $\Sigma\sigma$ with $\omega = 0$ first approached as an stability index, if the static friction coefficient increases, the index does also indicating more probability of instability, according to both approaches.

5 Conclusions

From the FE analysis, the following conclusions can be drawn:

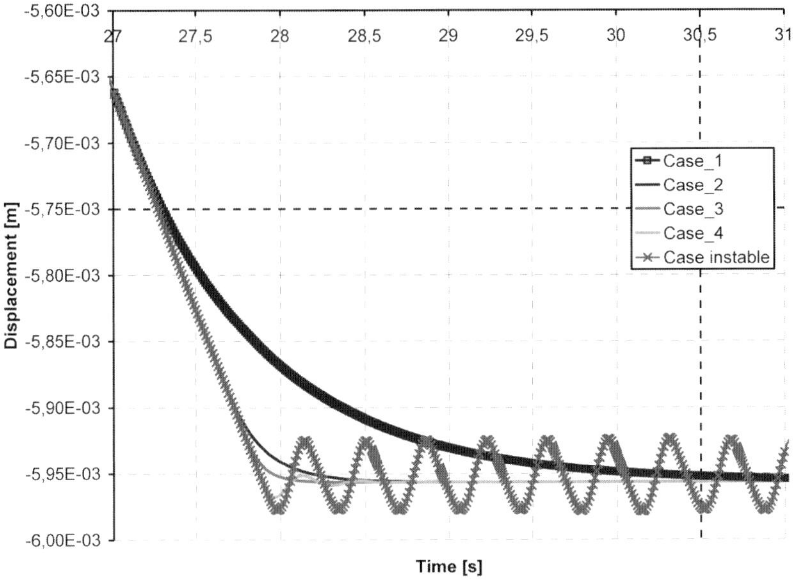

Fig. 16. FEA-CEA study 3.

- Previous analysis of friction mechanisms associated to rubber-metal sliding contacts establishes the existence of several parameters (temperature, sliding speed, contact pressure, etc.) that do influence their frictional behaviour.
- With the objective of introducing frictional behaviour of rubbers into FE modelling, the dependency of the friction coefficient with contact pressure and relative sliding speed (static-kinetic transition) has been successfully implemented by means of user subroutine (ABQ/Std) scripting.
- A 1D simplified "mass-spring" FE model has been created with the objective of analysing the phenomenon of "stick-slip". A parametrization method has been carried out by means of Python scripting. A real sliding system can be modelled through the input parameters of the "mass-spring" simplification.
- With the presently developed FEM and statistical-based tool, it is possible to determine which of the analysed system parameters ($p, \mu, v, \ldots$) do influence on "stick-slip" variables (frequency + amplitude). With the statistical post-processing (Minitab software), it is possible to determine how a single system parameter affects each "stick-slip" variable, which can be directly related to noise and vibrations of the system.
- Finally, if one desires to extract general conclusions of the dependencies of the phenomenon, it should be taken into account that both the amplitude and the period of "stick-slick" are, in fact, coupled in the presence of friction and, thus, valid conclusions about dependencies of the phenomenon are those which con-

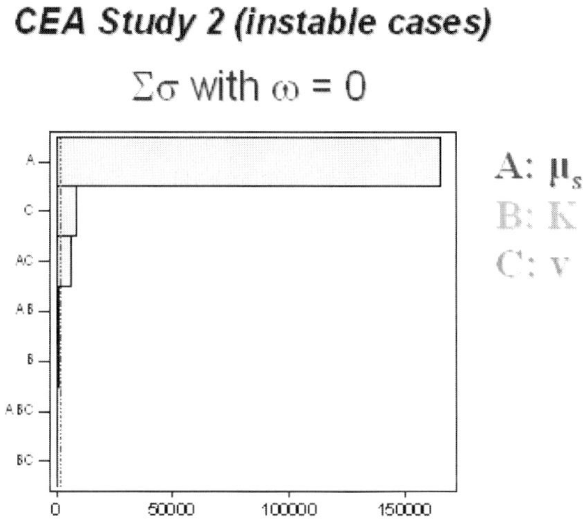

Fig. 17. Pareto chart for study 2, with $\Sigma\sigma$ as stability index.

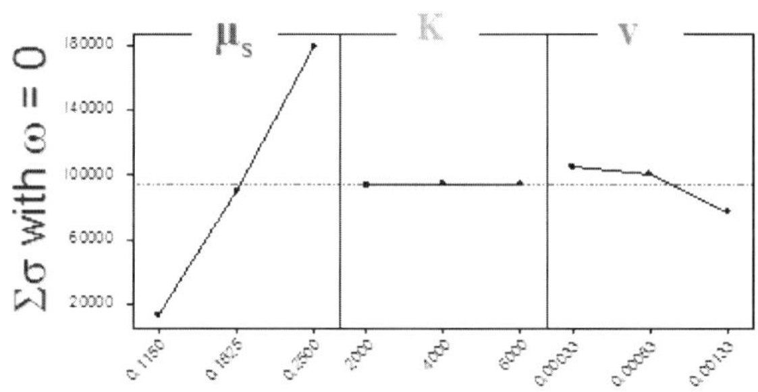

Fig. 18. Main effects plot for study 2, with $\Sigma\sigma$ as stability index.

sider how a given parameter is affecting both amplitude and frequency of "stick-slip" at the same time

From the complex eigenvalue analysis, it can be concluded that:

- Stability analysis by CEA is an appealing approach for "stick-slip" prediction by numerical simulation.
- It has been defined $\Sigma\sigma$ as first approach for stability index.
- Although it is an approximate solution it demands much less computational effort compared to transient FE analysis, allowing the simulation of complicated real systems.

- The main drawback is the difficulties of including non-linear effects coming from friction dependences and the lack of information concerning oscillation period/amplitude when they are within the scope of the analysis.

Acknowledgements

The present work is co-funded by Integrated Project "KRISTAL" (VI F.P.) and the "Torres Quevedo Program" from the Spanish Ministry of Education and Science.

References

1. F. Van De Velde, P. De Baets, The relation between friction force and relative speed during the slip-phase of a stick-slip cycle. *Wear*, 219:220–226, 1998.
2. F. Van De Velde, P. De Baets, J. Degrieck. The friction force during stick-slip with velocity reversal. *Wear*, 216:138–149, 1998.
3. S. Bouissou, J.P. Petit, M. Barquins. Normal load, slip rate and roughness influence on the polymethylmethacrylate dynamics of sliding. 1. Stable sliding to stick-slip transition. *Wear*, 214:156–164, 1998.
4. S. Bouissou, J.P. Petit, M. Barquins. Normal load, slip rate and roughness influence on the PMMA dynamics of sliding. 2. Characterisation of the stick-slip phenomenon. *Wear*, 215:137–145, 1998.
5. S. Bouissou, J.P. Petit, M. Barquins. Experimental evidence of contact loss during stick-slip: Possible implications for seismic behaviour. *Tectonophysics*, 295:341–350, 1998.
6. A.A. Batista, J.M. Carlson. *Bifurcations from steady sliding to stick slip in boundary lubrication. Physical Review E*, 57(5), 1998.
7. Hanbum Cho, J.R. Barber. Stability of the three-dimensional Coulomb friction law. *The Royal Society A*, 455:839–861, 1999.
8. T. Baumberger, P. Berthoud, C. Caroli. Physical analysis of the state- and rate-dependent friction law. II. Dynamic friction. *Physical Review B*, 60(6), 1999.
9. Christopher H. Sholz. Earthquakes and friction laws. *Nature*, 394, 1998.
10. H. Olsson, K.J. Astrom, C. Canudas de Wit, M. Gafvert, P. Lischinsky. Friction models and friction compensation. *Journal of European Control* 4(3), 2003.
11. O. Ronsin, K. Labastie Coeyrehourcq. State, rate and temperature-dependent sliding friction of elastomers. *The Royal Society A* 457:1277–1294, 2001.
12. J.J. Thomsen, A. Fidlin. Analytical approximations for stick-slip vibration amplitudes. *International Journal of Non-Linear Mechanics*, 38:389–403, 2003.
13. F. Xia. Modelling of a two-dimensional Coulomb friction oscillator. *Journal of Sound and Vibration*, 265:1063–1074, 2003.
14. E. Rabinowicz. *Friction and Wear of Materials*. Wiley-Interscience, 1995.
15. J. Neter, M.H. Knutner, C.J. Nachtscheim, W. Wasserman. *Applied Linear Statistical Models*, 4th Edition. McGraw-Hill, 1996.
16. D.C. Montgomery, A.E. Peck, G.G. Vining. *Introduction to Linear Regression*, 3rd Edition. Wiley, 2001.

17. D. Montgomery. *Diseño y Análisis de Experimentos*, 2nd Edition. Wiley, 2002.
18. S-W. Kung, K.B. Dunlap, R.S. Ballinger. Complex eigenvalue analysis for reducing low frequency brake squeal. In: *Proceedings SAE 2000 World Congress* (2000-01-0444).

Contact Mechanics of Wearing out Solids

Alfred Zmitrowicz

Institute of Fluid-Flow Machinery, Polish Academy of Sciences, ul. J. Fiszera 14, PL-80-952 Gdańsk, Poland
E-mail: azmit@imp.gda.pl

Abstract. Results of abrasive wear are identified as irreversible changes in bodies contours and as an increase of a clearance gap between contacting solids. Loose wear particles detached from the bodies transmit loads and displacements at the contact interface. In the contribution, descriptions of displacements and temperatures in the contacting solids are extended by taking into account the effects of the wear process. Illustrative examples demonstrate calculations of the abraded mass and fields of temperatures in pin-on-disc tests.

Key words: Friction, frictional heat, pin-on-disc test, wear, wear debris, wear depth, wear velocity.

1 Introduction

Wear is often identified with the gradual removal of material from rubbing surfaces of solids. Irreversible changes of shapes and sizes of the rubbing elements (i.e. changes in bodies contours) and an increase of a clearance gap between contacting solids are principal results of the wear process. The mechanism of wear involves formation of loose wear particles detached from the bodies. The wear particles are trapped and accumulated between the sliding surfaces. They form a thin, uniform and almost continuous intermediate layer which separates the bodies. The intermediate layer is called sometimes the third body [11, 12].

The main types of wear that are known, are as follows: adhesive, abrasive, oxidative, fatigue, fretting and erosion. In engineering, abrasive and fatigue wear are probably the most common cause of wear. Wear is the type of surface degradation or damage.

The first trials on numerical calculations of worn contours of the solids were given by Agelet de Saracibar and Chiumenti [1], Ko et al. [16], McColl et al. [18], Põdra and Andersson [22], Sfantos and Aliabadi [26], Strömberg [30], Telliskivi

Peter Wriggers and Udo Nackenhorst (eds.), IUTAM Symposium on Computational Contact Mechanics, 311–331.

[32]. Some researchers made use of fracture mechanics to study the wear phenomena, see [10, 20, 31]. A few phenomenological models for the third body have been proposed in the literature. Godet [11, 12], Berker and Van Arsdale [3] assumed the third body as a viscous fluid or a mixture. Elrod [6], Elrod and Brewe [7], Iordanoff et al. [13, 14] proposed models for the third body taking into account granular media. Zmitrowicz [35, 39] studied various continuum mechanics-based models for the interfacial layer, i.e. micro-polar thermoelastic solid, micro-polar fluid and thermoviscous fluid.

The main objective of this study is to obtain governing relations describing the solids subject to contact and wear in both formulations: the strong – based on general balance laws and constitutive equations, the weak one – based on variational principles. Classical variational formulations of the contact problems are extended by taking into account the effects of the wear process. The variational functionals and finite element approximations lead to equations useful in numerical procedures. Illustrative examples demonstrate calculations of the abraded mass and fields of temperatures in pin-on-disc tests for three types of materials.

2 Evolutions of Wearing out Surfaces

2.1 Wear Depth Profiles

A mechanical model of rubbing and wearing out solids consists of two contacting solids (A and B) and an interfacial layer of wear debris (S). Two quantities m_A and m_B describe rates of the mass flowing out from two contacting bodies (distinguished by subscripts A and B) due to the abrasion

$$m_A = -\rho_A \mathbf{v}^+ \cdot \mathbf{n}^+, \quad m_B = -\rho_B \mathbf{v}^- \cdot \mathbf{n}^-, \tag{1}$$

where ρ_A and ρ_B are mass densities, $\mathbf{n}^+$ and $\mathbf{n}^-$ are unit vectors normal to the wearing out surfaces Ω_{cA} and Ω_{cB}, see Figure 1. The velocities normal to the boundaries $(\Omega_{cA}, \Omega_{cB})$ define the wear process

$$\mathbf{v}^+ \cdot \mathbf{n}^+ = \upsilon^+(\mathbf{x}_A, t), \quad \mathbf{v}^- \cdot \mathbf{n}^- = \upsilon^-(\mathbf{x}_B, t). \tag{2}$$

$$\mathbf{x}_A \in \Omega_{cA}, \quad \mathbf{x}_B \in \Omega_{cB}, \quad t \in \langle 0, t_1 \rangle. \tag{3}$$

The contact regions Ω_{cA} and Ω_{cB} coincide. The wear velocities υ^+ and υ^- can be determined with the aid of wear laws.

An amount of the removed material can be estimated with the aid of a wear depth profile (a worn contour profile). This is a useful measure of the wear process, see [1, 9, 16, 18, 22, 26, 30, 40]. The wear depth profiles may have various shapes. Sometimes they are similar to pressure distributions in the contact area [9].

The depth of the material worn away from the bodies A and B in the time interval $\langle 0, t_1 \rangle$ is defined by the following integrals

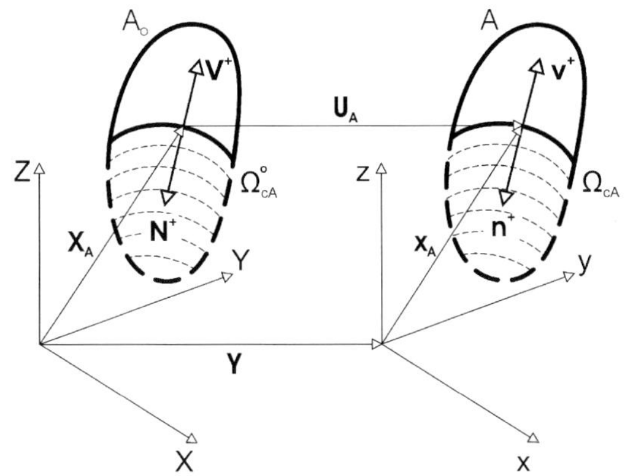

Fig. 1. Kinematics of the wearing out boundary Ω_{cA} of the solid body A in Lagrange and Euler descriptions.

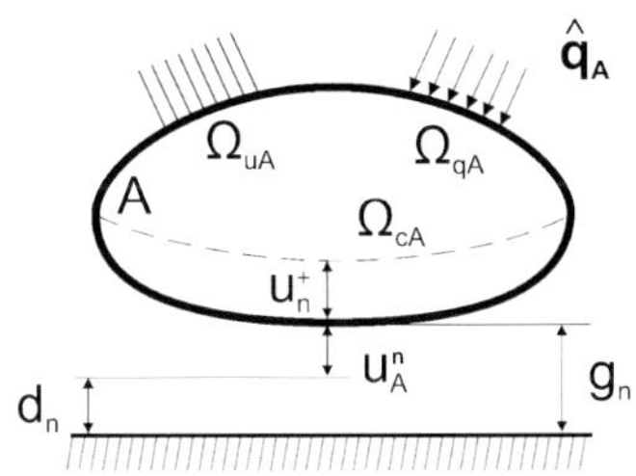

Fig. 2. Definitions of the wear depth profile (the worn contour profile), the initial gap and the total gap for the deformable body A in contact with a foundation.

$$u_n^+ = \int_0^{t_1} v^+(\mathbf{x}_A, t)\mathrm{d}t, \quad u_n^- = \int_0^{t_1} v^-(\mathbf{x}_B, t)\mathrm{d}t. \tag{4}$$

Equations (4) define a history of the wear process. Dimensions of the worn bodies are gradually reduced for the increasing sliding time. Wear is usually very slow process, but it is very steady and continuous.

A definition of a clearance gap between two wearing out bodies includes deformations of the bodies and evolutions of the wear depth profiles, see Figure 2. The total clearance gap (d_n) between two bodies consists of an initial gap (g_n), an increase of the gap as a result of the wear process of the bodies A and B (u_n^+, u_n^-) and deformations of the bodies in the normal direction to the contact (u_A^n, u_B^n), i.e.

$$d_n = g_n - u_A^n + u_n^+ + u_B^n - u_n^-. \tag{5}$$

2.2 Laws of Wear

In the subject literature, three stages in a development of wear equations have been classified: (a) empirical laws of wear, (b) phenomenological equations, (c) constitutive relations based on failure mechanisms of materials. Wear depends as much on the operating conditions (the normal pressure and the sliding action) as on physical properties of materials. Earlier analytical expressions for the prediction of wear connected a volume of the removed material with hardness only. To our days, various laws of wear have been proposed, and the great number of independent variables that influence the wear rate was used in the constitutive equations, see [2, 19]. In spite of that, at present, Archard law is a quantitatively simple calculation procedure of abrasive wear, see [1,9,16,18,22,25,26,28,30,32,33,40].

By analogy to Archard law, the wear velocities (2) can be defined as functions of the normal pressure and the sliding velocity, i.e.

$$v^{+} = -i_A|p_A^n|\,|\mathbf{V}_{AS}|, \quad v^{-} = -i_B|p_B^n|\,|\mathbf{V}_{BS}|, \tag{6}$$

where i_A and i_B are wear intensities of the bodies A and B, p_A^n and p_B^n are normal pressures, $\mathbf{V}_{AS}$ and $\mathbf{V}_{BS}$ are sliding velocities between the bodies and the intermediate layer S. In the tribology literature, the wear coefficients i_A and i_B are called the dimensional wear constants or the specific wear rates. Experimental measurements of the wear intensities belong to fundamental tests on mechanical properties of engineering materials.

The wear equations (6) satisfy the axiom of material objectivity, i.e. two different observers of the sliding at the contact region recognize the same wear velocities [35].

The constants in the phenomenological laws of wear (6) are restricted by two thermodynamic requirements [35, 41]:

(a) *Restrictions following from the Second Law of Thermodynamics.*
The internal energy spent at the wear process is always positive, and it satisfies the following inequality

$$m_A e_A = \rho_A i_A \left|p_A^n\right| |\mathbf{V}_{AS}|\, e_A \geq 0, \tag{7}$$

where e_A is the internal energy consumed by formation of a unit mass of wear debris. From (7) we obtain that the wear intensity coefficient is positive, i.e. $i_A \geq 0$ since all quantities in (7) are positive.

(b) *Constraints of the energy dissipated at the frictional contact.*
When two solids slide one against another, most of the energy dissipated in the friction process appears as frictional heat and the energy spent on the wear process (other forms of energy can be neglected). It can be assumed that 10% of the friction power is transformed into the wear process, the rest part of the friction power is converted into heat.

The following constraints of the energy dissipated at the contact between the body A and the layer S must be satisfied

$$\mathcal{P} = \mathbf{p}_A^t \cdot \mathbf{V}_{AS} = \underbrace{q_A^f + q_S^f}_{90\%} - \underbrace{m_A e_A}_{10\%}, \tag{8}$$

where $\mathcal{P}$ is the friction power, $\mathbf{p}_A^t$ is the friction force, q_A^f is the frictional heat flux entering into the body A, q_S^f is the heat supplied to the interfacial layer S. In general, Equation (8) does not decide which part of the friction power appears as frictional heat and the wear process energy.

2.3 Contact Constraints

In most cases, two unilateral contact constraints have to be satisfied at the contact region Ω_{cA}. The contacting bodies cannot penetrate each other and they are either separated or pressed on each other (no tensile normal forces)

$$d_n \geq 0, \quad p_A^n \leq 0. \tag{9}$$

The tangential component of the contact stresses is restricted by the following Coulomb law of isotropic friction

$$|\mathbf{p}_A^t| \leq \mu_A |p_A^n| \quad \text{(stick)}, \tag{10}$$

$$\mathbf{p}_A^t = -\mu_A |p_A^n| \frac{\mathbf{V}_{AS}}{|\mathbf{V}_{AS}|} \quad \text{(slip)}, \tag{11}$$

where μ_A is the friction coefficient.

The heat generated at or very close to the contact surface is transferred away into the bulk of rubbing solids, and it is conducted into surrounding. The frictional heat flux is given by

$$q_A^f = -w_A |p_A^n| \, |\mathbf{V}_{AS}|, \tag{12}$$

where w_A is the frictional heat intensity coefficient. Therefore, heat boundary conditions at the contact region Ω_{cA} have the following form

$$(\mathbf{k}_A \operatorname{grad} T_A) \cdot \mathbf{n}^+ = q_A^f, \tag{13}$$

where $\mathbf{k}_A$ is the thermal conductivity tensor, T_A is the temperature in the body A.

Assuming that 10% of the friction power transforms into wear of the body A, from (8) we get the following relation

$$-0.1 |\mathbf{p}_A^t| \, |\mathbf{V}_{AS}| = -m_A e_A. \tag{14}$$

After substituting the isotropic friction force (11) and the internal energy spent at the wear process (7) into Equation (14), we obtain a quantitative restriction on the wear intensity coefficient

$$i_A = 0.1 \frac{\mu_A}{\rho_A e_A} . \tag{15}$$

Rabinowicz [25] suggested that in estimations of e_A one should consider the sliding body “loaded to the limit”. It means that in micro-regions near the contact surface the accumulated strain energy reaches the critical value at which the yielding occurs. Most of this energy is transformed into production of the wear particles. Therefore, in rough calculations, wear can be assumed to result from the yielding of materials in the surface micro-regions [41].

3 Kinematics and Models of Wear Debris

3.1 Morphology and Physical Properties

After some time of sliding, the wear particles can be seen inside and outside of a wear track. The wear particles are generally detached mechanically by micro-stresses resulting from the applied load and sliding. There are numerous studies in the field of tribology including details about the size and nature of the wear particles, see [5, 21, 23, 24, 39]. The shape, size, edge details, surface roughness, number (concentration) and color are the main parameters characterizing morphological features of the wear particles. The particles do not have one particular morphology. They have the following shapes: flakes, chips, thin platelets, filings, powder-like particles, and others. Wear debris are produced in the form of macroscopic particles (size from a few to several micrometers) and microscopic particles (size from submicrometer to a few micrometers), see Figure 3. Small particles are much numerous than large ones. Recent methods of the particle morphology analysis are based on computer image analysis techniques, laser scanning microscopy, mathematics of fractals, neural networks, etc. [21, 23, 24].

Wear debris in the interface can be crushed into finer particles or compacted and agglomerated into large debris. It is often observed [5] that wear debris of flake and chip forms can be “rolled over” into spheres, cylinders or needles. The wear particles can slide and roll in contacting surfaces and in relation to each other. It takes part especially, when the contacting solids are in oscillating sliding. Each roll-like particle is subjected to opposing tangential forces at its top and bottom surfaces. The torque resulting from these forces causes the particle to roll. Usually, axes of roll particles are oriented perpendicular to the direction of sliding, indicating that they were generated by rolling and suggesting that the particle circulation is common in the contact, see Figure 4. The spherically shaped particle is a particularly intriguing type of wear debris observed e.g. in [5]. It is natural to assign to wear debris the models with micro-rotations.

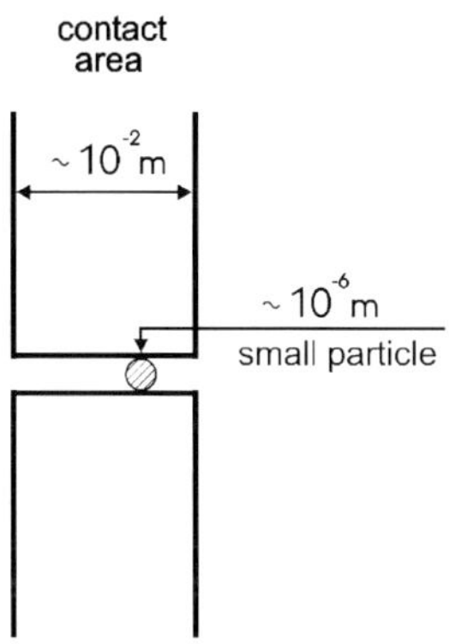

Fig. 3. Schematic view of a small particle at the contact area [12].

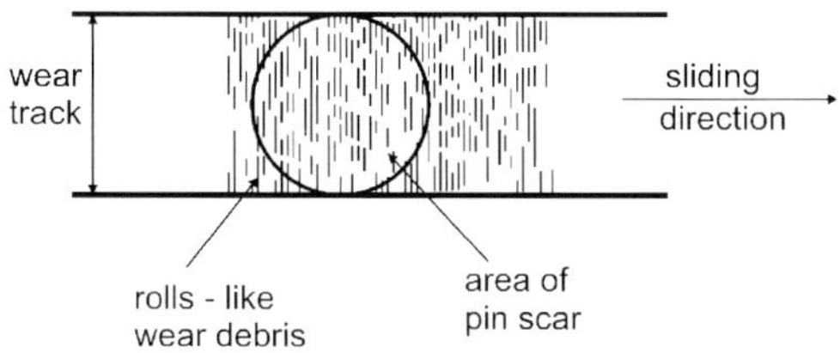

Fig. 4. Rolls-like wear debris observed in a wear track [15].

The material which was removed from the contacting surfaces is not in fact the same, either structurally or chemically, as the parent bodies material. Instead, it is a very fine-grained material which may be derived from both parts of the contacting system and may include components from the environment as well. Physical properties of the wear particles include composition, micro-structure, mass density, thermal expansion, thermal conductivity, specific heat, melting point, etc.

A number of the wear particles at the sliding interface can range from a single trapped particle to a continuous layer. Numerous experimental investigations and everyday praxis indicate the formation of the wear debris layer practically immediately after rubbing process starts, transforming the contact of two bodies into the contact of two bodies with the interfacial layer. The surfaces slide on each other separated by particles of the layer, see [5, 11–14]. The thickness of the layer is equal to a few micrometers.

The wear particles trapped and accumulated between sliding surfaces can be a source of significant mechanical phenomena. For example, the wear particles transmit translational and rotational displacements and heat at the contact interface. First of all, the presence of the wear particles between the sliding surfaces affects friction and wear phenomena very significantly [27]. The circulation of the wear particles is reflected by the friction coefficient which increases when particles are accumulated and decreases when particles are removed from the sliding interface. It has been reported that spherical and cylindrical particles roll over each other with resulting low friction. Therefore, wear debris can produce significant changes of the sliding conditions.

Various models can characterize: (a) quasi-solid, (b) quasi-fluid, and (c) granular-like behavior of the wear particles. It might be some promise in treating the interfacial layer of wear debris as a single continuum. The layer has certain properties which depend upon the physical and mechanical properties of the wear particles and the layer as a whole. For this propose one needs some equivalent mechanical properties to properly represent the layer, e.g. elastic modulus, Poisson ratio, shear modulus, hardness, etc. However, many problems of the constitutive laws for wear debris are still not solved.

3.2 Continuum Formulations

In the published literature, various continuum mechanics-based models for the interfacial layer have been proposed: (a) solid-like models, (b) fluid-like models, and (c) other models (e.g. mixtures of solid-liquid and solid-gas), see [3,5,12,14,35,39].

Advantages of the continuum approaches are: (a) the main stages of the wear process can be considered in this approach, i.e. the formation and circulation of wear debris, and the ejection from the contact area, (b) wear debris can undergo fragmentation and/or agglomeration, and (c) the third body is a thin layer, and the sliding occurs at the interface between the layer and the parent bodies.

Disadvantages of the continuum models are: (a) these models cannot be used of only a few wear particles are formed (a non-continuous interfacial medium), (b) difficulty to find quantitative values for physical constants, and (c) discontinuity in the velocity between the sliding bodies, i.e. an accommodation of the difference of speed of parent bodies cannot be realized in the bulk of the third body.

3.3 Discrete Formulations

Some authors postulated that mechanics of granular media should be applicable to define mechanical properties of the discontinuous and heterogenous third body, see [5–7, 13, 14]. Particles of the third body are modelled by assemblies of spheres or cylinders of micrometer size. Equations of translational and rotational motions are applied to each particle considered as an isolated rigid body.

Advantages of the granular material models are: (a) it is possible to simulate a non-continuous interfacial medium, (b) it is well known that the granular media can exhibit both solid-like and fluid-like behavior, and (c) the velocity accommodation takes place in the bulk of the third body by shearing.

Disadvantages of the granular material models are: (a) a number of particles in the layer is constant; formation of new particles and ejection of the particles from the contact region cannot be considered, (b) a choice of inter-particle forces is arbitrary, and (c) the layer has a finite thickness, and the size and shape of the particles have to be assumed.

4 Governing Equations

4.1 The Wearing Out Body A

Our model of wearing out solids consists of two contacting bodies (A and B) and the two-dimensional interfacial layer (S). The particles of the layer can translate and rotate independently. Differential forms of governing equations (motion, heat conduction and mass continuity) are constructed from general balance laws: mass, momentum, moment of momentum, energy and entropy, and from constitutive assumptions. Global and local balance equations are derived in [35, 38].

Deformations and displacements of the contacting body A are governed by the equilibrium equation which may be put in the following form

$$\operatorname{div} \boldsymbol{\sigma}_A - \bar{\mathbf{f}}_A = \mathbf{0}, \tag{16}$$

where $\boldsymbol{\sigma}_A$ is the Cauchy stress tensor, $\bar{\mathbf{f}}_A$ is the vector of body forces. The thermoelastic material and linearized theories are assumed for the body. Prescribed displacements and loads on Ω_{uA} and Ω_{qA} (respectively) are boundary conditions outside of the contact region, see Figure 2.

A heat conduction problem for the rubbing solid A is defined by the heat conduction equation as follows:

$$\rho_A c_A \frac{\partial T_A}{\partial t} = \operatorname{div}\left(\mathbf{k}_A \operatorname{grad} T_A\right) + Q_A, \tag{17}$$

where c_A is the specific heat, Q_A is the internal heat generation. Boundary conditions outside of the contact zone are as follows: there is the prescribed temperature (e.g. room temperature) on the boundary Ω_{sA}, the heat is convected into the surrounding (according to the Newton's rule) on the boundary Ω_{aA}.

The governing equations for the body B are similar to Equations (16) and (17). With the aid of the governing equations , material constitutive relations and boundary and initial conditions one can obtain fields of displacements and temperatures in the bodies A and B, i.e.

$$\mathbf{u}_A,\ \mathbf{u}_B,\ T_A,\ T_B. \tag{18}$$

Notice that Equations (16) and (17) are very classical and they are the same for rubbing and non-rubbing bodies [17, 33]. The fact that the bodies rub and wear out is included in adequate boundary conditions on Ω_{cA} and Ω_{cB}, see Equations (5), (9), (10) and (11). Tangential and normal components of the contact stresses on Ω_{cA} are given by

$$\mathbf{p}_A^t = (\mathbf{1} - \mathbf{n}^+ \otimes \mathbf{n}^+)(\boldsymbol{\sigma}_A \mathbf{n}^+) \quad \text{(stick)}, \tag{19}$$

$$\mathbf{p}_A^n = (\mathbf{n}^+ \otimes \mathbf{n}^+)(\boldsymbol{\sigma}_A \mathbf{n}^+) \equiv p_A^n \mathbf{n}^+, \tag{20}$$

where **1** is the unit tensor.

4.2 The Wear Debris Layer S

On a macroscopic scale, the layer which covers the contact zone can be regarded as infinitely thin. It makes possible to introduce simplifying assumptions: (a) to eliminate the dependence on the through-the-thickness coordinate, (b) to consider the layer as a two-dimensional material continuum.

A mass continuity equation and local balance equations of linear and angular momentum for the micro-polar thermoelastic layer S are given by

$$\frac{\partial \rho_S}{\partial t} + \text{div}_S \left(\rho_S \frac{\partial \mathbf{u}_S}{\partial t} \right) - m_A - m_B = 0, \tag{21}$$

$$\text{div}_S \, \boldsymbol{\sigma}_S - \bar{\mathbf{f}}_S - m_A \mathbf{V}_{AS} - m_B \mathbf{V}_{BS} = \mathbf{p}_A^t + \mathbf{p}_B^t, \tag{22}$$

$$\text{div}_S \, \boldsymbol{\mu} + (m_A + m_B) \mathbf{j}_S \frac{\partial \boldsymbol{\psi}}{\partial t} = \mathbf{c}, \tag{23}$$

where ρ_S is the mass density, $\mathbf{u}_S$, $\boldsymbol{\psi}$ are vectors of displacements and independent micro-rotations of wear debris, $\boldsymbol{\sigma}_S$ is the Cauchy stress tensor, $\bar{\mathbf{f}}_S$ is the body force, $\mathbf{p}_A^t$ and $\mathbf{p}_B^t$ are friction forces between the bodies A, B and the layer S, $\boldsymbol{\mu}$ is the couple stress tensor in the layer, $\mathbf{j}_S$ is the inertia tensor of wear debris, **c** is the couple of friction forces. Terms m_A and m_B define mass fluxes of wear debris supplied to the layer, $m_A \mathbf{V}_{AS}$ and $m_B \mathbf{V}_{BS}$ define momentum supplied to the layer by wear debris, and $(m_A + m_B)\mathbf{j}_S \partial \boldsymbol{\psi} / \partial t$ is the moment of momentum of supplied wear debris. The sliding velocities between the body A and the layer S and between the body B and the layer S are given by

$$\mathbf{V}_{AS} = (\mathbf{1} - \mathbf{n}^+ \otimes \mathbf{n}^+) \left(\frac{\partial \mathbf{u}_A}{\partial t} - \frac{\partial \mathbf{u}_S}{\partial t} \right), \tag{24}$$

$$\mathbf{V}_{BS} = (\mathbf{1} - \mathbf{n}^- \otimes \mathbf{n}^-) \left(\frac{\partial \mathbf{u}_B}{\partial t} - \frac{\partial \mathbf{u}_S}{\partial t} \right). \tag{25}$$

The couple of friction forces acting on upper and lower surfaces of an individual particle in S is defined as follows

$$\mathbf{c} = \zeta_A |p_A^n| \left(\mathbf{n}^+ \times \frac{\mathbf{V}_{AS}}{|\mathbf{V}_{AS}|} \right) + \zeta_B |p_B^n| \left(\mathbf{n}^- \times \frac{\mathbf{V}_{BS}}{|\mathbf{V}_{BS}|} \right), \tag{26}$$

where ζ_A, ζ_B are material constants.

A boundary line L of the layer S can be divided into parts, that are L_u, where boundary displacements can be given, L_q, where boundary forces are prescribed, L_ψ, where boundary micro-rotations can be defined and the part L_m, where boundary moments may occur. For example, L_u and L_ψ can be defined as boundaries where displacements and micro-rotations are restricted to be zero in the case of some constraints or foundations. L_q and L_m are boundaries where additional forces and moments are applied to the layer S, e.g. in the form of pressure and torque.

The heat conduction equation in the layer is as follows

$$\rho_S c_S \frac{\partial T_S}{\partial t} = \mathrm{div}_S \left(\mathbf{k}_S \,\mathrm{grad}_S\, T_S\right) - (m_A + m_B)\left[\epsilon_S - \frac{1}{2}\left(\mathbf{j}_S \frac{\partial \boldsymbol{\psi}}{\partial t}\right) \cdot \frac{\partial \boldsymbol{\psi}}{\partial t}\right] - \\ - \frac{1}{2} m_A \left(\mathbf{V}_{AS} \cdot \mathbf{V}_{AS}\right) - \frac{1}{2} m_B \left(\mathbf{V}_{BS} \cdot \mathbf{V}_{BS}\right) + q_S^f + q_S^c, \tag{27}$$

where c_S is the specific heat, ϵ_S is the internal energy, q_S^f is the frictional heat supplied to the layer, q_S^c is the heat conducted from the body A to the body B through the intermediate layer S. Terms with m_A and m_B in Equation (27) define changes of energy in the layer by wear debris supplied to the layer. Prescribed temperatures and heat fluxes into the surrounding are boundary conditions on the boundary line L.

Unknown quantities are: mass density, translations, micro-rotations and temperatures in the layer, i.e.

$$\rho_S,\ \mathbf{u}_S,\ \boldsymbol{\psi},\ T_S. \tag{28}$$

The wear debris layer S is characterized by the following constitutive equations independent of those characterizing the parent bodies A and B

$$\boldsymbol{\sigma}_S = \mathbf{E}_S \boldsymbol{\varepsilon}_S, \quad \boldsymbol{\varepsilon}_S(\mathbf{u}_S) = \frac{1}{2}\left[\mathrm{grad}_S\, \mathbf{u}_S + (\mathrm{grad}_S\, \mathbf{u}_S)^T\right], \tag{29}$$

$$\boldsymbol{\mu} = \mathbf{D}_S \boldsymbol{\kappa}, \quad \boldsymbol{\kappa}(\boldsymbol{\psi}) = \mathrm{grad}_S\, \boldsymbol{\psi}, \tag{30}$$

where $\boldsymbol{\varepsilon}_S$ is the strain tensor, $\boldsymbol{\kappa}$ is the couple strain measure in S, $\mathrm{grad}_S(\cdot)$ is the surface (two-dimensional) gradient operator, $\mathbf{E}_S$ and $\mathbf{D}_S$ are elasticity tensors with respect to displacements and micro-rotations respectively (see e.g. Eringen [8]).

5 The FE Approach

5.1 Variational Formulations

For numerical solutions of the contact problems with the aid of the finite element method, one needs a weak form of the governing equations. Variational formulations of displacements are derived with the aid of the principle of stationary total potential energy. The total potential energy of two contacting solids and the interfacial layer of particles with translational and rotational degrees of freedom [29] consists of the stored energy Π^{int} and the energy of external loads (i.e. external forces and contact forces) Π^{ext}, at the given step of deformation it achieves an extremum

$$\Pi = \Pi^{\text{int}} + \Pi^{\text{ext}} = \Pi(\mathbf{u}_A, \mathbf{u}_B, \mathbf{u}_S, \boldsymbol{\psi}) \longrightarrow \text{stationary}. \tag{31}$$

The functional Π is given in [36–38]. We search such $\mathbf{u}_A$, $\mathbf{u}_B$, $\mathbf{u}_S$ and $\boldsymbol{\psi}$ which guarantee the stationarity of Π at the given step of incremental approach and the iteration process, then the contact forces are assumed to be given. This is a simplified local weak formulation. The condition of stationarity is expressed by the first variation of the functional (31) equal to zero

$$\delta\Pi\left(\mathbf{u}_A, \mathbf{u}_B, \mathbf{u}_S, \boldsymbol{\psi}\right) = 0, \tag{32}$$

$$\delta\Pi = \left[\frac{\partial\Pi}{\partial\mathbf{u}_A}, \delta\mathbf{u}_A\right]_A + \left[\frac{\partial\Pi}{\partial\mathbf{u}_B}, \delta\mathbf{u}_B\right]_B + \left[\frac{\partial\Pi}{\partial\mathbf{u}_S}, \delta\mathbf{u}_S\right]_S + \left[\frac{\partial\Pi}{\partial\boldsymbol{\psi}}, \delta\boldsymbol{\psi}\right]_S, \tag{33}$$

$$\frac{\partial\Pi}{\partial\mathbf{u}_A} = \mathbf{0}, \quad \frac{\partial\Pi}{\partial\mathbf{u}_B} = \mathbf{0}, \quad \frac{\partial\Pi}{\partial\mathbf{u}_S} = \mathbf{0}, \quad \frac{\partial\Pi}{\partial\boldsymbol{\psi}} = \mathbf{0}, \tag{34}$$

where [.,.], etc., mean scalar products (represented by integrals over the corresponding volumes and surface parts respectively) and $\partial\Pi/\partial\mathbf{u}_A$, etc., are gradients of the functional Π, see [4].

5.2 Finite Elements Approximations

Discretized forms of the variational functionals (approximated with the aid of the finite element method) lead to equations useful in numerical calculation procedures. The discretization of the bodies and the layer by means of finite elements reduces the problem to nodal parameters. Finite element approximations of displacement, strain and stress fields in an element of the thermoelastic body A are given by

$$\mathbf{u}_A\left(\mathbf{x}_A\right) = \mathbf{N}_A\left(\mathbf{x}_A\right)\mathbf{u}_A^e, \tag{35}$$

$$\boldsymbol{\varepsilon}_A\left(\mathbf{x}_A\right) = \mathbf{B}_A\left(\mathbf{x}_A\right)\mathbf{u}_A^e, \quad \boldsymbol{\sigma}_A\left(\mathbf{x}_A\right) = \mathbf{E}_A\mathbf{B}_A\left(\mathbf{x}_A\right)\mathbf{u}_A^e, \tag{36}$$

where $\mathbf{N}_A$ is the matrix of shape functions, $\mathbf{B}_A$ is the differential operator matrix of the shape functions, $\mathbf{u}_A^e$ is the vector of displacements at nodes of the finite element, $\mathbf{E}_A$ is the elasticity matrix. Furthermore, we approximate the local fields in the micropolar layer S, i.e. displacements, micro-rotations, two measures of strains and two measures of stresses

$$\mathbf{u}_S, \ \boldsymbol{\psi}, \ \boldsymbol{\varepsilon}_S, \ \boldsymbol{\kappa}, \ \boldsymbol{\sigma}_S, \ \boldsymbol{\mu}. \tag{37}$$

5.3 Displacements of the Body A and the Layer S

The stationarity condition of the total potential energy for any finite element of the body A leads to the equilibrium equation (and boundary conditions)

$$\frac{\partial \Pi^e}{\partial \mathbf{u}_A^e} = \mathbf{K}_{A1}^e \mathbf{u}_A^e + \widetilde{\mathbf{q}}_A^e + \widetilde{\mathbf{p}}_A^e, \tag{38}$$

where $\mathbf{K}_{A1}^e$ is the local stiffness matrix, $\widetilde{\mathbf{q}}_A^e$ is the local vector of external and body loads, $\widetilde{\mathbf{p}}_A^e$ is the local vector of contact loads. For the whole body A, the equilibrium equation is an algebraic equation system

$$\mathbf{K}_{A1}\mathbf{u}_A + \widetilde{\mathbf{q}}_A + \widetilde{\mathbf{p}}_A = \mathbf{0}, \tag{39}$$

where $\mathbf{K}_{A1}$ is the global stiffness matrix, $\widetilde{\mathbf{q}}_A, \widetilde{\mathbf{p}}_A$ are global vectors of loads.

Displacements in any finite element of the layer S are defined by the following equation

$$\frac{\partial \Pi^e}{\partial \mathbf{u}_S^e} = \mathbf{K}_{S1}^e \mathbf{u}_S^e + \widetilde{\mathbf{q}}_S^e + \widetilde{\mathbf{p}}_S^e + \widetilde{\mathbf{m}}_1^e, \tag{40}$$

where $\mathbf{K}_{S1}^e$ is the local stiffness matrix of the layer's element, $\mathbf{u}_S^e$ is the vector of nodal displacements, $\widetilde{\mathbf{q}}_S^e$ is the local vector of loads in the layer, $\widetilde{\mathbf{p}}_S^e$ is the local vector of contact loads, $\widetilde{\mathbf{m}}_1^e$ is the momentum of wear debris supplied to the layer. Micro-rotations in the finite element of the layer S are given by

$$\frac{\partial \Pi^e}{\partial \boldsymbol{\psi}^e} = \mathbf{K}_{S2}^e \boldsymbol{\psi}^e + \widetilde{\mathbf{a}}^e + \widetilde{\mathbf{c}}^e + \widetilde{\mathbf{m}}_2^e, \tag{41}$$

where $\mathbf{K}_{S2}^e$ is the local stiffness matrix with respect to the rotations, $\boldsymbol{\psi}^e$ is the vector of rotations at nodes of the finite element, $\widetilde{\mathbf{a}}^e$ is the local vector of boundary moments, $\widetilde{\mathbf{c}}^e$ is the local vector of frictional moments, $\widetilde{\mathbf{m}}_2^e$ is the moment of momentum of wear debris.

5.4 Temperatures of the Body A and the Layer S

The heat conduction problem in the bodies A and B and in the layer S can be formulated with the aid of the variational approach, see [34]. In this case, the heat conduction problem is equivalent to the minimization of some functional which describes fields of temperatures in bodies (T_A, T_B) and in layer (T_S) and boundary conditions, i.e.

$$\widetilde{\Pi} = \widetilde{\Pi}(T_A, T_B, T_S), \tag{42}$$

$$\frac{\partial \widetilde{\Pi}}{\partial T_A} = 0, \quad \frac{\partial \widetilde{\Pi}}{\partial T_B} = 0, \quad \frac{\partial \widetilde{\Pi}}{\partial T_S} = 0. \tag{43}$$

The functional $\widetilde{\Pi}$ is given in [37]. By analogy to the stress analysis, the fields of temperatures are approximated with the aid of finite elements. Then, the heat conduction equation in any finite element of the body A is given by

$$\frac{\partial \widetilde{\Pi}^e}{\partial \mathbf{T}_A^e} = \mathbf{C}_A^e \dot{\mathbf{T}}_A^e + \mathbf{K}_{A2}^e \mathbf{T}_A^e + \widetilde{\mathbf{Q}}_A^e + \widetilde{\mathbf{q}}_{f1}^e, \tag{44}$$

where $\mathbf{T}_A^e$ is the vector of nodal temperatures at the finite element, $\mathbf{C}_A^e$ is the local thermal capacity matrix, $\mathbf{K}_{A2}^e$ is the local conductivity matrix, $\widetilde{\mathbf{Q}}_A^e$ is the local vector of thermal loads, $\widetilde{\mathbf{q}}_{f1}^e$ is the frictional heat vector.

Temperatures in any element of the layer S are defined with the aid of the following equation

$$\frac{\partial \widetilde{\Pi}^e}{\partial \mathbf{T}_S^e} = \mathbf{C}_S^e \dot{\mathbf{T}}_S^e + \mathbf{K}_{S3}^e \mathbf{T}_S^e + \widetilde{\mathbf{Q}}_S^e + \widetilde{\mathbf{q}}_{f2}^e + \widetilde{\mathbf{m}}_3^e, \tag{45}$$

where $\widetilde{\mathbf{q}}_{f2}^e$ is the vector of frictional and conducted heat, $\widetilde{\mathbf{m}}_3^e$ is the energy of wear debris supplied to the layer.

5.5 The Mass Density Field in the Layer S

The mass density field in the layer S, see Equation (21), can be approximated with the aid of the finite elements. During the sliding process, the wear particles can be accumulated between the sliding surfaces, or eliminated out of the contact region. The particles of the intermediate layer S leave the contact zone through the boundary line L due to the relative motion of the contacting bodies or due to different motions of wear debris and the contacting bodies. The mass efflux per unit time and unit length of the boundary L can be governed by an arbitrary assumed relation.

5.6 Solution algorithms

A common way to solve nonlinear contact problems is to adopt step-by-step incremental procedures. Following a loading history, contact constraints are satisfied at any increment step with the aid of iteration techniques. The contact problem is treated as a sequence of conservative problems solved successively (step-by-step). There is the energy dissipation, therefore the total potential energy is corrected at every step of the deformation history. The contact forces are calculated at every step of the iteration process with the aid of a regularization technique. Nowadays, formulations of contact problems (i.e. the problems with constraints on solution variables) are more refined by applications of the Lagrange multiplier method, penalty method and their generalizations. Then, the contact constraints are calculated together with displacements of the bodies in process of solving the problem [17, 33].

Terms with velocities in the equilibrium equations are replaced by increments. They are treated as definitions of applied loads (forces, torques) at the given step

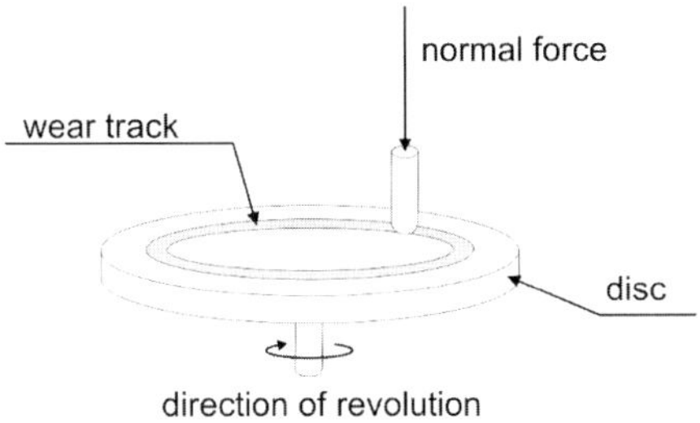

Fig. 5. The pin-on-disc test device.

of the deformation history. They give contributions to the equilibrium of forces and torques and to the external work. Large sliding of the deformable body can be composed with rigid body displacements and deformations.

In the wear process, dimensions of the wearing out bodies are gradually reduced. The shapes and sizes of the worn bodies should be calculated together with fields of stresses and strains. An updating process of a geometry of the wearing out solids can be given with the aid of the classical methods: re-meshing technique, repositioning the boundary surface nodes, etc.

6 Illustrative Examples

The pin-on-disc testing machine is a typical device used to study friction and wear of materials. A stationary pin is pressed against a rotating disc, see Figure 5. Pałżewicz and Jabłoński in the Institute of Fluid-Flow Machinery (Gdańsk) investigated a short duration contact between the pin and the steel disc under high pressure and high sliding speed conditions. It was an experimental simulation of dry friction in sliding bearings when a lubricating film is broken, and a part of the load is carried by an unlubricated contact.

Illustrative examples demonstrate calculations of the abraded mass and fields of temperature in the pin for three types of materials: steel, aluminium (Al) and tin (Sn). Aluminium and tin are typical components of bearing metals, e.g. AlSn20, AlSn20Cu1. The friction coefficients were obtained by experimental testing in the Institute of Fluid-Flow Machinery. Other material constants of the model were estimated with the aid of the subject literature, and they are collected in Table 1.

The pin is described by a one-dimensional continuum. In the model, the contact region, the normal pressure and the sliding velocity are known in advance. The normal pressure is given by $p_n = 31.5$ MPa. The disc moves with the sliding velocity $V = 3.548$ m/s with respect to the pin. The wear velocity of the pin is defined by v^+, see Equation (6). The depth of the material removed u_n^+ in the time of sliding $t \in \langle 0, 4\ \text{s})$ is defined by Equation (4). Abraded mass from the pin is given by

Table 1. Material constants used in the calculations of the pin-on-disc test.

Material	Friction coefficient μ_p	Wear intensity i_p (MPa^{-1})	Mass density ρ_p (kgm^{-3})	Thermal diffusivity K (m^{2}s^{-1})	Coefficient of heat transfer b (s^{-1})
Steel	0.1	$0.03 \cdot 10^{-6}$	$7.9 \cdot 10^3$	$13.8 \cdot 10^{-6}$	$2.26 \cdot 10^{-3}$
Al	0.07	$0.5 \cdot 10^{-6}$	$2.7 \cdot 10^3$	$85.9 \cdot 10^{-6}$	$3.46 \cdot 10^{-3}$
Sn	0.026	$2.0 \cdot 10^{-6}$	$7.3 \cdot 10^3$	$39.8 \cdot 10^{-6}$	$4.96 \cdot 10^{-3}$

$$m_{\text{pin}} = \rho_p \int_{A_p} u_n^+ dS = \rho_p i_p p_n V t A_p, \tag{46}$$

where ρ_p is the mass density of the pin, i_p is the wear intensity of the pin, $A_p = 200 \cdot 10^{-6}$ m^2 is the cross-section area of the pin.

Figure 6 shows the removed mass from the pin with respect to the sliding time. The least abraded mass is for the steel pin. For the short duration contact, the temperature does not change significantly the wear intensity. Therefore, the coefficients i_p deal with the fixed room temperature 20°C.

The values of the removed mass from the pin are very small numbers, but the abrasion effects are significant, and they can be observed by a visual inspection of the contact surfaces. In engineering, the human eye is a sensitive instrument to detect the abrasion effects. The surfaces are smoothed, and a surface roughness decreases compared with the initial surfaces.

The heat generated at, or very close to the contact, is transferred away into the bulk of of the pin and the disc, and is conducted into the surrounding. The equation of heat conduction in the pin with included heat transfer into the surrounding has the following form

$$\frac{\partial T}{\partial t} = K \frac{\partial^2 T}{\partial y^2} - b(T - T_a), \tag{47}$$

$$T = T(y, t), \quad y \in \langle 0, \infty), \quad t \in \langle 0, t_1 \rangle, \tag{48}$$

where T is the temperature of the pin, y is the coordinate along the pin, $T_a = 20$°C is the room temperature, K is the thermal diffusivity coefficient, b is the coefficient of Newton's heat transfer rule. The initial and boundary conditions of the heat conduction problem are given by

$$T(y, t = 0) = T_a, \quad T(y \to \infty, t) = T_a, \tag{49}$$

$$k \frac{\partial T}{\partial y}(y = 0, t) = q_f = w_p p_n V, \tag{50}$$

where k is the thermal conductivity coefficient, q_f is the frictional heat flux entering into the pin, see Equation (12). It was assumed that 90% of the friction power transforms into heat, see Equation (8). After substituting the isotropic friction force (11)

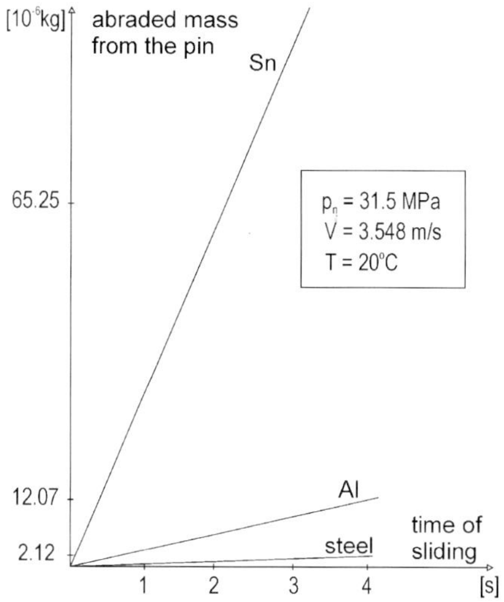

Fig. 6. The progress of the abraded mass from the pin after 0 to 4 seconds of sliding for three types of materials: steel, aluminum and tin.

into the energy constraints (8) we obtain a quantitative restriction on the frictional heat intensity coefficient w_p. The total frictional heat is divided into two parts, a part of heat which is transferred into the pin (40%) and a part of heat transferred into the disc (60%).

Figure 7 presents temperature plots in the pin with respect to the length coordinate of the pin, i.e. with respect to the distance from the contact point. The temperatures strongly decrease along the pin axis since the heat is conducted into the surrounding and the time of sliding is short. The highest temperatures occur at the contact point.

Figure 8 illustrates contact temperatures produced by friction, i.e.

$$T_{\text{contact}} = T(y = 0, t). \tag{51}$$

The contact temperatures increase with respect to the sliding time. The temperature curves correspond to the short time of sliding.

The contact temperature in the steel pin reaches very high value. In the experiments, sparking and associated color changes (red color) of the contact surface have been observed in the steel pin.

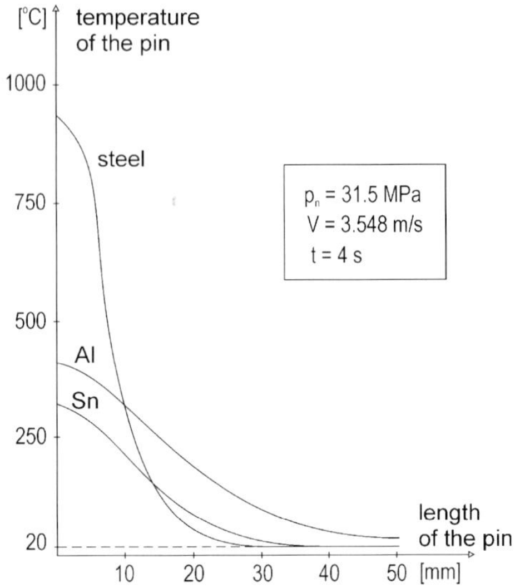

Fig. 7. Temperature distributions in the pin plotted with respect to the distance from the contact point to the top of the pin after 4 seconds of sliding.

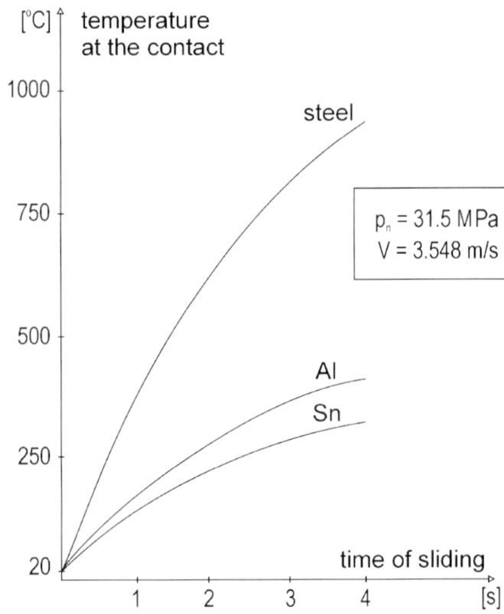

Fig. 8. The progress of the contact temperatures for the time of sliding from 0 to 4 seconds.

7 Conclusions

(a) Irreversible changes in shapes of the bodies and the increase of gaps between the contacting solids are principal results of abrasive wear. The sizes of the wearing out bodies are gradually reduced in the process of abrasion.

(b) The amount of the removed material can be estimated with the aid of the wear depth and the wear depth profile. They are useful measures of the wear process, and they can be described with the aid of the wear laws. The definition of the gap includes deformations of the bodies and evolutions of the wear profiles.
(c) The measures of wear (i.e. the wear depth, the reduced dimensions, the removed mass, etc.) have non-zero values as long as an observable amount of the material is removed. Usually, but not always (see Section 6), a significant period of the sliding time must be taken into account. Due to this, rather long simulation times are needed.
(d) In this study, classical strong and weak formulations of the contact problems are extended by including the interfacial layer of wear debris with translational and rotational degrees of freedom. The wear particles transmit loads, displacements and heat at the contact interface.
(e) The presence of wear debris implies modifications of friction and wear processes as sliding progress. The practical importance of the wear particles depends on a tribological system under investigations. It is extremely important in contacts where formation of loose particles is permanent. For example, the wear particles are always present inside the contact region in bearings, in fretting processes (i.e. between contacting surfaces subjected to vibrations or cyclic stressing) and in prostheses of human joints.

Acknowledgement

The author thanks the Alexander von Humboldt Foundation for financial support.

References

1. Agelet de Saracibar C, Chiumenti M (1999) On the numerical modeling of frictional wear phenomena, *Comput. Methods Appl. Mech. Engrg.*, 177:401-426.
2. Bahadur S (1978) Wear research and development, *ASME Journal of Lubrication Technology*, 100:449-454.
3. Berker A, Van Arsdale WE (1992) Phenomenological models of third body reology. In: Dowson D et al. (Eds), *Wear Particles: From the Cradle to the Grave*. Elsevier, Amsterdam.
4. Bufler H (1979) Zur Struktur der Gleichungen der Elastokinetik und Elastostabilität, *Z. Angew. Math. Mech.*, 59:73-78.
5. Dowson D, Taylor CM, Childs THC, Godet M, Dalmaz G (Eds) (1992) *Wear Particles: From the Cradle to the Grave*. Elsevier, Amsterdam.
6. Elrod HG (1988) Granular flow as a tribological mechanism - a first look. In: Dowson D et al. (Eds), *Interface Dynamics*. Elsevier, Amsterdam.
7. Elrod HG, Brewe DE (1992) Numerical experiments with flows of elongated granules. In: Dowson D et al. (Eds) *Wear Particles: From the Cradle to the Grave*. Elsevier, Amsterdam.

8. Eringen C (1975/76) *Continuum Physics*, Vol.1-4. Academic Press, New York.
9. Fouvry S, Kapsa P, Zahouani H, Vincent L (1997) Wear analysis in fretting of hard coatings through a dissipated energy concept, *Wear*, 203-204:393-403.
10. Glodež S, Ren Z, Flašker J (1998) Simulation of surface pitting due to contact loading, *Int. J. Numer. Meth. Eng.*, 43:33-50.
11. Godet M (1984) The third-body approach: a mechanical view of wear, *Wear*, 100:437-452.
12. Godet M (1988) Modeling of friction and wear. In: Ling FF, Pian CHT (Eds), *Approaches to Modeling Wear*. Springer, Berlin.
13. Iordanoff I, Seve B, Berthier Y (2002) Solid third body analysis using a discrete approach: influence of adhesion and particle size on macroscopic properties, *ASME Journal of Tribology*, 124:530-538
14. Iordanoff I, Berthier Y, Descartes S, Heshmat H (2002) A review od recent approaches for modelling solid third bodies, *ASME Journal of Tribology*, 124:725-735.
15. Kalker JJ (1992) Modification of two-body contact conditions to account for the third body. In: Dowson D et al. (Eds), *Wear Particles: From the Cradle to the Grave*. Elsevier, Amsterdam
16. Ko DC, Kim DH, Kim BM (2002) Finite element analysis for the wear of Ti-N coated punch in the piercing process, *Wear*, 252:859-869.
17. Laursen TA (2002) *Computational Contact and Impact Mechanics: Fundamantals of Modeling Interfacial Phenomena in Nonlinear Finite Element Analysis*. Springer, Berlin.
18. McColl IR, Ding J, Lee SB (2004) Finite element simulation and experimental validation of freeting wear, *Wear*, 256:1114-1127.
19. Meng HC, Ludema KC (1995) Wear models and predictive equations: their form and content, *Wear*, 181-183:443-457.
20. Ohmae N, Tsukizoe T (1980) Analysis of a wear process using the finite element method, *Wear*, 61:333-339.
21. Peng Z, Kirk TB (1999) The study of three-dimensional analysis techniques and automatic classification systems for wear particles, *ASME Journal of Tribology*, 121:169-176.
22. Põdra P, Andersson S (1999) Simulating sliding wear with finite element method, *Tribology International*, 32:71-81.
23. Podsiadlo P, Stachowiak GW (2000) Scale-invariant analysis of wear particle surface morphology I: Theoretical background computer implementation and technique testing; II: Fractal dimension; III: Pattern recognition, *Wear*, 242:160-179, 180-188, 189-201.
24. Podsiadlo P, Stachowiak GW (2005) Development of advanced quantitative analysis methods for wear particle characterization and classification to aid tribological system diagnosis, *Tribology International*, 38:887-897.
25. Rabinowicz E (1995) *Friction and Wear of Materials*, second edition. John Wiley & Sons, New York, pp.161-162.
26. Sfantos GK, Aliabadi MH (2006) Application of BEM and optimization technique to wear problems, *Int. J. Solids Structures*, 43:3626-3642.
27. Sherrington P, Hayhurst P (2001) Simultaneous observation of the evolution of debris density and friction coefficient in dry sliding steel contacts, *Wear*, 249:182-187.
28. Shillor M, Sofonea M, Telega JJ (2004) Quasistatic viscoelastic contact with friction and wear diffusion, *Quart. J. Appl. Math.*, 62:379-399.

29. Steinmann P, Stein E (1997) A unifying treatise of variational principles for two types of micropolar continua, *Acta Mechanica*, 121:215-232.
30. Strömberg N (1999) A Newton method for three-dimensional fretting problems, *Int. J. Solids Structures*, 36:2075-2090.
31. Suh NP (1986) *Tribophysics*. Prentice-Hall, New Jersey.
32. Telliskivi T (2004) Simulation of wear in a rolling-sliding contact by a semi-Winkler model and the Archard's wear law, *Wear*, 256:817-831.
33. Wriggers P (2002) *Computational Contact Mechanics*. John Wiley & Sons, Chichester.
34. Zienkiewicz OC (1971) *The Finite Element Method in Engineering Science*. McGraw-Hill, London.
35. Zmitrowicz A (1987) A thermodynamical model of contact, friction and wear: I governing equations; II constitutive equations for materials and linearized theories; III constitutive equations for friction, wear and frictional heat, *Wear*, 114:135-168, 169-197, 199-221
36. Zmitrowicz A (2000) Variational approach to contact, friction and wear problems, *Z. Angew. Math. Mech.*, 80(suppl.2):S463-S464
37. Zmitrowicz A (2000) Strong and weak formulations of contact problems with various laws of friction, frictional heat and wear, *Machine Dynamics Problems*, 24:209-222
38. Zmitrowicz A (2002) Wear profiles and wear debris taken into account in variational approaches to contact mechanics. In: Mang HA, Rammerstorfer FG, Eberhardsteiner J (Eds), *Proceedings of Fifth World Congress on Computational Mechanics*. Vienna University of Technology, http://wccm.tuwien.ac.at
39. Zmitrowicz A (2005) Wear debris: a review of proprties and constitutive models, *Journal of Theoretical and Applied Mechanics*, 43:3–35
40. Zmitrowicz A (2006) Wear patterns and laws of wear - a review, *Journal of Theoretical and Applied Mechanics*, 44:219–253
41. Zmitrowicz A (2006) Models of kinematics dependent anisotropic and heterogenous friction, *Int. J. Solids Structures*, 43:4407-4451

Modelling of Grain Boundaries Contact in Metals

N.M. Vlasov and I.I. Fedik

Scientific Research Institute, Scientific Industrial Association "Luch", Zheleznodorozhnaya 24, Podolsk Moscow Region, 142100 Russia
E-mail: iifedik@luch.podolsk.ru

Abstract. Internal stresses occur within a material in the presence of non-uniform deformation. The main types of the internal stresses are the thermal and residual ones and fields of structural defects as well. These stresses have an essential effect on the diffusion processes kinetics. In this case change of the strength material properties takes place. The properties degradation is accompanied by damaging and failing the material. The physical mechanisms underlying changes of properties include, for example, decreasing of surface fracture energy, stress corrosion cracking, and hydrogen embrittlement. The diffusion process is described by a non-stationary equation of a parabolic type under both initial and boundary conditions. The purpose of this paper is simulating the material damage as a result of running the diffusion processes. Triple grain boundaries are considered as structural defects. They serve as stress concentrators under external loadings. This is caused by the orientation dependence of elastic and thermophysical characteristics of the contiguous grain material. The dilatation field of considered defects depends logarithmically on the radial coordinate. Such a dependence enables one to obtain an exact analytical solution for the task on hydrogen segregation kinetics. Analytical relations for the field of atomic hydrogen concentration near the triple grain boundaries are given. If the concentration of hydrogen atoms exceeds the solubility limit at a given temperature, hydride phases are formed in some metals (e.g., Zr). Hydride growth kinetics in the stress field of structural defects under study is considered. The changes of the volume hydride are accompanied by microcrack formation along the grain boundaries.

Key words: Internal stresses, diffusion kinetics, damage simulation.

1 Introduction

The strength of structural components in modern structures depends on the level and character of internal stresses that are introduced by non-uniform temperature distribution (thermal stresses) [1, 2], irreversible deformation (residual stresses), and structural defects that may be present [3]. The stresses change the strength properties as a result of diffusion processes present. The physical mechanisms underlying

Peter Wriggers and Udo Nackenhorst (eds.), IUTAM Symposium on Computational Contact Mechanics, 333–339.

changes of properties include, for example, decreasing of surface fracture energy, stress corrosion cracking, and hydrogen embrittlement [4, 5]. These mechanisms also include diffusion of interstitial impurities (e.g., hydrogen and oxygen) in the internal stress field. Hydrogen is dominant among these interstitial impurities. That dominance is caused by the high diffusion mobility of hydrogen atoms in a wide range of temperature. At room temperature, for example, the diffusion coefficient of hydrogen atoms is greater by several orders of magnitude than the diffusion coefficients of other substitutional and/or interstitial impurities. Triple grain boundaries are considered the most pertinent structural defects. They serve as stress concentrators under dynamic and temperature loadings. This is caused by the orientation dependence of thermo- physical and elastic characteristics of contiguous grain materials. Considered defects contact directly the grain boundaries and play a dominant role in damaging them. A possible mechanism of the formation of microcracks along grain boundaries can be described as follows. When the breaking of grain boundaries decreases, cavities of cylindrical shape are formed around triple grain boundaries, which are linear defects. Hydrogen atoms diffuse into the cavities, where H_2 forms rapidly and pressurizes the cavities greatly. Microcracks are formed under the internal pressure along the grain boundaries. The pressure of molecular hydrogen within the cavities results in the embrittlement of metal exposed to a hydrogen medium. In certain metals (e.g, Zr), hydride phases are formed near the triple grain boundaries. The change of the hydride volume (e.g., by 9–12% for Zr) results in microcracks developing along grain boundaries. The purpose of this study was to analyze the mechanism(s) of the embrittlement of metals exposed to a hydrogen medium resulting from the hydrogen pressure within the cavities and by the formation of hydride phases in metals with non-uniform properties.

2 Hydrogen Pressure in Cavities

The nuclei of cylindrically-shaped cavities are formed near triple grain boundaries when braking of grain-boundary slippage occurs. Hydrogen atoms diffuse to cavity boundaries and then pass into the metal per se. The pressure of molecular hydrogen within cavities increases and results in the opening of microcracks along contiguous grain boundaries. The kinetics of the process is described by a parabolic equation under corresponding initial and boundary conditions [6–8]:

$$\frac{1}{D}\frac{\partial C}{\partial t} = \Delta C + \frac{\nabla(C\nabla V)}{kT}, \qquad r_0 < r < R,$$
$$C(r,0) = C_0, \quad C(r_0,t) = 0, \quad C(R,t) = C_0, \tag{1}$$

where D is the coefficient of hydrogen diffusion, k is the Boltzmann constant, T is temperature (K), r_0 is the internal cavity radius, R is the outer radius of cavity surrounding, C_0 is the average concentration of hydrogen atoms, and V is the potential of hydrogen atom interaction with triple grain boundaries. Initially the concentration of hydrogen atoms is equal to an average value. Such a concentration exists on the outer cavity boundary as well. It is caused by the fact that, when the hydrogen atoms migrate into the cavity near triple grain boundaries with a positive dilatation, hydrogen from neighbouring triple grain boundaries with a negative dilatation moves to its outer boundary ($r = R$). A boundary condition as $r = r_0$ means that hydrogen atoms quickly form molecular hydrogen on the internal cavity boundary. Consequently, the concentration of atomic hydrogen is zero. Interaction potential V depends logarithmically on coordinates

$$V = -A \ln \frac{r}{R}, \qquad A = \frac{\mu\omega(1+\nu)}{3\pi(1-\nu)}\delta\upsilon, \tag{2}$$

where μ is the shear module, ν is Poisson's ratio, ω is the rotation vector module of triple grain boundaries, and $\delta\upsilon$ is the crystal volume change due to placing hydrogen atoms. The relations (see Equation (2)) are given by unessential constants. The latter does not influence the process of atomic hydrogen diffusion, because the diffusion flux is proportional to ∇V ($\nabla V = 0$, where V is a harmonic function). The $\delta\upsilon$ is constant for all materials and equal to 3×10^{-30} m^3 [2]. For triple grain boundaries with positive dilatation (tension stress), ω is taken to be positive. The negative value of V ($\delta\upsilon > 0, \omega < 0$) corresponds to the hydrogen atom attraction to the triple grain boundary line.

Equation (1) is simplified when Equation (2) is taken into consideration,

$$\begin{gathered}\frac{1}{D}\frac{\partial C}{\partial t} = \frac{\partial^2 C}{\partial r^2} + \frac{1+\alpha}{r}\frac{\partial C}{\partial r}, \quad r_0 < r < R, \quad \alpha = \frac{A}{kT} < 0,\\ C(r,0) = C_0, \quad C(r_0,t) = 0, \quad C(R,t) = C_0. \end{gathered} \tag{3}$$

Non-dimensional task parameter α characterizes a ratio of binding energy of a hydrogen atom with the triple grain boundaries to thermal motion energy. In the case of $|\alpha| \ll 1$, the stress field of triple grain boundaries is the small perturbation of the basic diffusion flux of hydrogen atoms. For $|\alpha| \gg 1$, the main contribution to the diffusion flux of hydrogen atoms is given by the stress field of triple grain boundaries. When $|\alpha| \approx 1$, diffusion fluxes are comparable at the expense of concentration gradients and potential V. Thus for $\alpha = -1$, the field of atomic hydrogen concentration near the cylindrical shape cavity is derived from the relation

$$\begin{gathered}\frac{1}{D}\frac{\partial C}{\partial t} = \frac{\partial^2 C}{\partial r^2}, \quad r_0 < r < R,\\ C(r,0) = C_0, \quad C(r_0,t) = 0, \quad C(R,t) = C_0, \end{gathered} \tag{4}$$

which becomes

$$\frac{C_0 - C}{C_0} = \frac{R-r}{R-r_0} - \frac{2}{\pi}\sum_{n=1}^{\infty} \frac{\sin \frac{\pi n(r-r_0)}{(R-r_0)}}{n} \exp\left[-\frac{\pi^2 n^2 D t}{(R-r_0)}\right]. \tag{5}$$

Thus, the stress field of triple grain boundaries transforms the task from a cylindrical symmetry into a plain one. This means that the cylindrically symmetrical field of hydrogen atom concentration is formed according to the plain symmetry law. The change of such a symmetry increases the formation rate of the concentration profile from hydrogen atoms that follows directly from diffusion Equation (3).

As the number of hydrogen atoms is much less than the number of vacancies in the cavity, the laws of an ideal gas are used to determine the pressure of molecular hydrogen, p. If the cavity size does not change, the diffusion flux of hydrogen atoms is equal to

$$|\mathbf{j}| = \frac{DC_0}{R - r_0}\left\{1 + 2\sum_{n=1}^{\infty} \exp\left[-\frac{\pi^2 n^2 Dt}{(R - r_0)^2}\right]\right\}. \tag{6}$$

The pressure of molecular hydrogen in the cavity for time τ can be obtained by the following simple transformations

$$p = nkT = \frac{DC_0 N_0 kT}{(R - r_0)r_0}\left\{\tau + \frac{2(R - r_0)^2}{\pi^2 D}\sum_{n=1}^{\infty}\frac{1 - \exp\left[-\frac{\pi^2 n^2 D\tau}{(R-r_0)^2}\right]}{n^2}\right\}, \tag{7}$$

where n is the number of hydrogen molecules in a volume unit, and N_0 is the number of atoms in a volume unit. The number N_0 converts the non-dimensional hydrogen atom concentration into a dimensional one (at/m^3). At $D\tau/(R - r_0)^2 \gg 1$, Equation (7) becomes

$$p = \frac{DC_0 N_0 kT}{(R - r_0)r_0}\left\{\tau + \frac{(R - r_0)^2}{3D}\right\},$$

$$\sum_{n=1}^{\infty}\frac{1}{n^2} = \frac{\pi^2}{6}. \tag{8}$$

For $D \approx 10^{-9}$ m^2/s, $R/r_0 \approx 10^2$, $r_0 \approx 2 \times 10^{-6}$ m, $C_0 \approx 10^{-4}$ at.%, $kT \approx 4 \times 10^{-21}$ J, $N_0 \approx 10^{29}$ at/m^3, $\tau \approx 10^3$ s, we obtain $p \approx 100$ MPa. Such a pressure is able to "open" the grain boundaries contiguous with the cavity down to the formation of microcracks (stress concentrators). Thus the pressure of molecular hydrogen in the cavities near triple grain boundaries can cause the embrittlement of the metal. The above estimation for the pressure of molecular hydrogen clearly has a random character, since many pertinent factors were not taken into consideration. These factors could include the degree of hydrogen dissociation on the cavity boundary, the effect of pressure on the hydrogen ingress into the cavity, the drop of the volume concentration of atomic hydrogen as its migration proceeds into the cavity, and so forth.

3 Hydride Phase Formation

The hydrogen embrittlement of metals is related to the formation of hydride phases as well. A hydride nucleus forms when the concentration of hydrogen atoms near triple grain boundaries with the positive dilatation reaches a solubility limit at a given temperature. The further growth of hydrides is limited by the diffusion supply of hydrogen atoms. The hydride formation is the phase transfer of a second series, which occurs without forming nuclei. This holds true for metals that are free of defects. However, local segregations of hydrogen atoms are formed near structural defects and convert into hydride at the moment of reaching the solubility limit. Further growth of such a nucleus is caused by the diffusion of hydrogen atoms. In this case, the concentration of hydrogen atoms on the transient boundary changes dramatically: $C = C_1$ for a new phase and $C = C_2$ in the surrounding matrix ($C_1 > C_2$, $C_2 < C_0$, where C_0 is the average concentration of hydrogen atoms). Physically this means that a hydride boundary instantly captures atomic hydrogen from the solution and supplies it to the new phase with a higher hydrogen concentration. The influence of a stress field is that, in addition to the concentration gradient, the hydrogen atoms migrate due to the stress field gradient of triple grain boundaries. Because of this, the velocity of the hydride phase boundary increases. The relationship defining the hydride growth kinetics around triple grain boundaries can be written as [3]

$$\begin{aligned}
&\frac{1}{D}\frac{\partial C}{\partial t} = \frac{\partial^2 C}{\partial r^2} + \frac{1+\alpha}{r}\frac{\partial C}{\partial r},\\
&C(R_1,t) = C_2, \quad C(r,0) = C_0\ (r \geq R_0), \quad C(\infty,t) = C_0,\\
&(C_1 - C_2)\frac{dR_1}{dt} = D\left(\left|\frac{dC}{dr}\right| + \left|\frac{C\alpha}{r}\right|\right)_{r=R_1},
\end{aligned} \tag{9}$$

where R_0 is the radius of hydride phase nucleus, and R_1 is the current hydride radius. The distance between the lines of triple grain boundaries is assumed to exceed the typical size of hydride phase considerably. This allows for consideration of the new phase growth in the unlimited matrix. Changing the hydride size obeys the law $R_1(t) = \beta\sqrt{Dt}$, where β is a non-dimensional parameter, the value of which is taken from the mass balance equation on an interphase boundary. In relation to "a stationary interphase boundary", a quadratic equation was obtained for determining the parameter β

$$\beta^2 - \frac{2\beta}{\sqrt{\pi}}\left|\frac{C_2 - C_0}{C_1 - C_2}\right| - \left|\frac{2C_2}{C_1 - C_2}\right| = 0. \tag{10}$$

When $\alpha = 0$ in Equation (9), under invariable initial and boundary conditions we obtained

$$\frac{1}{D}\frac{\partial C}{\partial t} = \frac{\partial^2 C}{\partial r^2} + \frac{1}{r}\frac{\partial C}{\partial r},$$

$$C(R_1, t) = C_2, \quad C(r, 0) = C_0 \ (r \geq R_0), \quad C(\infty, t) = C_0,$$

$$(C_1 - C_2)\frac{dR_1}{dt} = D\left(\frac{dC}{dr}\right)_{r=R_1}. \tag{11}$$

In this case, the growth of hydride results from the hydrogen atom concentration gradient. A transcendental equation can be obtained from a mass balance equation on the interphase boundary for determining parameter β_1, according to the expression $\beta_1\sqrt{Dt}$

$$\beta_1 = \frac{2}{\sqrt{\pi}}\left|\frac{C_2 - C_0}{C_1 - C_2}\right| \frac{K_1\left(\beta_1\frac{\sqrt{\pi}}{2}\right)}{K_0\left(\beta_1\frac{\sqrt{\pi}}{2}\right)}, \tag{12}$$

where $K_0(x)$ and $K_1(x)$ are modified Bessel functions of the second series of zeroth and first orders respectively. Assume $C_0 \approx 2.0 \times 10^{-4}$, $C_2 = 10^{-4}$, $C_1 = 3.0 \times 10^{-4}$ (at.%) without limiting community. This gives $\beta = 1.3$ and $\beta_1 = 0.8$ for the non-dimensial parameters. The stress field near triple grain boundaries clearly accelerates the growth of the hydride phase. The essential change of volume hydride wedges off the grain boundaries, forming microcracks. The hydrogen embrittlement of metal is observed at macroscopic scales as a result of the formation and growth of hydrides.

4 Conclusion

Internal stresses have an essential effect on the character of diffusion processes and change the strength of the metals. The kinetics of the processes is described by a parabolic equation under corresponding initial and boundary conditions. The formation of microcracks along the grain boundaries occurs when hydrogen atoms interact with the metal. A possible mechanism of hydrogen embrittlement is as follows. Under some conditions (for example, under braking of grain-boundary slippage), cylindrically-shaped cavities are formed near triple grain boundaries. Molecular hydrogen creates an essential pressure within the cavities, followed by formation of microcracks along grain boundaries. In some metals, hydride phases are formed near triple grain boundaries. The change of the hydride volume also results in the formation of microcracks along grain boundaries. Because of this, structural defects like triple grain boundaries, when interacting with hydrogen contents, cause the formation of microcracks along grain boundaries. The embrittlement of metal is observed macroscopically when interacting with hydrogen. The results of theoretical analysis are employed to example the hydrogen embrittlement of metals with non-uniform properties.

References

1. Melan, E., Parkus, H., *Wärmespannungen infolge stationärer Temperaturfelder*. Springer-Verlag, Wien, 1953.
2. Parkus, H., *Instationäre Wärmespannungen*. Springer-Verlag, Wien, 1959.
3. Teodosiu, C., *Elastic models of Crystal Defects*. Springer, Heidelberg, 1982.
4. Smith, E., The stress in a zirconium alloy due to the hydride precipitation misfit strains. *Journal of Materials Science*, 32:1127, 1997.
5. Varias, A.G., Massih, A.R., Simulation of hydrogen embrittlement in zirconium alloys under stress and temperature gradients. *Journal of Nuclear Materials*, 279:273, 2000.
6. Vlasov, N.M., Gontar, A.S., Zaznoba, V.A., Decomposition of solid solution in an alloy under high plastic strain. *Technical Physics*, 46:569, 2001.
7. Vlasov, N.M., Fedik, I.I., Hydrogen segregation in the of threefold junctions of grain boundaries. *International Journal of Hydrogen Energy*, 22:921, 2002.
8. Vlasov, N.M., Fedik, I.I., Structural and impurity traps for hydrogen atoms. *International Journal of Hydrogen Energy*, 31:265, 2006.

Segregation of Particulate Material Using the Discrete Element Method

Hashem Alkhaldi and Peter Eberhard

Institute of Engineering and Computational Mechanics, University of Stuttgart, Pfaffenwaldring 9, 70569 Stuttgart, Germany
E-mail: {alkhaldi,eberhard}@itm.uni-stuttgart.de

Abstract. Particulate or granular materials are an integral part of our environment. Due to their fascinating physical properties and their wide variety of applications in industrial and technological processes, they are the subject of intensive studies and scientific research. The related studies are often based on numerical simulations. It is still considered as challenging to investigate computationally complicated phenomena of dense granular systems and, therefore, parallel computation techniques should be applied. In this study, we investigate particle segregation of dry granular mixtures using a commonly used machine in granular processing: a tumbling screening machine. Screeners are generally considered as sifting units which are rotated as material is fed into their interior. The system under consideration can be modelled as a multibody system which consists basically of the machine itself and the system of the mixed particles. Fine particles usually fall through the sieve openings while oversized particles are ejected out through certain outlets located around the machine body The concept of the discrete element method (DEM) that considers the motion of each single particle individually is applied in this study. Dynamical values of particle positions, velocities and orientations are tracked at each time step of the simulation. Particle-to-particle and particle-to-wall/sieve collisions will appear under the tumbling action of the rotating structure. The soft particle method, which is one of the most common discrete element simulation techniques, is applied. In this method, the normal and frictional forces between particles themselves and particles and machine walls are calculated according to a penalty method. Results from numerical computations are shown and interpreted.

Key words: Discrete element method, granular material, screening, segregation, sorting machine.

1 Introduction

Particle separation phenomena are important in granular media studies [1, 12]. Screeners and separators are used in a large number of industrial applications requiring separation and classification of powders or other bulk materials by particle

Peter Wriggers and Udo Nackenhorst (eds.), IUTAM Symposium on Computational Contact Mechanics, 341–355.

size as well as separation of particles by density, magnetic properties or electrical characteristics. The operation of particle separation is divided into two main categories, continuous and batch operations. In continuous operation, the particles are continuously fed into the separation unit during the whole separation process. This type of particle separation is usually called *screening*. On the other hand, batch operation is used if the particulate material is charged only once. This kind of batch separation is commonly described by the term *sieving*.

Jansens and Glastonburys work in 1967 is among the earliest research in studying particle screening phenomena [6]. They analyzed the dynamics of a screening process and studied the factors that affect the screening performance. Over a certain range of operating variables, the screening efficiency of two types of particles over a vibrating screen has been observed in [15]. The variables include flow rate, deck angle, angular velocity and mesh size. The results show that the separation process is sensitive to the operating variables.

The discrete element method (DEM) was early proposed by Cundall and Strack [5] and was adapted in [7, 8] to be used for the analysis of the internal dynamics of tumbling mills. Analytical works, e.g. [9, 10], consider particles with or without adhesive and frictional forces using the penalty approach. Since MD simulations are very time consuming, many researchers use supercomputers and parallel programming to simulate large systems of particles on longer time scales, e.g. [11]. Therefore, a parallel computation has the potential to alleviate current computer-based limitations, allowing much larger granular systems with greater physical reality to be analyzed. Replicated data and spatial decomposition algorithms are among the most frequently used parallel techniques in molecular dynamics. Load balancing is also an important issue that should be taken into consideration when running parallel computations. It refers to the practice of distributing work among different processors such that all processes are kept busy all of the time.

Screeners are sifting units which are rotated as material is fed into their interior. The finer particles should fall through the sieve opening and oversized particles are thrown out through certain outlets, see [2]. Screening round particles of different sizes using a rotary round deck separating machine is the main problem of interest in this study. To reach our goal of better understanding the processes, a tumbler screening machine has been modeled. This interesting technical process of particle segregation is simulated by our discrete element simulation program. The efficiency of the sorting process is determined by counting the number of both right-sorted and undesired-gangue particles of the whole process at the different levels of the machine. Different simulations are conducted to study the effect of some parameters on the machine performance, e.g. the machine speed and the feeding rate of particles. Continuous feeding processes with different flow rates as well as batch sieving with a limited number of particles are analyzed and compared.

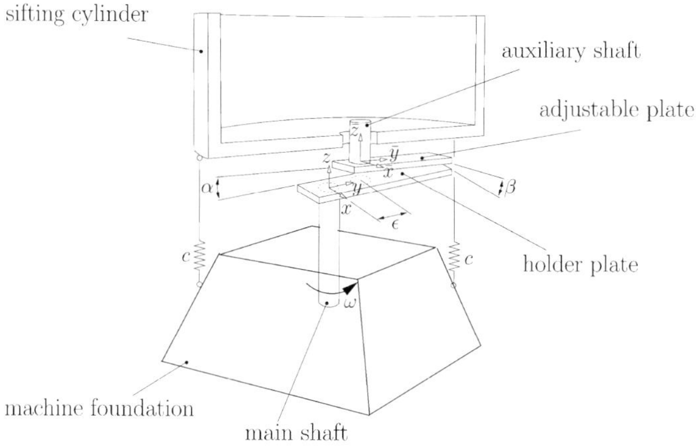

Fig. 1. Main parts of the TSM machine.

2 Description of the Tumbler Screening Machine

The investigated tumbler screening machine (TSM) consists of several parts (see Figure 1):

(a) Machine foundation which is heavy enough to hold the machine structure and contains the driving motor.
(b) Holder and adjustable plates with angles α and β to create the tumbling motion during rotation.
(c) Main and auxiliary rotating shafts with a certain margin of eccentricity in between to hold the entire body of the machine and to transmit the motion to the main sifting unit.
(d) Sifting meshes which represent the separating units and consist of a group of potentially different-sized sieves along the successive decks of the machine.
(e) Sifting vessel which is the main cylinder in which the bulk material is charged. This vertical rotating vessel can be designed as a uniform or stepped multi level oblique barrel with exits distributed at each sorting level and located at the outer periphery around the body of the machine, see Figure 2.
(f) Feeding container which is used in the simulation of continuous screening through which the TSM is charged by the particle flow. The size of the exit nozzle of the container determines the particle flow rate. The machines are charged from the top and in the middle of the highest level of the sifting unit.

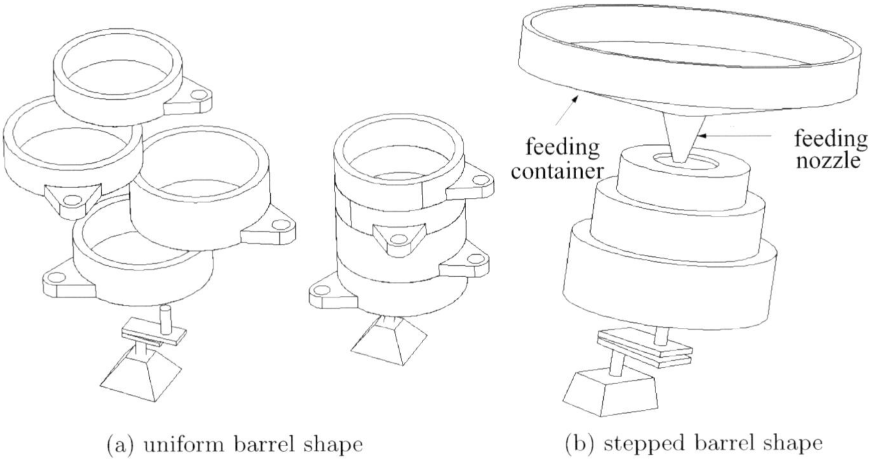

Fig. 2. Stepped and uniform TSM models.

3 Machine and Particle Movement

The machine is modelled as a multibody system, see e.g. [13]. It consists basically of the machine itself with different layers of sifters and the particulate material to be screened and classified. The basic motion of the tumbler screening machine is gyratory. The angular velocity ω is assumed to be constant during the separation process.

The TSM can be used for materials with particle sizes between micrometers and millimeters, such as those used in the chemical, pharmaceutical, food and plastics industries. The complicated three-dimensional screening motion of the machine is mainly independent of the material load. This screening motion is adjustable to improve the separation process. Experiments are necessary to find optimal process parameters. As the screening material is continuously fed into the center of the top screen, it is distributed over the entire screening surface from the center towards the periphery. The particles should travel in a spiral motion due to the combination of the tangential and radial inclination and motion of the screen and to the friction between the particles and the surface area of the mesh.

In the center, many fine particles will pass through the mesh holes. Towards the outside, the horizontal and vertical accelerations are increasing, causing the particles with near mesh size to be separated. The remaining oversized particles are carried to the outlets. This is repeated on every screen deck. After the granular materials have been tumbled in the rotating vessel and then segregated depending on their size, the particles are classified by outlets located in the walls.

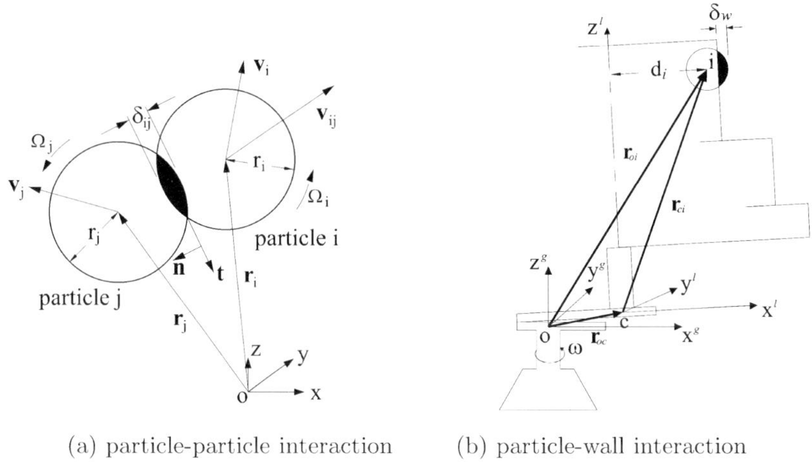

Fig. 3. Vector description of particle-to-particle and particle-to-wall contacts.

4 System Equations of Motion

Each particle has six degrees of freedom in translation and rotation. Considering particle i, the Newton–Euler equations of motion which are later solved numerically are

$$m_i \mathbf{a}_i = \mathbf{F}_i, \qquad \mathbf{I}_i \boldsymbol{\alpha}_i = \mathbf{M}_i, \tag{1}$$

where m_i and $\mathbf{I}_i$ are the mass and the inertia tensor of particle i, $\mathbf{F}_i$ and $\mathbf{M_i}$ are the force and torque vectors, $\mathbf{a}_i$ and $\boldsymbol{\alpha}_i$ are the linear and angular accelerations. Considering two bodies i and j in an N particle system, see Fig. 3a, the force $\mathbf{F}_i$ and torque $\mathbf{M}_i$ acting on particle i can be calculated as

$$\mathbf{F}_i = \sum_{j=1,\ j\neq i}^{N} \mathbf{F}_{ij} + m_i \mathbf{g}, \tag{2}$$

$$\mathbf{M}_i = \sum_{j=1,\ j\neq i}^{N} \mathbf{M}_{ij} = \sum_{j=1,\ j\neq i}^{N} \mathbf{r}_i \times \mathbf{F}_{ij}\ , \qquad i = 1, \ldots, N, \tag{3}$$

where $\mathbf{r}_i$ is the vector from the center to a point on the surface of particle i and $\mathbf{g}$ is the gravity vector.

The contact calculations are based on the soft-particle model. Applying the penalty method, the normal and frictional forces between the colliding particles can be determined. This model assumes that the contact forces result from an unphysical overlap between the bodies in contact, see [2, 3]. As a result of collisions, the particles will dissipate energy due to the normal and frictional contact losses. The normal force $\mathbf{F}_{ij}$ between two colliding particles i and j can be formulated as

$$\mathbf{F}_{ij} = (k_p \delta_{ij} + c_p \dot{\delta}_{ij})\mathbf{n}. \tag{4}$$

Substitution of $\mathbf{F}_{ij}$ in Equations (2) and (3) will yield the resultant contact force and torque acting on particle i due to the particle-particle interaction, where

$$\delta_{ij} = (r_i + r_j) - (\mathbf{r}_{oi} - \mathbf{r}_{oj})^T \mathbf{n}, \qquad \mathbf{n} = \frac{\mathbf{r}_{oi} - \mathbf{r}_{oj}}{|\mathbf{r}_{oi} - \mathbf{r}_{oj}|}, \tag{5}$$

k_p, c_p are the spring stiffness and damping coefficient of the penalty approach, $\mathbf{n}$ is the normal unit vector between the two particles, δ_{ij} and $\dot{\delta}_{ij}$ are the overlap and the relative velocity between the two colliding particles i and j. The system of forces and torques $(\mathbf{F}_i, \mathbf{M}_i)$ which acts on each individual particle is calculated in the global coordinate system.

In order to investigate the particle-to-mesh and particle-to-wall contacts, an inertial system $K_g\{o, x^g, y^g, z^g\}$ and a moving coordinate system $K_l\{c, x^l, y^l, z^l\}$ are defined. The body fixed coordinate system K_l is rotating with the tumbling cylinder with its origin located at the base of the auxiliary shaft on the inclined adjustable plate, see Figure 3b. The position of any particle i inside the rotating vessel can be expressed by

$$\mathbf{r}^g_{ci} = \mathbf{r}^g_{oi} - \mathbf{r}^g_{oc} \quad \text{or} \quad \mathbf{r}^l_{ci} = \mathbf{R}_{lg}(\mathbf{r}^g_{oi} - \mathbf{r}^g_{oc}), \tag{6}$$

where $\mathbf{r}^g_{oc}$ is the translation vector from point o to point c given in the coordinate system K_g and $\mathbf{R}_{lg}$ is the rotation matrix between the coordinate systems. The particle velocity can be expressed relative to the rotating coordinate system by differentiating Equation (6) once with respect to time

$$\dot{\mathbf{r}}^l_{ci} = \mathbf{R}_{lg}(\dot{\mathbf{r}}^g_{oi} - \dot{\mathbf{r}}^g_{oc}) + \dot{\mathbf{R}}_{lg}(\mathbf{r}^g_{oi} - \mathbf{r}^g_{oc}). \tag{7}$$

Substituting $\mathbf{v}^l_i = \dot{\mathbf{r}}^l_{ci}$ and $\mathbf{v}^g_i = \dot{\mathbf{r}}^g_{oi}$ in Equation (7) for particle velocity we get

$$\mathbf{v}^l_i = \mathbf{R}_{lg}(\mathbf{v}^g_i - \dot{\mathbf{r}}^g_{oc}) + \dot{\mathbf{R}}_{lg}(\mathbf{r}^g_{oi} - \mathbf{r}^g_{oc}), \tag{8}$$

where $\dot{\mathbf{r}}^g_{oc}$ and $\dot{\mathbf{R}}_{lg}$ represent the rate of change of the translation vector and the rotation matrix with respect to time.

5 Parallelizing the Problem

A spatial decomposition method for short-range parallel direct simulations can shorten the computation times. In this method a part of the physical simulation domain is assigned to each processor. This processor computes the forces on particles

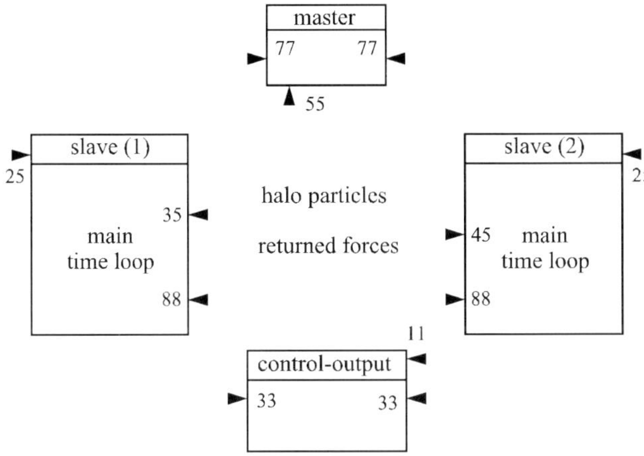

Fig. 4. Message exchange between master, slaves and control-output codes.

in its domain, determines forces from neighboring domains, and updates the positions and velocities of all particles within its box at each time step. Each processor tracks particles as they enter and exit its sub-domain. As the simulation progresses, processors exchange particles as they move from one sub-domain to another. In this case, particles are reassigned to new processors as they move through the physical domain.

In our work, the existing molecular dynamics serial code was rearranged and modified to work in parallel using a message passing library, see Figure 4. The parallel code is divided into three main parts:

- *Master program*: This part is responsible for generating the different slaves and control-output codes and for distributing the original data to them. It collects also the output data for each simulation time step and redistributes them to the different slaves according to the fixed predefined physical boundaries assigned to the processors.
- *Slaves program*: The slaves, which are generated according to the selected number of processors, receive the data from the master program at each simulation time step. The neighborhood-list is generated and the contact forces are then computed. The positions and velocities of all particles are updated and sent directly to the master program in order to redistribute them again at the next time step. Another copy of this updated data is sent also to the control-output program to collect and arrange them in the suitable form of the final output file required for animation. The slaves should also communicate with each other and exchange data belonging to those particles, which lie in the so-called *halo region* at the domain boundaries.

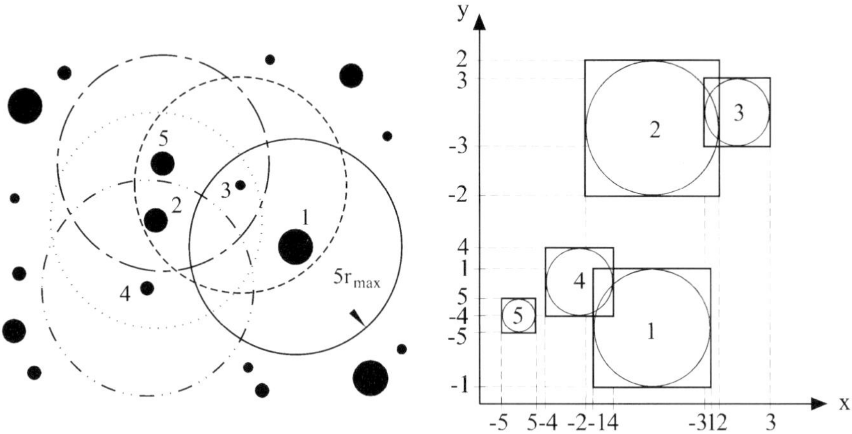

(a) Verlet particle domains (b) bounding boxes

Fig. 5. Verlet list and linked linear lists for neighborhood computations.

- *Control-output program*: It collects the data, which is sent from the different slaves at each output time step during simulation. These outputs are arranged in the final output file for animation.

Performance of the parallel program is of paramount importance in parallel computation. Measuring the performance is usually done by calculating the *speedup* of the parallel simulation which is a measure of how much faster the program runs on the parallel machine than it does on a serial one, i.e.

$$\text{Speedup} = \frac{\text{serial execution time}}{\text{parallel execution time}} \,. \tag{9}$$

6 Neighbor List Computations

There are some basic techniques used to speed up simulations of particulate systems. Representatives of these techniques are the *Verlet* approach and the *linked linear list* approach. For each method the neighboring particles are stored in a neighbor data-structure after the pre-sorting has been done. Once the neighbor list is built, examining it for possible interactions is much faster than checking all particles in the system. Since the update of these lists can be optimized, the essential calculation operation for collision detection can be reduced from $O(N^2)$ to an order proportional to the size of the system, i.e. $O(N)$, where N is the number of particles in the system.

- *Verlet approach*: As shown in Figure 5a a circular domain (spheres in the spatial case) is chosen. The radius of these circles is considered as the cutoff radius of the

Verlet approach, see [6]. This radius is often chosen as five times the maximum particle radius $r_{\max}$ in the system, i.e. the Verlet radius is $r_v = 5r_{\max}$. The optimal radius depends on the density of the whole system, the particle radii distribution and on the speed of the particles. Particles within these domains are considered as a neighbors to the original particle i in the center of these Verlet circles and added to the neighbor list. Refreshing the neighbor list e.g. every 100th time step is quite acceptable for relatively dense systems.

- *Linked linear list approach*: The idea of using this method is to enclose the particles within *bounding boxes*. The edges of the bounding box are aligned to the system axes x, y, and z (see Figure 5b) and are projected orthogonally to them. The begin and end of the bounding boxes are stored in sequence. The overlap of the bounding boxes should be checked at each time step according to some intersection rules, see [14]. The advantage of this method is that to update a nearly sorted list is a computationally cheap operation.

7 Particle-Mesh Contact Detection

The particle transportation through the mesh depends directly on the position of the specified particle with respect to the hole boundaries. In reality, the mesh is built of a combination of intersecting wires over which the particles will have a contact. It happens that the particle contacts with the mesh wires and is reflected in different directions around the hole. This will depend on the exact point on the wire where the particle hits. Taking into consideration all those precise contacts with the mesh wires will lead to many calculations and infeasible computational efforts. In order to reduce these computations while still reaching physical results, another less computationally expensive approach is suggested. This approach depends on defining a reduced mesh hole which is only used in simulation, to consider all contact possibilities and adjust the number of passed and reflected particles in a realistic way. This method assumes that the undersized particles pass through the mesh if they are small enough and their centers are over the reduced hole, otherwise they are reflected. The reduction in the mesh hole diameter can be adjusted to agree with experimental observations which are currently under investigation.

The particle-mesh contact should be detected in both directions x^l and y^l. The particle will be considered to be reflected if it is in contact with the mesh surface and its center is over the solid part and not over a reduced hole, otherwise it will fall through the mesh if it is small enough. At any level k, consider the points a and b on the borders of two successive holes on the mesh surface (see Figure 6)

$$\begin{aligned} x_a^l &= w + j(2w + g), \\ x_b^l &= x_a^l + g, \qquad j = -(n_k - 1), \ldots, j = (n_k - 2), \end{aligned} \tag{10}$$

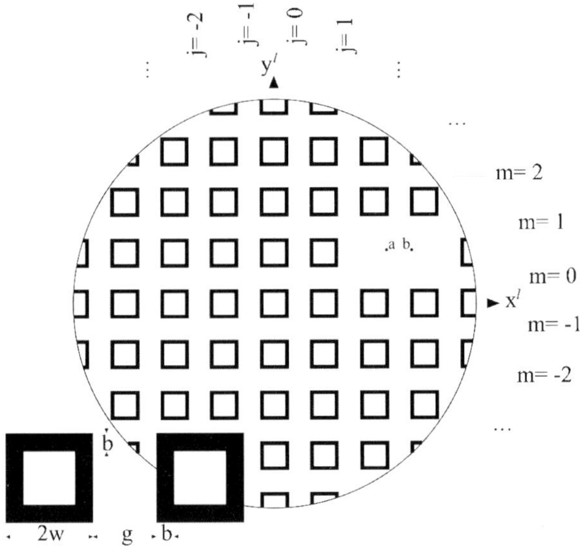

Fig. 6. Contact detection with mesh surface.

where j is an integer counter over the number of the gap aisles in the mesh surface along the x^l direction and n_k is the number of holes along the radius of the mesh plate. To check the particle position with respect to the mesh holes in x^l direction, a flag integer u_x is defined as

$$u_x = \begin{cases} 0 & \text{if } (x_a^l - b) \leq x_{ci}^l \leq (x_b^l + b), \\ 1 & \text{else,} \end{cases} \tag{11}$$

where x_{ci}^l is the x position of point i in the local coordinate system K_l and b is the adjustable hole clearance. This parameter is defined to be used as a simulation parameter to adjust the size of the mesh openings. Reduced holes with adjustable clearance b help in controlling the falling particles through the mesh holes during the simulation process. The adjustable parameter b is selected to adjust the simulation results in order to be in a good agreement with experimental observations. Similarly, the y^l direction is investigated yielding u_y. In order to determine the contact forces with the mesh surface, it is checked if the particle is in direct contact with the solid part of the mesh or if it is over a hole, i.e.

$$u_x u_y = \begin{cases} 1 & \text{particle potentially falls through sieve,} \\ 0 & \text{particle is reflected from sieve.} \end{cases} \tag{12}$$

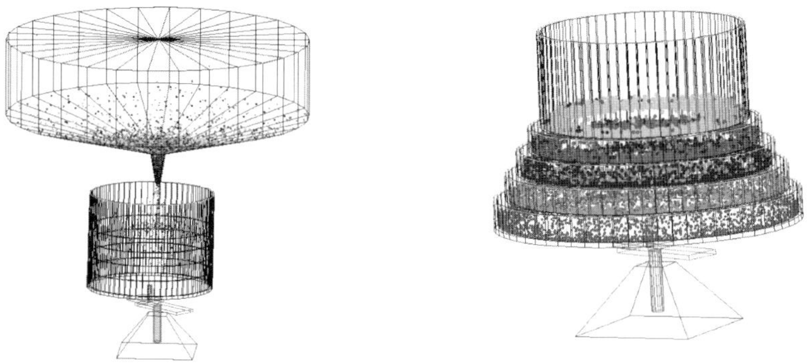

(a) continuous screening (b) batch sieving

Fig. 7. Particle segregation in continuous and batch modes of 10499 round particles.

8 Numerical Results and Discusion

Different simulations are performed to investigate screening phenomena in the rotating tumbling machine. Particle distribution and sifting rates of the separated particles are studied. In this study, the particles are investigated for both batch and continuous feeding in a uniform and a stepped multi level oblique vertical cylinder, see Figure 7. Some of the undersized particles are unintentionally forced to travel outside the machine together with the sorted ones. Those undesired particles, which are called gangue particles, reduce the efficiency of the machine. For an efficient separation process, the good selection of proper operating conditions reduces the number of these gangue particles. Mathematically, to measure machine efficiency, the characteristic value c can be defined as

$$c = \frac{100}{m} \sum_{k=1}^{m} c_k, \qquad c_k = \frac{s_k}{s_k + g_k}, \tag{13}$$

where c_k is the individual efficiency of layer k of the machine, s_k and g_k are the number of the oversized sorted and the undersized gangue particles at level k of the machine, respectively, m is the total number of barrels in the machine where $k = 1$ is the lowest level and $k = m$ is the highest one.

8.1 Influence of Machine Velocity

The screening machine can be run in different angular velocities. This angular speed has a great influence on the sifting rate of the mixed particles at each of the different sorting levels of the machine. The angular velocity ω should not be too high in order to avoid too fast motion of particles. Very high velocities of the particles will decrease

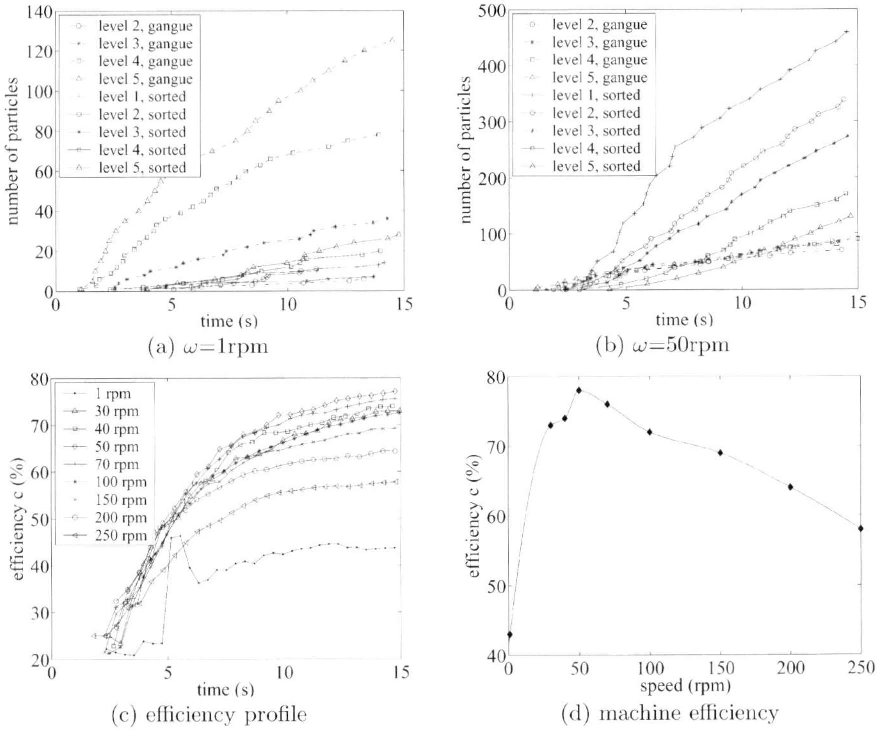

Fig. 8. Influence of rotational speed on the particle sifting rates at different sorting levels of the machine in continuous screening operation, $\alpha = 1°$, $\beta = 0.65°$, $w/g = 1$, $b = 0.8w$ and feeding rate 147 particles per second.

the separation rate of the mixture and reduce the chance for undersized particles to fall down through the mesh openings. On the other side the rotational speed should not be too slow since low separation rates and a big number of undesired gangue particles can then be expected due to the insufficient mixing of particles.

In order to catch the best separation performance, the optimal speed of the machine that maximizes the number of the sorted particles and minimizes the gangue ones, should be found. In order to do this, we simulate the problem for different angular velocities and measure the efficiency of each set of geometrical, contact and material parameters. It is observed that the sorting under low rotational velocity leads to bad separation rates, e.g. low efficiency of $c = 40\%$ is recorded for this case. The number of undesired gangue particles is even much higher than those of desired sorted ones in most of the machine levels, see Figure 8a.

On the other hand, a relatively high velocity of 50 rpm leads to better results and higher efficiency of about $c = 78\%$, see Figure 8b. Going further, to very high velo-

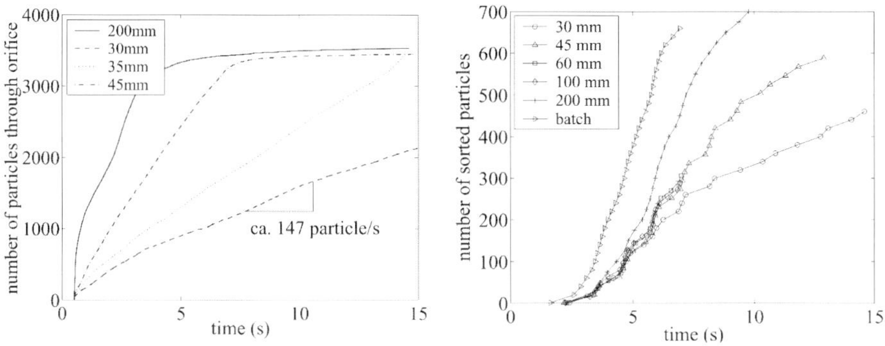

(a) flow rate for different orifice radii (b) lowest sorting level

Fig. 9. Influence of the feeding rate on the sorted number of particles in batch sieving and continuous screening operation modes, $\alpha = 1°$, $\beta = 0.65°$, $w/g = 1$, $b = 0.8w$ and $\omega = 50$ rpm.

cities would decrease the efficiency dramatically, see Figure 8c. The efficiency starts low and reaches its steady state in about $12 \div 20$ s. It is observed here that intermediate speed of about 50 rpm gives the best screening rates during the continuous screening process, see Figure 8d.

8.2 Influence of Particles Feeding

The effect of the feeding rate of the mixed particulate material on the machine efficiency is also investigated. Increasing the size of the output nozzle of the feeding container will increase the feeding rate of the particle flow. The feeding rate of the particles is not stationary during the whole simulation process, see Figure 9a. For a small orifice size, e.g. 30 mm, the flow rate will stay stationary at 147 particle/s over the whole simulation after a few seconds. In continuous screening, it is observed that low feeding rates with nozzle radius of 30 mm will lead to a bad screening performance. There, the sorted particles travel slowly to the machine exits together with some of the undesired gangue ones.

To study the machine efficiency, the number of sorted particles in the lowest level of the sifting unit is recorded, see Figure 9b. The lowest-level-particles usually experienced to many collisions and face different obstacles during their travel. It is observed that very low feeding rates are not recommended in the case of continuous screening. Increasing the rate of the feeding material will increase the machine efficiency.

9 Conclusions

In this paper particle screening phenomena in a rotary tumbling screening machine are studied. The discrete element method is applied and used as a simulation tool for the separation process. This method proves its ability to be a powerful numerical modeling tool for solving problems in granular media. For specific geometrical and contact parameters, sifting rates and machine efficiency are recorded. Particles are simulated in uniform and stepped models of tumbling cylinders. For both continuous screening and batch sieving, it was found that the segregation process is very sensitive to the rotational speed of the machine. This speed should be selected carefully to maximize the number of sorted particles and to improve the sifting rates for the different machine levels. Too high and too low speeds will lead to a bad screening performance. Furthermore, the particle feeding rates have a great influence on the machine efficiency. For the same operating conditions, batch sieving often shows better results compared to continuous screening.

Parallel computation can be efficiently used to speed up the MD simulations of dense granular systems. Improving the accuracy of the simulation requires to be more realistic in implementing the contact forces and the associated contact parameters of the dynamical system of the granular system. Physical contacts inside the TSM require some more detailed investigations. These parameters can be obtained from special experiments and standard tests. For better understanding of the particle sorting and transportation between the different layers of the machine, experimental studies have to be performed.

Acknowledgements

By an agreement between the Jordanian Higher Council for Science and Technology (HCST) and the German Academic Exchange Service (DAAD), the first author had the chance to work on this project at the Institute of Engineering and Computational Mechanics, University of Stuttgart. The financial support of the DAAD is highly appreciated.

References

1. Alexander A, Shinbrot T, Muzzio F (2001) Granular segregation in the double-cone blender: Transitions and mechanisms, *Physics of Fluids*, 13:578–587.
2. Alkhaldi H, Eberhard P (2006) Computation of screening phenomena in a vertical tumbling cylinder, *Proceedings in Applied Mathematics and Mechanics (PAMM)*, submitted for publication.
3. Alkhaldi H, Eberhard P (2006) Particle screening phenomena in an oblique multi-level tumbling reservoir – A numerical study using discrete element simulation, *Granular Matter*, submitted for publication.

4. Allen M, Tildesley D (1987) *Computer Simulation of Liquids*, Clarendon, Oxford.
5. Cundall P, Strack O (1979) A discrete numerical model for granular assemblies, *Geotechnique*, 29:47–65.
6. Jansen M, Glastonbury J (1967) The size separation of particles by screening, *Powder Technology*, 1:334–343.
7. Mishra B, Rajamani R (1992) The discrete element method for the simulation of ball mills, *Applied Mathematical Modelling*, 16:598–604.
8. Mishra B, Thornton C (2002) An improved contact model for ball mill simulation by the discrete element method, *Advanced Powder Technology*, 13:25–41.
9. Muth B (2001) Simulation von Kontaktvorgängen einfacher Körper mit Methoden der Molekulardynamik (in German), DIPL-87, Institute B of Mechanics, University of Stuttgart, Germany.
10. Muth B, Müller M, Eberhard P, Luding S (2004) Contacts between many bodies, *Machine Dynamics Problems*, 28:101–114.
11. Plimpton S (1995) Fast parallel algorithms for short-range molecular dynamics, *Journal of Computational Physics*, 117:1–19.
12. Rhodes M (2005) *Introduction to Particle Technology*, Wiley-Interscience, Chichester.
13. Schiehlen W, Eberhard P (2004) *Technische Dynamik* (in German), Teubner, Stuttgart.
14. Schinner A (1999) Fast algorithms for the simulation of polygonal particles, *Granular Matter*, 2:35–43.
15. Standish N, Bharadwaj A, Hariri-Akbari G (1986) A study of the effect of operating variables on the efficiency of a vibrating screen, *Powder Technology*, 48:161–172.

Unilateral Contact and Damage Analysis in Masonry Arches

M. Campo[1], G.A. Drosopoulos[2], J.R. Fernández[1] and G.E. Stavroulakis[3]

[1]*Departamento de Matemática Aplicada, Facultade de Matemáticas, Universidade de Santiago de Compostela, 15782 Santiago de Compostela, Spain*
E-mail: {macampo, jramon}@usc.es
[2]*Department of Material Science and Technology, University of Ioannina, GR-45100 Ioannina, Greece*
E-mail: me01122@cc.uoi.gr
[3]*Department of Production Engineering and Management, Technical University of Crete, GR-73132 Chania, Greece and Department of Civil Engineering, Technical University of Braunschweig, Germany*
E-mail: gestavr@dpem.tuc.gr

Abstract. Two unilateral models, the first with contact interfaces and the second with continuous damage material (Fremond's model), are applied on the nonlinear analysis and collapse of masonry arches. The results are compared with the predictions of the classical Heyman theory.

Key words: Unilateral analysis, Fremond damage model, masonry arches, limit analysis, collapse analysis, Heyman theory.

1 Introduction

In the present work the ultimate failure load and the collapse mechanism of a stone arch bridge are estimated by the usage of (a) a multi-part elastic model with unilateral contact frictional interfaces and (b) a continuous damage model. According to the contact model, contact interfaces simulating potential cracks are considered and their opening or sliding indicates crack initiation. The second model deals with the material damage consisting in the development of internal microcracks that affect the load bearing capacity of the material. Satisfactory comparison of the results is reported here.

Peter Wriggers and Udo Nackenhorst (eds.), IUTAM Symposium on Computational Contact Mechanics, 357–363.

2 Description of the Unilateral Contact and the Damage Model

A stone bridge consists of stone blocks and the mortar joints. Blocks have high strength in compression and low strength in tension while mortar has generally low strength. Heyman [1] introduced the collapse mechanism method while several computational methods for the evaluation of the limit load of a masonry arch followed. In this paper two models are used for the calculation of the failure load and mechanism. Their predictions on a bridge with a concentrated force applied at the quarter span of the arch, which is probably the worst position of the live load [1], are compared.

2.1 The Unilateral Contact-Friction Model

The elastic model of the bridge is divided by a number of interfaces perpendicular to the center line of the arch. Unilateral contact law governs the behavior in the normal direction of an interface, indicating that no tension forces can be transmitted in this direction. The behavior in the tangential direction takes into account that sliding may or may not occur.

The unilateral contact-friction problem is a nonlinear problem in which the nonlinearity is restricted to certain interfaces or the boundary. Consequently, the equations of equilibrium are nonlinear even if the material obeys a linear elastic law or a small displacement assumption is considered. For the frictional contact problem the Virtual Work equation is written in a general form as

$$\int_V \boldsymbol{\tau} : \delta\boldsymbol{\epsilon} dV = \int_S \delta\mathbf{u} \cdot \mathbf{t} dS + \int_V \delta\mathbf{u} \cdot \mathbf{f} dV + \int_{S'} \delta\mathbf{u} \cdot \mathbf{t}^n dS' + \int_{S'} \delta\mathbf{u} \cdot \mathbf{t}^t dS', \quad (1)$$

where $\boldsymbol{\tau}$ is the stress matrix, $\delta\boldsymbol{\epsilon}$ is the virtual strain matrix; $\delta\mathbf{u}$ is the virtual displacement matrix, $\mathbf{t}$ and $\mathbf{f}$ are surface and body force vectors respectively, and $\mathbf{t}^n$ and $\mathbf{t}^t$ are the normal and tangential traction vectors on the actual contact boundary S'. The contact constraint is enforced in the Virtual Work equation, with Lagrange multipliers representing the contact pressures. Lagrange multipliers are also used to enforce sticking conditions in the tangential direction. The set of the nonlinear equations is solved by the Newton–Raphson incremental iterative procedure.

In the simplest case of frictionless contact, the problem can be formulated as a parametric linear complementarity problem (LCP) [2]

$$\mathbf{Ku} + \mathbf{N}^T\mathbf{r} = \mathbf{P}_o + \lambda\mathbf{P}, \quad (2a)$$

$$\mathbf{Nu} - \mathbf{g} \leq 0, \quad (2b)$$

$$\mathbf{r} \geq 0, \quad (2c)$$

$$(\mathbf{Nu} - \mathbf{g})^T\mathbf{r} = 0. \quad (2d)$$

Equation (2a) expresses equilibrium equations of the unilateral contact problem without friction, $\mathbf{K}$ is the stiffness matrix and $\mathbf{u}$ is the displacement vector. $\mathbf{P}_o$ denotes the self-weight of the structure and $\mathbf{P}$ represents the live load multiplied by the scalar parameter λ. $\mathbf{N}$ is an appropriate geometric transformation matrix and vector $\mathbf{g}$ contains the initial gaps for the description of the unilateral contact joints. Relations (2b), (2c), (2d) represent the constraints of the unilateral contact problem for the whole discretized structure, i.e. no-penetration, no-tension and either closed or open (complementarity). The enforcement of the constraints can be achieved by using Lagrange multipliers. Thus, $\mathbf{r}$ is the vector of Lagrange multipliers corresponding to the inequality constraints and is equal to the corresponding contact pressure ($-\mathbf{t}^n$).

The calculation of the ultimate load is based on the exploitation of the solvability conditions for linear complementarity problems and variational inequalities. Every part of the bridge between two interfaces may develop rigid body displacements, which must be compatible with the constraints of the unilateral contact problem, otherwise no solution exists. In other words an equilibrium configuration may or may not exist, depending on the geometry of the structure and the direction of the applied loading. Collapse occurs at the load step of the load incrementation where solvability is first violated. A more detailed description of the unilateral contact-friction model can be found in [3].

2.2 The Damage Model

In this case, a linear elastic model is considered where the damage of the material has been taken into account. The damage is caused by the opening of internal micro-cracks which may lead to the decrease in the load carrying capacity of the material, and, eventually, to the possible failure of the system. The material is assumed to be elastic with constitutive law

$$\tau = \zeta \mathbf{K}(\delta\epsilon), \tag{3}$$

where ζ is the scalar *damage field*. Following the ideas of Frémond and Nedjar [4, 5], the damage field is defined as an internal variable and measures the fractional decrease in the strength of the material, taking the values as the ratio

$$\zeta = \zeta(\mathbf{x}, t) = \frac{E_{\text{eff}}}{E} \tag{4}$$

between the effective modulus of elasticity of the material E_{eff} and that of the damage-free material E. Due to this definition, the variable ζ varies between 0 and 1. When $\zeta = 1$ the material is damage-free, when $\zeta = 0$ the material is completely damaged and for $0 < \zeta < 1$ it is partially damaged. Following the derivation in [4, 5] the evolution of the microfractures responsible for the damage is described by the differencial inclusion

$$\dot{\zeta} - \kappa \Delta\zeta + \partial\psi_{[\zeta_*,1]}(\zeta) \ni \phi(\delta\epsilon, \zeta), \tag{5}$$

where κ is the damage diffusion constant, $\partial\psi_{[\zeta_*,1]}$ is the subdifferential of the indicator function of the interval $[\zeta_*, 1]$ (ζ_* is supposed to be a lower bound for the damage) and ϕ is the damage source function, which contributes to the evolution of the damage depending on the mechanical stresses, and in this case has the form

$$\phi(\delta\boldsymbol{\epsilon}, \zeta) = \lambda_1 \left(\frac{1-\zeta}{\zeta}\right) - \frac{1}{2}\lambda_2\, \delta\boldsymbol{\epsilon}^+ \cdot \delta\boldsymbol{\epsilon}^+ + \lambda_3. \tag{6}$$

Here $\delta\boldsymbol{\epsilon}^+$ is the positive part of the strain tensor, that is, only tension is supposed to contribute to the development of damage and $\lambda_1, \lambda_2, \lambda_3$ are process parameters. We notice that the existence of such lower limit ζ_* is justified since when damage tends to zero, microcracks are too dense and modelling the material as elastic ceases to make sense. The body is supposed to initially occupy a domain Ω whose boundary $\partial\Omega = \Gamma$ is divided into two disjoint parts: Γ_D, where the body is supposed to be fixed (homogeneus Dirichlet condition), and Γ_N, where surface tractions ($\boldsymbol{f}_N$) are considered (Neumann boundary condition). In order to analyze and solve numerically this model, a variational formulation is obtained, consisting in an evolutionary variational equation for the mechanical behaviour and a nonlinear variational inequality for the damage evolution:

$$(\zeta \mathbf{K}\boldsymbol{u}(t), \delta\boldsymbol{\epsilon}(\boldsymbol{w}))_{L^2(\Omega)^{2\times 2}} = (\boldsymbol{f}_B(t), \boldsymbol{w})_V + (\boldsymbol{f}_N(t), \boldsymbol{w})_{L^2(\Gamma_N)^2}, \quad \forall \boldsymbol{w} \in V, \tag{7}$$

$$\begin{aligned}(\dot{\zeta}(t), \xi - \zeta(t))_{L^2(\Omega)} &+ (\nabla\zeta(t), \nabla(\xi - \zeta(t)))_{L^2(\Omega)^2} \\ &\geq (\phi(\delta\boldsymbol{\epsilon}(t), \zeta(t)),\ \xi - \zeta(t))_{L^2(\Omega)}, \quad \forall \xi \in \mathcal{K},\end{aligned} \tag{8}$$

where

$$V = \{\boldsymbol{v} \in [H^1(\Omega)]^2; \quad \boldsymbol{v} = \boldsymbol{0} \quad \text{on} \quad \Gamma_D\},$$
$$\mathcal{K} = \{\xi \in H^1(\Omega); \quad \zeta_* \leq \xi \leq 1 \quad \text{a.e. in} \quad \Omega\},$$

are the sets of admissible displacements and damage fields, respectively. Moreover, $(.,.)_X$ denotes the inner product on the variational space X. Details about the variational analysis of this model can be found in [6]. By using approximate problems, the theory of set-valued pseudomonotone operators and a priori estimates, the existence of a unique solution is obtained with the following regularity,

$$\boldsymbol{u} \in L^\infty(0, T; V), \quad \zeta \in L^2(0, T; H^2(\Omega)) \cap H^1(0, T; L^2(\Omega)).$$

Its numerical resolution is performed in [7], where error estimates on the approximate solutions are also provided and, as a consequence, under some regularity conditions the linear convergence of the algorithm is achieved. Collapse occurs at the loading step for which the whole cross-section of the bridge reaches the lower bound for the damage ζ_*.

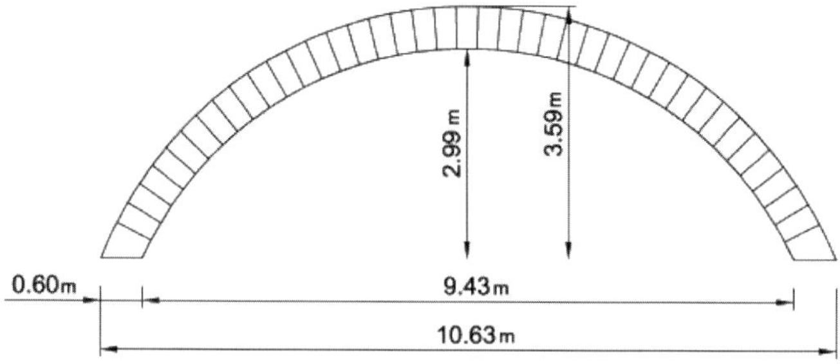

Fig. 1. Geometry of the masonry bridge – Forty contact interfaces.

3 An Application on a Real Scale Masonry Bridge

A plane stress model of a stone arch bridge without fill is considered [8], as shown in Figure 1. Loading includes self-weight and a concentrated load at the quarter span of the bridge. Young's modulus is 5 GPa, Poisson's ratio 0.3 and density 2200 kg/m^3.

3.1 General

For the contact analysis, the finite element model consists of 3036 quadrilateral, four-node, bilinear finite elements with two translational degrees of freedom per node. A typical finite element length is 0.05 m. A load increment equal to 0.02 kN is considered in the iterative incremental procedure. The friction coefficient is chosen equal to 0.6. This value is high enough to prevent sliding. Therefore a direct comparison of the results with the ones provided by the damage model is possible. In addition, a relatively large number of interfaces equal to 40 has been considered for the arch as the exact number of interfaces along the bridge's geometry tends to be meaningless in case many interfaces are used [3].

The damage model consists of 8400 triangular Lagrange finite elements with two translational degrees of freedom per node, and a total of 4515 nodes. Furthermore, the following data for the damage model have been used: $k = 0.001$ s (time step size), $\kappa = 10^{-2}$, $\zeta_* = 0.01$, $\lambda_1 = 2.d - 3$, $\lambda_2 = 5 \times 10^9$, $\lambda_3 = 7 \times 10^{-4}$.

3.2 Results

The failure load for the contact model of 40 interfaces is 87.14 kN. This limit load compares well with the one obtained by the damage model, which is equal to 90 kN. In Figure 2 the force-displacement diagrams of the two methods are compared. Moreover, both methods predict the same, well-known, mechanism of collapse. A four hinges mechanism arise in case of a quarter span load. The same conclusion arises from both experimental research [9] and the classical collapse mechanism

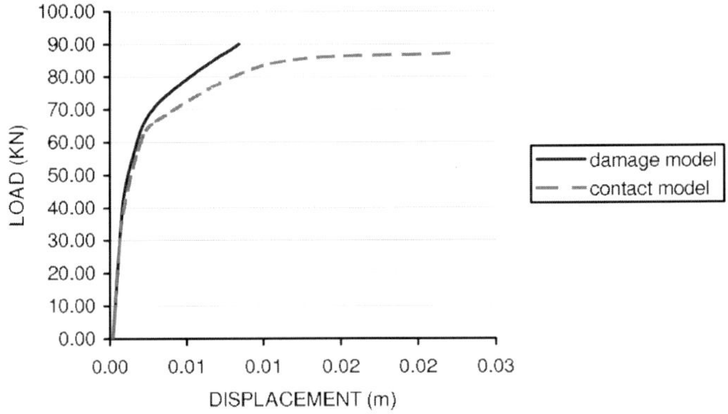

Fig. 2. Force-displacement diagram.

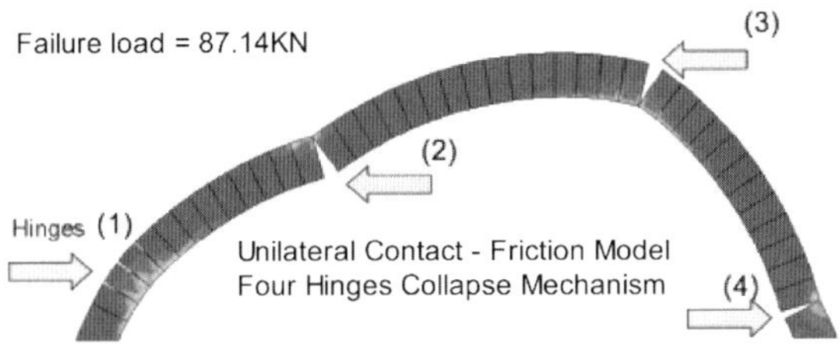

Fig. 3. Collapse mechanism from the unilateral contact-friction model.

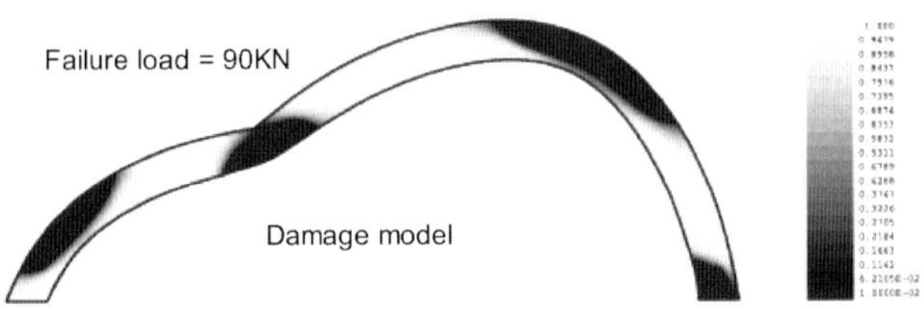

Fig. 4. Collapse mechanism from the damage model.

method of Heyman [1]. The above results are schematically shown in Figures 3 and 4. In Figure 4 also the damage field over the deformed configuration (deformations multiplied by a factor 50) is plotted.

4 Conclusions

The collapse mechanism of both contact and damage models at failure coincides with the one predicted by the classical collapse mechanism method of Heyman. The failure load obtained by the two methods is almost identical as well. However, there is a divergence in the force-displacement diagrams where the damage model has an ascending branch at failure. Further investigation, including for example parameter identification, will possibly lead to better comparison results.

Acknowledgement

The results presented here have been obtained within the research project "Smart Systems. New Materials, Adaptive Systems and Their Nonlinearities. Modelling, Control and Numerical Simulation", supported by the European Union with contract number HPRN-CT-2002-00284.

References

1. Heyman J (1982) *The Masonry Arch*. Ellis Horwood Series in Engineering Science, England.
2. Haslinger J, Miettinen M, Panagiotopoulos PD (1999) *Finite Element Method for Hemivariational Inequalities. Theory, Methods and Applications*. Kluwer Academic Publishers, Dordrecht.
3. Drosopoulos GA, Stavroulakis GE, Massalas CV (2006) Limit analysis of a single span masonry bridge with unilateral frictional contact interfaces. *Eng Struct*, 28:1864–1873.
4. Frémond M, Nedjar B (1995) Damage in concrete: The unilateral phenomenon. *Nuclear Eng Design*, 156:323–335.
5. Frémond M, Nedjar B (1996) Damage, gradient of damage and principle of virtual work. *Internat J Solids Struct*, 33:1083–1103.
6. Kuttler KL, Shillor M (2006) Quasistatic evolution of damage in an elastic body. *Nonlinear Anal Real World Appl*, 7:674–699.
7. Campo M, Fernández JR, Kuttler KL, Shillor M (to appear) Quasistatic evolution of damage in an elastic body: Numerical analysis and computational experiments. *Appl Numer Math*.
8. Page J (1989) Load tests to collapse on two arch bridges at Strathmashie and Barlae. TRRL Research Report 201, Department of Transport, Crowthorne (England).
9. Page J (1993) *Masonry Arch Bridges. TRL State of the Art Review*. HMSO, London.

Discontinuous Numerical Analysis of Masonry Vaults

R. Bravo[1] and J.L. Pérez-Aparicio[2]

[1]*Department of Structural Mechanics and Hydraulic Engineering, University of Granada, E-18001 Granada, Spain*
E-mail: rbravo@ugr.es
[2]*Department of Continuum and Structural Mechanics, Polytechnic University of Valencia, E-46022 Valencia, Spain*
E-mail: jopeap@upvnet.upv.es

Abstract. In this article the behavior of buried vaults (arches, tunnels, bridges, etc.) is studied with Discontinuous Deformation analysis, a numerical method that allows for the discontinuity of the structure and bases its physics on contact and friction among blocks. Two vault geometries are studied, the first semicircular and the second similar to that of oval arches. The considered loads are the weight of the structure, those related with lateral filling, with embankment and with concentrated (through a short distribution) forces at the peak. These loads are transformed to side forces with standard contact algorithm (penalty, Coulomb friction) and to point forces applied to the center of gravity with simple formulae from Soil Mechanics.

The analysis results are compared with experimental ones taken from the literature with, for most cases, very good agreement given the uncertainties on the geometry and material properties and given the intrinsic quality dispersion of masonry structures. The comparison is done for collapse load as a function of the number of joints, safety coefficients and limit point loads, including several modes of failure. The geometry of collapse is also compared with that of the experiments, obtaining here very good agreement.

Key words: Masonry, discontinuous numerical methods, contact, friction, experimental, oval arch, vault, collapse, hinge, filling, embankment, point load.

1 Introduction

One of the earliest applications of Discontinuous Methods was the failure analysis of masonry structures. Due to the fact that these structures are discontinuous, Continuum Mechanics does not apply, and Discrete Deformation Analysis (DDA) is particularly well suited. DDA is well suited not only because it is discontinuous but also because it is based on a displacement formulation more compatible with the targeted type of problems. Other methods such as Discrete Element (DEM) were designed as

Peter Wriggers and Udo Nackenhorst (eds.), IUTAM Symposium on Computational Contact Mechanics, 365–380.

explicit dynamic, which requires very small time steps and artificial damping, stability and convergence having problems for static or quasi-static cases [2]. The basis of the method is well known (see [3]) and will not be repeated here.

Masonry structures are composed of blocks whose stability is achieved by contact and friction among each other, directly or indirectly through a thin interface of elastoplastic material. This interface is often a layer of mortar with more or less constant thickness, which is simulated by the parameters of numerical contact: friction, cohesion, etc., without the addition of new unknowns.

In this article, the field experiments of two-dimensional vaults made of masonry (cut stone) at real scale are described. They are taken from the literature [5] and were performed by the CBTP Solen for the *Société National des Chemins de Fer* in 1980. In this source, the stability of these vaults with different geometries is tested against a variable number of blocks and different applied loads:

- Case 1. Filling thrust plus convoy weight. Collapse as a function of the number of blocks.
- Case 2. Filling plus embankment. Influence of the embankment thickness.
- Case 3. Filling plus point load. Influence of the concentrated load.

Filling refers to the soil at the sides of the vault and embankment to the horizontal layers of soil placed on top of the vault, either to stabilize it (as, for instance, on a bridge) or to constitute the terrain load (as in tunnels). The point load could represent the weight transmitted to an arch in an architectonic construction.

The results of the experiments are compared to those from the numerical simulation. Given the lack of statistics on the experiments (only one test was performed for each configuration) and the uncertainties in the data (dimensions, materials, etc.), most of the results are very close. These comparisons allow us to validate the applicability of discontinuous numerical methods for more or less real curved masonry structures under several types of soil effects.

2 Limit Calculation and Analogy with Contact Algorithms

Two basic concepts are introduced: friction and cohesion. The first F_f is a force related to the normal F_n with a coefficient $\mu = \tan\phi$ (where ϕ is the angle of friction), or in physical terms the mechanical clinging between two pieces due to the asperities in their surfaces. The cohesion, F_c, is the magnitude of the friction force when the normal is null, that is, the force necessary to break the joint. The total friction force is therefore $F_f = \mu F_n + F_c$.

Due to the fact that the tensile strength of masonry structures is negligible, special methods called "limit calculation" have existed for some time. They are based on the concept of the "breaking joint" from Coulomb (1773), whose basic hypotheses are:

- Lack of tractions in the structure, so that all sections are under compression.

- The tangential forces in the section must be lower than the resistance given by the friction.

These two hypotheses are analogous to two of the standard contact algorithm conditions: those of Kuhn–Tucker and of Coulomb friction respectively, see [4].

Although DDA can handle deformable bodies, in this work and for the sake of simplicity the blocks are considered rigid, that is, there will be no computation of block stresses and consequently no elastic failure on compression, as happens in some particular situations in the following cases. The method can also be dynamic. However, since we are interested only in the initial failure of the structure, the static situation is considered sufficient.

3 Case 1: Buried Vault

Experimental and numerical tests for this section were performed for a model of a vault under the action of filling terrain and forced loads. The geometry (Figure 1), corresponds to a structural arch whose thickness increases from the peak to the abutments (vertical lower blocks), considered to be supported by on the rigid foundations. The arch is assembled by dry masonry, that is, direct contact between the stone blocks without mortar between them. Stability is thus achieved through the curvature of the structure and the contact and friction generated by the rugosity of the block. To facilitate the collapse, the vault has an even number of blocks; there is no keystone.

3.1 Experimental Tests

The experimental testing program consists of breaking real scale masonry vaults with 7, 15, 25 and 59 joints. These vaults were covered on the outside with soil from the abutment to the peak of the vault. In addition, there are two hydraulic jacks symmetric to the keystone at 1.5 m with a surface of $1.5 \times 1.5\ \mathrm{m}^2$, which can apply concentrated vertical load, see Figure 1.

A load intensity q_0 is defined as a parameter for all external loading, from the jacks and from the thrust of the lateral filling. The intensity is equal to $q_0 = P_v/l$, where $l = 1.5$ is the length (in the plane of study) of the jack and P_v is the point load of the jack.

A vault with 199 joints is also reported in the reference, but this configuration has not been included here since the critical loads and failure mechanisms are nearly the same as for a vault with 59 joints.

3.2 Numerical Computation with DDA

The next step is the numerical modeling of the vaults with DDA and the comparison of the results obtained with those from the experiments. The data required are:

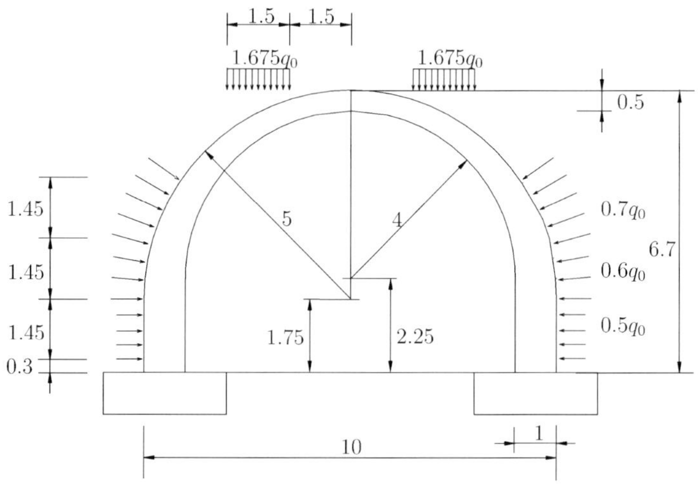

Fig. 1. Buried vault: geometry and loads.

- *Geometry*. The shape of the vault was estimated from the reference, and the dimensions were taken directly. The abutment is modeled as a single rectangular block for all vaults, but the circular part is divided into $N + 1$ equal divisions, where N represents the number of joints. The foundations are represented by a rigid block of rectangular shape that is prevented from moving.

The restraints for the foundation blocks are the only essential boundary conditions of the model. In the absence of loads, the rigid body motions of structure are prevented by contact and friction between blocks, according to the Kuhn–Tucker conditions (see [4]).

- *Elastic properties*. The blocks are considered very rigid, with a Young's modulus of $E = 1 \times 10^{12}$ N/m^2 and a standard Poisson's coefficient of $\mu = 0.25$.

The elastic modulus was chosen so that the value would be high enough to enforce rigidity but not interfere with computer precision.

- *Contact parameters*. From [6] a friction angle for dry masonry of $\phi = 30°$ and a cohesion of $c = 0$ N/m^2 were obtained. For the penalty parameter the usual value of $\epsilon_N = 1 \times 10^7$ N/m was used, preventing penetrations between blocks.
- *Applied loads*. As mentioned before, the load consists of vertical concentrated load from jacks and lateral, hydrostatic-like (perpendicular to the extrados or external circular line) from the filling. The weight of the structure is applied directly to the center of gravity of each block, using a density of $\gamma_s = 2500$ kg/m^3. The filling is considered to be gravel with a density of $\gamma_f = 800$ kg/m^3, a friction angle of $\phi = 30°$, and no cohesion.

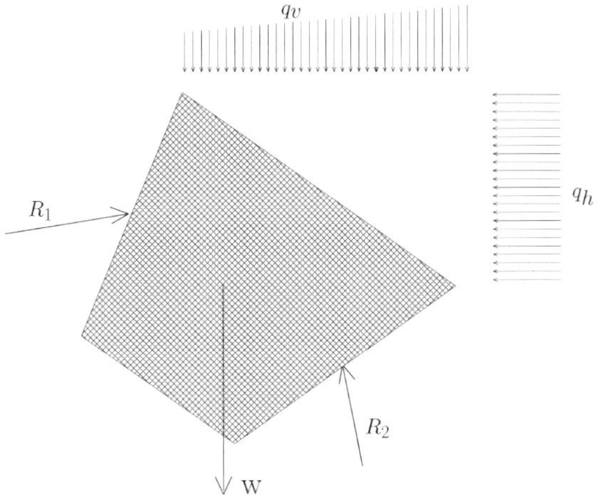

Fig. 2. External loads on a single block.

The contact forces act on the interface sides through the contact algorithms. Notice that they are not perpendicular to the side due to the tangential friction component.

Figure 2 shows a schema of the loads on each block. In addition to the vertical active loads (present only in a few blocks by the peak) that contribute to q_v, the lateral loads from the filling contribute to q_v and to q_h, see also Equation (2). Both distributed loads are applied to the side in the extrados but projected to the block center of gravity, see Equation (1), along with the weight W.

$$F_v = \int_{\theta_1}^{\theta_2} q\ R \sin\alpha\ \mathrm{d}\alpha = -q\ R\,(\cos\theta_2 - \cos\theta_1),$$

$$F_h = \int_{\theta_1}^{\theta_2} q\ R \cos\alpha\ \mathrm{d}\alpha = q\ R\,(\sin\theta_2 - \sin\theta_1), \tag{1}$$

where q is the distributed load (either vertical or horizontal), R the external radii and θ_1, θ_2 the angles which limit the polar load.

3.3 Numerical Results

As the jack point load is the main unknown of the problem, several runs were made to find this critical load. Starting from a value slightly lower than the experimental, the load is increased by 5 kN until collapse is observed. The structure collapses when several symmetric hinges form close to the abutments, at the peak and at other locations and contact is lost in one or more blocks. The positions of these hinges are the other important unknowns of the computations.

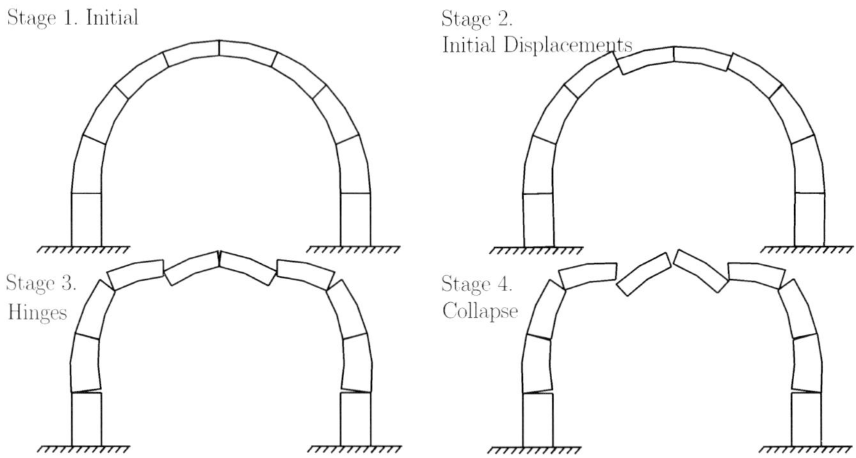

Fig. 3. Collapse sequence for a vault with 7 joints.

In Figure 3, a collapse sequence for a 7-joint vault is presented. A particular load intensity q_0 is kept constant during the computation, and successive DDA static iterations give the equilibrium geometries. The first stage represents the intact structure. The second shows the initial displacements. Given the high relative load of the jack, these occur mainly for the upper two blocks with a significant sliding between them and their neighbors. In the third stage, sliding is replaced by rotations, and symmetric hinges appear by the abutments and the peak and in the first and second blocks (from the top, any direction). Here it is very clear that the arch "opens" to the sides due mostly to the lower hinges. Finally in the fourth stage the two upper blocks separate from the neighbors and drop due to lack of friction restraint. As mentioned before, the computation is static in the sense that no inertial forces are considered. The observed displacements are static and due to the applied loads. Also, as is well know to architects, the hinges alternate: Along the circular line of the vault, an external hinge follows an internal one or vice versa (as in the following examples).

In Figure 4, initial collapse situations are shown for vaults with 15 and 25 joints. In the first one, we can see a block displacement pattern similar to Figure 3, including the length of the top blocks that descend and the position of the hinges. Some asymmetry can be seen in both configurations, due to the fact that the contact algorithms produce little bifurcation during the simulation. This is important in the corner-to-corner contact. Far from being detrimental this effect is often very useful in obtaining average results, which are more realistic than those from the purely deterministic simulations.

Finally in Figure 5 the collapse sequence of a vault with a large number of joints (59) is shown. Here, symmetry was forced by analyzing only half of the vault and

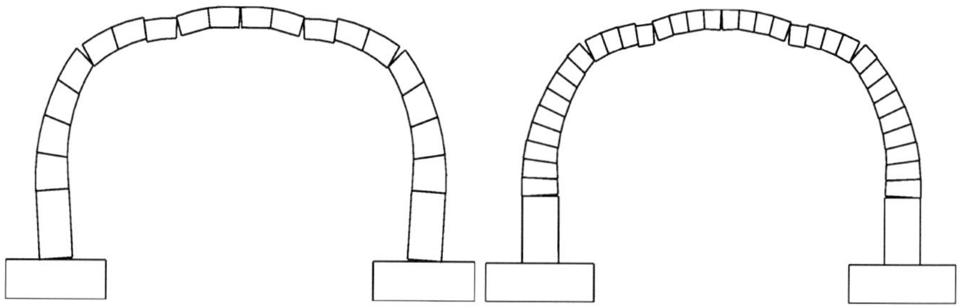

Fig. 4. Final collapse for vaults with 15 and 25 joints.

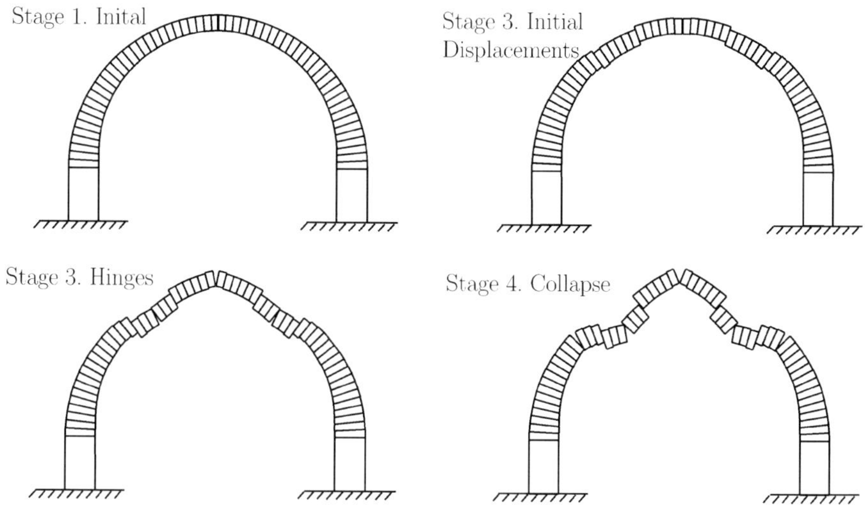

Fig. 5. Collapse stages for a 59-joint vault.

prescribing zero friction for the top block, in contact with a rigid vertical surface. In this configuration the effect of the vertical loads is mostly local at both right and left of the peak, where blocks along the jack length descend at each side. Due to the high number of blocks (and therefore contacts) no hinges are very clear (except for the peak) until the last stage: friction between the contact sides of the blocks prevent excessive movement until the collapse. Only during the last moments of collapse are both hinges and sliding evident. Due to the discontinuity of the jack loads, the vault opens in the middle, while the global vertical movement of the top more or less compensates the inward horizontal thrust. The whole vault behaves more like a chain than an arch. Therefore, the effects are mostly local. For instance, the vault does not expand laterally, since the upper blocks descend rather than rotate.

Table 1. Experimental and numerical results. Critical load as a function of joint number.

No. of joints	Critical Load Experimental (kN)	Critical Load DDA (kN)	Error %
7	250	280	12.2
15	206	210	1.6
25	206	205	–0.8
59	205	205	0.1

In Table 1 experimental and numerical results corresponding to the critical point loads for the different configurations are presented.

Notice that when the number of joints increases the critical load tends to an asymptotic value. This result is well known in architecture and represents the fact that, when there are many blocks and consequently more joints, there are more weak points that permit the turning or sliding of the blocks. Other authors, such as Drosopoulosa et al. [7] have reported this tendency with discontinuous numerical and special finite element methods, but without validating their results with experiments. We can also see that the error is reduced substantially with a large number of joints. This is due to the fact that the solution is less sensitive to uncertainties in the data when multiple possible failure modes are provided to DDA. As in any converging numerical method, the solution is better when the number of elements is higher. Finally the implemented kinematics are of first order, meaning that for large blocks some precision in the movements are lost.

4 Case 2: Influence of Embankment Thickness in the Stability of an Oval Vault

In this section a numerical simulation of the second experimental test form [5] is presented. Again the geometry and loads are estimated from the reference, see Figure 6. The vault geometry is now oval. The lateral thrust from the filling still holds but instead of the jack vertical load, a distributed load from an embankment of variable thickness h_e is considered.

This type of oval can be unstable for two reasons (see Figure 7):

- The lateral thrust of the filling dominates over the vertical loads. This occurs when the thickness of the embankment is low. Collapse occurs by elevation of the vault's peak and the formation of hinges, see Figure 7a.
- The vertical forces are prevalent, which happens when the thickness is high. The collapse occurs by descent of the vault's peak and the formation of hinges, see Figure 7b.

There is thus an optimal range in the thickness of the embankment that stabilizes the vault.

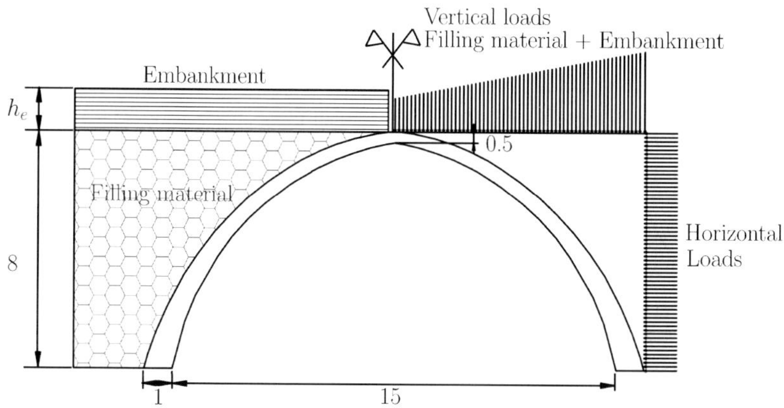

Fig. 6. Scheme of an oval vault and external loads.

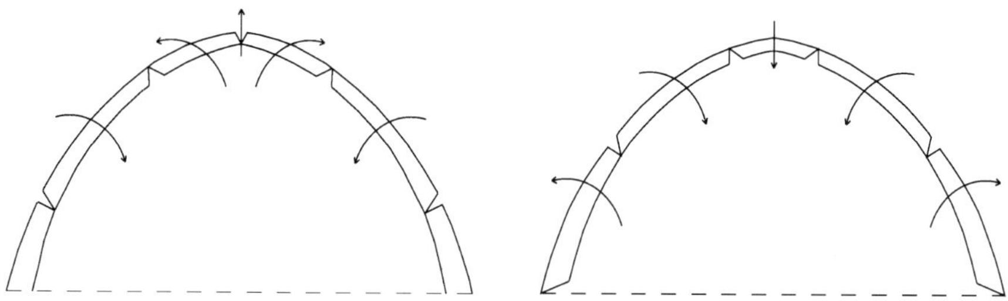

Fig. 7. Collapse schemes. Left, lateral loads prevalent. Right, vertical loads prevalent.

4.1 Experimental Tests

The experimental vault is assembled with 10 blocks (consequently 9 joints) and is supported on the foundations only; that is, the lower blocks can rotate. The situation is very similar to the previous case, except for the geometry and the vertical external loads, which produce different failure modes.

Starting from a situation of ruin due to lateral thrust (lower limit), the oval is tested by increasing the embankment thickness until new collapse is caused by excessive weight (upper limit). As mentioned above, for the lower limit the peak rises and a central hinge appears, while for the upper limit the peak does not open and descends (Figure 7). In both situations lateral hinges open alternately (toward the intrados or the extrados), causing rotations that are also alternating and with different directions for the two limits. For the upper limit, the rotations at the base cause the lower blocks partially to lose contact with the base.

Figure 9 shows the experimental trend of the security factor C_u (from 1 to ∞) of the vault versus the embankment thickness h_e with a continuous line. The lower limit zone starts at $h_e = 0.32$ m with a $C_u = 1$. After that point, a linear distribution up to $h_e = 0.9$ m with strong slope is observed, meaning that a small increase in thickness greatly increases the structure's safety. After this, there is a flat and short zone (up to $h_e = 1.9$ m) corresponding to failure by compression of the blocks. When the embankment thickness increases even more, there is a progressive decrease in safety due to excessive weight until failure is reached again for $h_e = 11.9$ m (upper limit).

4.1.1 Numerical Computation with DDA

Through DDA, the safety coefficients of the oval vault have been calculated. The own weight is computed directly from its volume and the previous material. For the filling, the same values are also used.

A density of $\gamma = 2000$ N/m^2 is given by Delbecq [5] for the embankment. Taking into account the reported compaction, an increase of 10% is considered for low thicknesses, with a final value $\gamma_e = 2200$ N/m^2. For higher thicknesses ($h_e > 1.9$ m) the value from the reference is used, due to the fact compaction is much less effective. As the load is completely vertical, no friction or cohesion considerations are necessary here.

Figure 8 represents the loads of a single block. For this case, beside the contact and weight point loads, the actions of the filling and the embankment are represented. These actions are calculated with Equation (2) from Lambe and Whiteman [1], so that the vertical and horizontal intensity loads are:

$$q_h = \frac{2}{3}(1 - \sin\phi)Hg\gamma_f,$$

$$q_v = \frac{1}{2}(h_l + h_r)\gamma_f g + h_e g \gamma_e, \tag{2}$$

where g is the gravity acceleration.

These intensity loads are transformed into point loads simply by multiplying them by the horizontal and vertical projections of the external side of the block. The point loads are then applied to the center of gravity.

With the complete DDA model, by trial and error, the lower and upper limits which stabilize the vault are computed, giving values of $h_e = 0.5$ m and $h_e = 12$ m, very close to the experimental ones. In Figure 9 the distribution of the safety coefficient C_u (limit load divided by actual load) is represented, with a continuous line for experimental results and with ticks for the numerical. Given the uncertainties in he data, the general agreement is very good except for the transition between limits. In this zone the resultant of the forces is very close to the center of gravity (inside the central core of the section) for which only compressions are present, almost no friction and no turning moment of the block. In this situation the only possible failure is by compression, but as mentioned before the blocks are considered rigid. Therefore, elastic computation is not reasonable. DDA predicts a very large

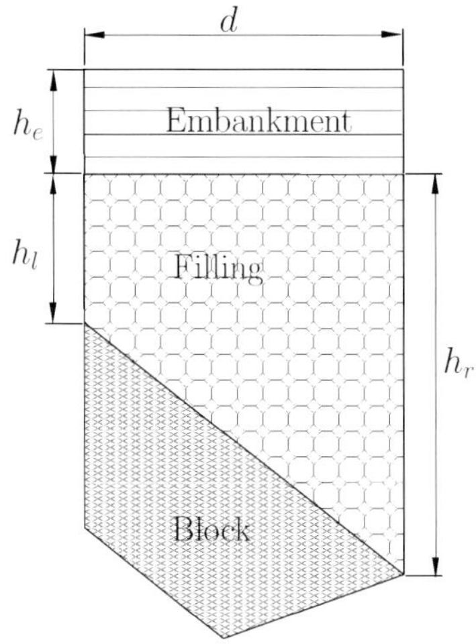

Fig. 8. External loads on a single block.

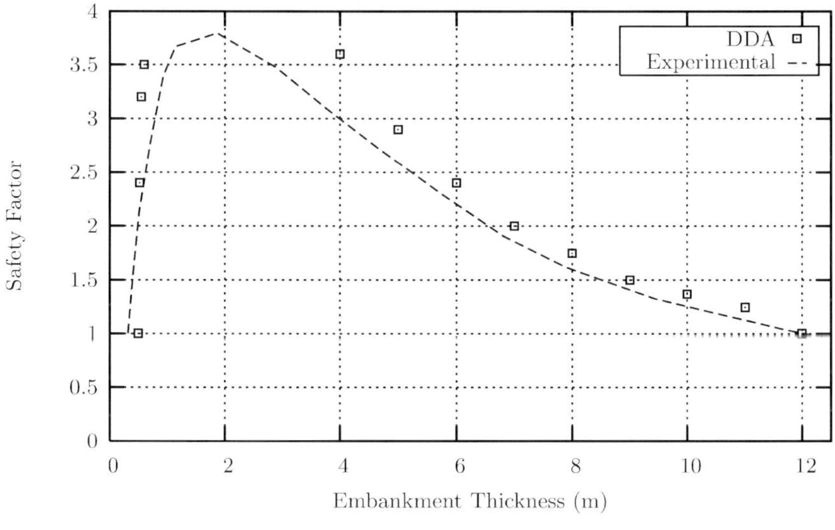

Fig. 9. Safety coefficient versus embankment thickness.

and unreasonable C_u in this zone, which is evident in the trend of the numerical data. Ongoing work is focusing on how to consider the blocks elastic and apply standard failure criteria for the stone compression.

The second comparison between experimental and numerical results is related to the failure modes of the oval vault. Figure 10 shows the collapse of the oval vault,

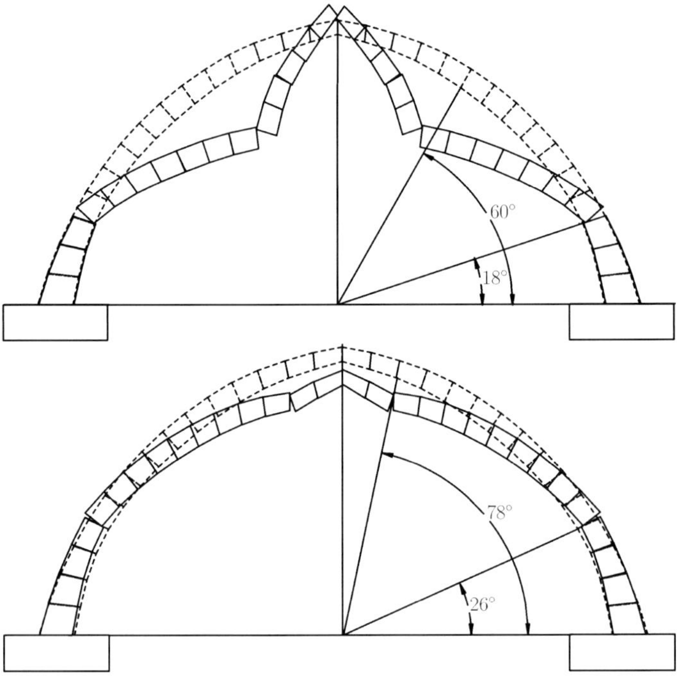

Fig. 10. Collapse of the oval vault for the lower limit (top figure) by raising of the peak and for the upper limit (bottom figure) by descent of the peak.

where it can be seen that to increase the precision of the method 3 times more blocks (30) than have been used in the experiment.

For the lower limit of embankment thickness (top figure), it is evident that the lateral forces are prevalent with the result of that the peak rises. Due to symmetry a hinge tends to open in the peak to the extrados. In the arch, two other hinges (one on each side) open alternately, the first one from the top to the intrados. For the upper limit (bottom figure) DDA predicts just the contrary situation: due to the weight of the embankment, the prevalent load is the vertical: the peak descends and two alternate hinges (different than those from the earlier result) open in the arch.

The angles of these hinges have been recorded experimentally and numerically and are shown in Table 2. The angles refer to the initial configuration, that is, at the initial stages of the collapse. It is worth noting that the results are very close, with the exception of the intermediate angle of the upper limit. However, a simulation with only the 10 real blocks would give much lower precision, since the DDA would not have enough choices to find the solution.

Table 2. Hinge angles for embankment thickness.

Limit	Numerical			Experimental		
Lower	18°	60°	90°	19°	64°	90°
Upper	0°	26°	78°	0°	37°	78°

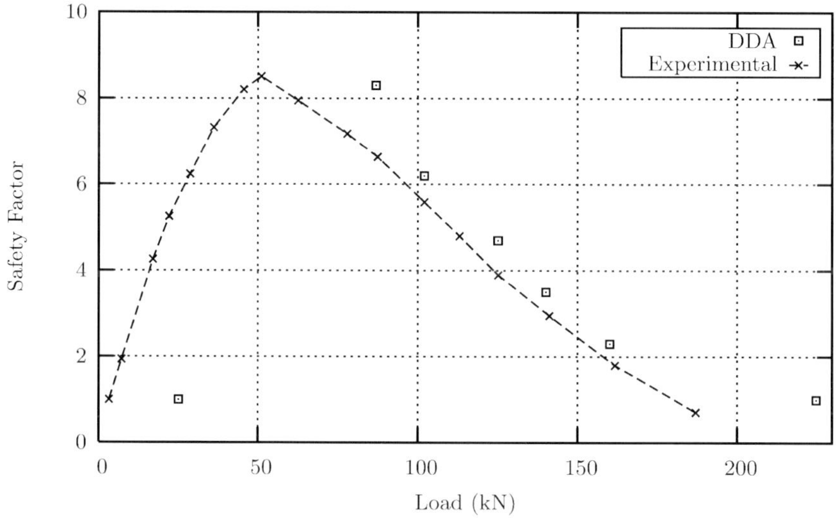

Fig. 11. Safety coefficient versus equivalent concentrated load.

5 Case 3: Influence of Concentrated Load in the Stability of an Oval Vault

Here the oval vault from the previous section is studied under the same conditions except for the vertical load, which is now concentrated at the peak of the vault, through a rectangle of width 2 m and variable height with a load intensity of q_p. The length of the load distribution covers approximately one block at each side of the peak. The embankment is now of a small constant thickness of $h_e = 0.2$ m, for which it is considered compacted, with density $\gamma_e = 2200$ N/m^2. The rest of material properties and geometry are the same as before.

The objective is again to study the safety coefficient, but now versus the concentrated load, which is equivalent to the load intensity. Figure 11 shows the relationship, which is similar to the previous one: there are a lower and upper limit of the load under which the oval vault collapses, although with different modes. In this figure, there is no transition zone related to failure by compression.

By trial and error, we find that DDA predicts a lower limit of 25 kN and an upper of 225 kN with safety coefficients of 1. While the second is acceptable compared with

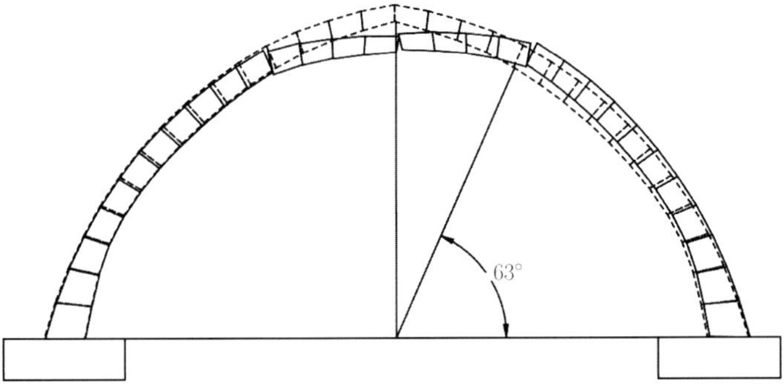

Fig. 12. Collapse of the oval vault for the upper limit by descent of the peak.

the experimental of 186 kN, the first is very different. For any value slightly higher than that of the lower limit, DDA prediction goes to unreasonably high values of C_u. On the other hand, the agreement is good in the upper limit zone. Uncertainties in the data could play a role in the differences, and the problem is probably very sensitive to one or several parameters. For instance, since this is a failure related with punching, an increase of the friction angle yields better results and more stability in the lower limit zone, but no physical bases for this increase could be found. Notice that the punching is not produced at the border of the concentrated load distribution but 3 more blocks to each side of the peak, that is, it is not a local failure. Ongoing work is being directed to solving this shortcoming.

In Figure 12, the failure mode corresponding to the upper limit is shown. The lower limit is not represented, since it is exactly equal to the one in the previous case. This is due to the fact that the failure is caused by lateral loads which are equal in both cases, while the vertical loads play a marginal role. On the other hand the simulation of the upper zone is acceptable; the closer to the upper limit the better.

As in the previous case, the positions of the hinges are compared in Table 3. In spite of the error in the computation of the safety factor in the lower limit zone, the simulation is very good, with a maximum error of 10%. The failure mode for the upper limit is somewhat different than that in Case 2: 4 blocks at each side of the peak descend together, and only one hinge opens to the extrados. In this hinge there is a combination of friction and rotation, but the rest of the arch is unaffected by the vertical loads, moving just slightly inwards due to the lateral loads from the filling.

For both embankment thickness and concentrated load, it is evident that in the oval vault collapse friction plays a very small role and most movement comes from the rotation of series of blocks. Due to the inclination of the vault sides, no much lateral movement occurs except slight motion for the upper limit, which creates a hinge in the base. Note also that the qualitative motions (movement of the peak, direction of the rotations) are the same as those given by the experiments, see Figure 7.

Table 3. Hinge angles for concentrated load.

Limit	Numerical			Experimental		
Lower	18°	60°	90°	19°	64°	90°
Upper	0°	63°	90°	0°	57°	90°

6 Conclusions

Several experiments taken from the literature have been simulated using a discontinuous numerical method, Discrete Deformation Analysis (DDA). This kind of method is better suited than continuous ones, as it is much closer to the physics of the problem, and the development, implementation and computational time are smaller. Three cases for two different vaults are presented. The first case is related to the resistance of a circular vault versus the number of joints, which as is well known is asymptotic. The other two, with the resistance of an oval vault under distributed and concentrated load. Some of the parameters of the problem (e.g. geometry) were taken from the reference but others had to be estimated from other sources or by best guess.

The failure modes are very well simulated by the numerical method in all cases, as well as the limit parameters for the first two, but the collapse load in third case has discrepancies in one of the behavior zones. In addition, a failure mechanism must be implemented, one related not to the stability of the structure but to the individual failure of the block under elastic compression. For this failure mechanism, elasticity in the blocks and dynamics in the method must also be considered. Further work is currently being conducted to solve these shortcomings and to apply DDA to configurations not tested in the experiments, such as effects of foundation settlement and different positions of the concentrated loads, and seismic behavior. When the numerical tool is completely finished and validated, it could be an important aid for the optimal design of real vaults and arches, where currently elevated safety factors of 3 or more are used.

References

1. T.W. Lambe and R.V. Whitman. *Soil Mechanics*. J. Wiley & Sons, 1969.
2. A. Munjiza. *Combined Finite-Discrete Element Method*, J. Wiley & Sons, 2004.
3. G.H. Shi. *Discontinuous deformation analysis – A new model for the statics and dynamics of block systems*, Thesis (Ph.D.), University of California, Berkeley, 1988.
4. P. Wriggers. *Computational Contact Mechanics*. J. Wiley & Sons, 2002.
5. J.M. Delbecq. *Analyse de la Stabilité des Ponts en Maçonnerie par la Théorie du Calcul à la Rupture*. Tesis, Ecole Nationale des Ponts et Chaussées, Paris, France, 1983.
6. Martín-Caro Álamo. *Análisis Estructural de Puentes de Arco de Fábrica. Criterios de Comprobación*. Tesis Doctoral, UPM, Madrid, 2001.

7. G.A. Drosopoulosa, G.E. Stavroulakis and C.V. Massalasa. Limit analysis of a single span masonry bridge with unilateral frictional contact interfaces. *Engineering Structures*, 28:1864–1873, 2006.

Author Index

Subject Index

IUTAM Bookseries

1. P. Eberhard (ed.): *IUTAM Symposium on Multiscale Problems in Multibody System Contacts*. 2007 ISBN 978-1-4020-5980-3
2. H. Hu and E. Kreuzer (eds): *IUTAM Symposium on Dynamics and Control of Nonlinear Systems with Uncertainty*. 2007 ISBN 978-1-4020-6331-2
3. P. Wriggers and U. Nackenhorst (eds): *IUTAM Symposium on Computational Methods in Contact Mechanics*. 2007 ISBN 978-4020-6404-3